PARTICLES AND THE UNIVERSE

PARTICLES AND THE UNIVERSE

Proceedings of the International Symposium on
Particles and the Universe,
held at Thessaloniki, Greece, June 24-29, 1985

edited by

G. LAZARIDES
Aristotle University of Thessaloniki
Thessaloniki, Greece

Q. SHAFI
Bartol Research Foundation
University of Delaware
Delaware, U.S.A.

1986

NORTH-HOLLAND PHYSICS PUBLISHING
AMSTERDAM • OXFORD • NEW YORK • TOKYO

© Elsevier Science Publishers B.V., 1986

All rights reserved. No part of this publication may be reproduced, stored in a retrieval system, or transmitted, in any form or by any means, electronic, mechanical, photocopying, recording or otherwise, without the prior permission of the publisher, Elsevier Science Publishers B.V. (North-Holland Physics Publishing Division), P.O. Box 103, 1000 AC Amsterdam, The Netherlands.

Special regulations for readers in the USA: This publication has been registered with the Copyright Clearance Center Inc. (CCC), Salem, Massachusetts. Information can be obtained from the CCC about conditions under which photocopies of parts of this publication may be made in the USA.

All other copyright questions, including photocopying outside of the USA, should be referred to the publisher.

ISBN: 0 444 87005 9

Published by:
North-Holland Physics Publishing
a division of
Elsevier Science Publishers B.V.
P.O. Box 103
1000 AC Amsterdam
The Netherlands

Sole distributors for the U.S.A. and Canada:
Elsevier Science Publishing Company, Inc.
52 Vanderbilt Avenue
New York, N.Y. 10017
U.S.A.

PREFACE

This book contains most of the contributions to the International Symposium on Particles and the Universe held at the Aristotle University of Thessaloniki, Greece from June 24 to 29, 1985. The meeting was organized scientifically by G. Lazarides (Thessaloniki), A. Salam (Trieste) and Q. Shafi (Bartol). The local organization was due to S. Abatzidis, M. Contadakis, N. Charalambakis, T. Christidis, N. Ganoulis, M. Hatzis, G. Lazarides (chairman), J. Magriotis, N. Papadakis, S. Papargyris, J. Paschalis, A. Trochides and M. Zigomalas.

The Symposium was sponsored mainly by the Aristotle University of Thessaloniki, the Ministry of Northern Greece and the Greek Ministry of Research and Technology. It was also sponsored by the Greek Ministry of Commerce and the Municipality of Thessaloniki.

The meeting was linked to the celebration of the 2300th anniversary of the foundation of the city of Thessaloniki, the metropolis of Northern Greece; the seal displayed on the cover marks this celebration. It was intended to emphasize the fact that Northern Greece, together with Ionia, were the birth places of natural philosophy. The Cosmologists of Ionia Thales, Anaximandros, Anaximenes, Heracletos, and the thinkers of Northern Greece, Leucippos, Democritos and Aristotle founded the materialistic thinking in contrast to the idealists of the Greek south.

Thessaloniki   G. LAZARIDES
February 1986   Q. SHAFI

TABLE OF CONTENTS

PREFACE   v

Heterotic string theory
    D.J. GROSS   1

Low energy superstring theory
    B.A. OVRUT   21

Renormalization group analysis of superstring models
    C.R. NAPPI   29

Superstrings and preons
    J.C. PATI   39

Scalar sector of gauge theories and the quest for
a unified theory
    M.A.B. BÉG   61

Composite Higgs and composite fermions
    H.M. GEORGI   79

Phenomenology of colour exotic fermions
    D. LÜST and G. ZOUPANOS   93

The problem of families
    G. SENJANOVIĆ   103

The neutrino mass, lepton flavor and lepton number
non-conservation
    J.D. VERGADOS   115

Heavy and light Dirac neutrinos and a techniphoton
    B. HOLDOM   133

Fermion masses in GUTs with intermediate scales
C. PANAGIOTAKOPOULOS 139

Quantum cosmology
A. VILENKIN 143

The dynamics of new inflation
R. BRANDENBERGER 151

Geometric effects in cosmological phase transitions
B.L. HU 169

Cosmic strings and galaxy formation
T.W.B. KIBBLE 177

Recent developments in the cosmic string theory
of galaxy formation
N. TUROK 189

"Invisible" axion detectors
P. SIKIVIE 201

Cosmic axion decay to photons in the microwave
and infrared regions
T.J. WEILER 215

Superconducting strings and membranes
G. LAZARIDES 223

Quark and lepton masses in a six dimensional
SO(12) model
C. WETTERICH 231

Attractor in a higher-dimensional cosmology
K. MAEDA 239

Searching for Cygnets
E.W. KOLB 247

Fractional statistics, exceptional preons, scalar dark matter, lepton number violation, neutrino masses, and hidden gauge structure
   A. ZEE   257

Non-abelian interactions in Rubakov-Callan effect
   K.S. NARAIN   263

Jets, $W^{\pm}$ and Z^o production in UA2
   G. GOGGI   271

Magnetic monopoles from Stanford to MACRO
   W.P. TROWER   287

LIST OF PARTICIPANTS   293

AUTHOR INDEX   295

HETEROTIC STRING THEORY

David J. GROSS*

Joseph Henry Laboratories, Princeton University, Princeton,
New Jersey 08544, USA

1. INTRODUCTION

High energy physics is, at present, in an unusual state. It has been clear for some time that we have succeeded in achieving many of the original goals of particle physics. We have constructed theories of the strong, weak and elecromagnetic interactions and have understood the basic constituents of matter and their interactions. The "standard model" has been remarkably successful and seems to be an accurate and complete description of physics, at least at energies below a Tev. Indeed, as we have heard in the experimental talks at this meeting, there are at the moment no significant experimental results that cannot be explained by the color gauge theory of the strong interactions (QCD) and the electroweak gauge theory. New experiments continue to confirm the predictions of these theories and no new phenomenon have appeared.

This success has not left us sanguine. Our present theories contain too many arbitrary parameters and unexplained patterns to be complete. They do not satisfactorily explain the dynamics of chiral symmetry breaking or of CP symmetry breaking. The strong and electro- weak interactions cry out for unification. Finally we must ultimately face up to including quantum gravity within the thoery. However, we theorists are in the unfortunate situation of having to address these questions without the aid of experimental clues. Furthermore extrapolation of present theory and early attempts at unification suggest that the natural scale of unification is 10^{16} Gev or greater, tantalizingly close to the Planck mass scale of 10^{19} Gev. It seems very likely that the next major advance in unification will include gravity. I do not mean to suggest that new physics will not appear in the range of Tev energies. Almost all attempts at unification do, in fact, predict a multitude of new particles and effects that could show up in the Tev domain (Higgs particles, Supersymmetric partners, etc.), whose discovery and exploration is of the utmost importance. But the truly new threshold might lie in the totally inaccessible Planckian domain.

---

*Research supported in part by NSF Grant PHY80-19754.

In this unfortunate circumstance, when theorists are not provided with new experimental clues and paradoxes, they are forced to adopt new strategies. Given the lessons of the past decades it is no surprise that much of exploratory particle theory is devoted to the search for new symmetries. However it is not enough simply to dream up new symmetries, one must also explain why these symmetries are not apparent, why they have been heretofore hidden from our view. This often requires both the discovery of new and hidden degrees of freedom as well as mechanisms for the dynamical breaking of the symmetry.

Some of this effort is based on straightforward extrapolations of established symmetries and dynamics, as in the search for grand unified theories (SU_5, SO_{10}, E_6,...), or in the development of a predictive theory of dynamical chiral gauge symmetry breaking (technicolor, preons, ...). Ultimately more promising, however, are the suggestions for radically new symmetries and degrees of freedom.

First there is supersymmetry, a radical and beautiful extension of space-time symmetries to include fermionic charges. This symmetry principle has the potential to drastically reduce the number of free parameters. Most of all it offers an explanation for the existence of fermionic matter, quarks and leptons, as compelling as the argument that the existence of gauge mesons follows from local gauge symmetry.

An even greater enlargement of symmetry, and of hidden degrees of freedom is envisaged in the attempts to revive the idea of Kaluza and Klein, wherein space itself contains new, hidden, dimensions. These new degrees of freedom are hidden from us due to the spontaneous compactification of the new spatial dimensions, which partially breaks many of the space-time symmetries of the larger manifold. Although strange at first, the notion of extra spatial dimensions is quite reasonable when viewed this way. The number of spatial dimensions is clearly an experimental question. Since we would expect the compact dimensions to have sizes of order the Planck length there clearly would be no way to directly observe many (say six) extra dimensions. The existence of such extra dimensions is not without consequence. The unbroken isometries of the hidden, compact, dimensions can yield a gravitational explanation for the emergence of gauge symmetries (and, in supergravity theories, the existence of fermionic matter). A combination of supergravity and Kaluza-Klein thus has the potential of providing a truly unified theory of gravity and matter, which can provide an explanation of the known low energy gauge theory of matter and predict its full particle content.

Attempts to utilize these new symmetries in the context of ordinary QFT, however, have reached an impasse. The problems one encounters are most severe if one attempts to be very ambitious and contemplate a unified theory of pure supergravity (in, say, 11 dimensions), which would yield the observed low energy gauge group and fermionic spectrum upon compactification. First of all we do not have a satisfactory quantum theory of gravity, even at the perturbative level. Einstein's theory of gravity, as well as its supersymmetric extensions, is nonrenormalizable. We know that that means that there must be new physics at the Planck length. We are clearly treading on thin ice if we attempt to use this potentially inconsistent theory as the basis for unification.

Even if we ignore this issue, and focus on the low energy structure of such theories, it appears to be impossible to construct realistic theories without a great loss of predictive power. The primary obstacle is the existence of chiral fermions (i.e. the fact that the weak interactions are V-A in structure). In order to generate the observed spectrum of chiral quarks and leptons it appears to be necessary to retreat from the most ambitious Kaluza-Klein program, which would uniquely determine the low energy gauge group as isometries of some compact space and introduce gauge fields by hand. Furthermore the supergravity theories ubiquitously produce a world which would have an intolerably large cosmological constant. Finally no realistic and compelling model has emerged. This brings us to string theories which offer a way out of this impasse.

2. STRING THEORIES

String theories offer a way of realizing the potential of supersymmetry, Kaluza-Klein and much more. They represent a radical departure from ordinary quantum field theory, but in the direction of increased symmetry and structure. They are based on an enormous increase in the number of degrees of freedom, since in addition to fermionic coordinates and extra dimensions, the basic entities are extended one dimensional objects instead of points. Correspondingly the symmetry group is greatly enlarged, in a way that we are only beginning to comprehend. At the very least this extended symmetry contains the largest group of symmetries that can be contemplated within the framework of point field theories--those of ten-dimensional supergravity and super Yang-Mills theory.

The origin of these symmetries can be traced back to the geometrical invariance of the dynamics of propagating strings. Traditionally string theories are constructed by the first quantization of a classical relativistic one dimensional object, whose motion is determined by requiring that the invariant area of the world sheet it sweeps out in space-time is extremized.

In this picture the dynamical degrees of freedom of the string are its coordinates, $X_\mu(\sigma,\tau)$ (plus fermionic coordinates in the superstring), which describe its position in space time. The symmetries of the resulting theory are all consequences of the reparametrization invariance of σ,τ parameters which label the world sheet. As a consequence of these symmetries one finds that the free string contains massless gauge bosons. The closed string automatically contains a massless spin two meson, which can be identified as the graviton whereas the open string, which has ends to which charges can be attached, yields massless vector mesons which can be identified as Yang-Mills gauge bosons.

String theories are inherently theories of gravity. Unlike ordinary quantum field theory we do not have the option of turning off gravity. The gravitational, or closed string, sector of the theory must always be present for consistency even if one starts by considering only open strings, since these can join at their ends to form closed strings. One could even imagine discovering the graviton in the attempt to construct string theories of matter. In fact this was the course of events for the dual resonance models where the graviton (then called the Pomeron) was discovered as a bound state of open strings. Most exciting is that string theories provide for the first time a consistent, even finite, theory of gravity. The problem of ultraviolet divergences is bypassed in string theories which contain no short distance infinities. This is not too surprising considering the extended nature of strings, which softens their interactions. Alternatively one notes that interactions are introduced into string theory by allowing the string coordinates, which are two dimensional fields, to propagate on world sheets with nontrivial topology that describe strings splitting and joining. From this first quantized point of view one does not introduce an interaction at all, one just adds handles or holes to the world sheet of the free string. As long as reparametrization invariance is maintained there are simply no possible counterterms. In fact all the divergences that have ever appeared in string theories can be traced to infrared divergences that are a consequence of vacuum instability. All string theories contain a massless partner of the graviton called the dilaton. If one constructs a string theory about a trial vacuum state in which the dilaton has a nonvanishing vacuum expectation value, then infrared infinities will occur due to massless dilaton tadpoles. These divergences however are just a sign of the instability of the original trial vacuum. This is the source of the divergences that occur in one loop diagrams in the old bosonic string theories (the Veneziano model). Superstring theories have vanishing dilaton tadpoles, at least to one-loop order. Therefore both the superstring and the heterotic string are explicitly finite to one loop order and there are strong arguments that this persists to all orders!

String theories, as befits unified theories of physics, are incredibly unique. In principle they contain no freely adjustable parameters and all physical quantities should be calculable in terms of h, c, and m_{planck}. In practice we are not yet in the position to exploit this enormous predictive power. The fine structure constant α, for example, appears in the theory in the form $\alpha \exp(-D)$, where D is the aforementioned dilaton field. Now the value of this field is undetermined to all orders in perturbation theory (it has a "flat potential"). Thus we are free to choose its value, thereby choosing one of an infinite number of degenerate vacuum states, and thus to adjust α as desired. Ultimately we might believe that string dynamics will determine the value of D uniquely, presumably by a nonperturbative mechanism, and thereby eliminate the nonuniqueness of the choice of vacuum state. In that case all dimensionless parameters will be calculable. Even more, string theories determine in a rather unique fashion the gauge group of the world and fix the number of space-time dimensions to be ten.

Finally and most importantly, string theories lead to phenomenologically attractive unified theories, which could very well describe the real world.

3. CONSISTENT STRING THEORIES

The number of consistent string theories is extremely small, the number of phenomenologically attractive theories even smaller. First there are the closed superstrings, of which there are two consistent versions. These are theories which contain only closed strings which have no ends to which to attach charges and are thus inherently neutral objects. At low energies, compared to the mass scale of the theory which we can identify as the Planck mass, we only see the massless states of the theory which are those of ten dimensional supergravity. One version of this theory is non-chiral and of no interest since it could never reproduce the observed chiral nature of low energy physics. The other version is chiral. One might then worry that it would suffer from anomalies, which is indeed the fate of almost all chiral supergravity theories in ten dimensions. Remarkably the particular supergravity theory contained within the chiral superstring is the unique anomaly free theory in ten dimensions. It however contains no gauge interactions in ten dimensions and could only produce such as a consequence of compactification. This approach raises the same problems of reproducing chiral fermions that plagued field theoretic Kaluza-Klein models and has not attracted much attention.

Open string theories, on the other hand, allow the introduction of gauge groups by the time honored method of attaching charges to the ends of the strings. String theories of this type can be constructed which yield, at low energies, N=1 supergravity with any SO(N) or Sp(2N) Yang-Mills group. These however, in addition to being somewhat arbitrary, were suspected to be anomalous. The discovery by Green and Schwarz, last summer, that for a particular gauge group--SO_{32}--the would be anomalies cancel, greatly increased the phenomenological prospects of unified string theories.

The anomaly cancellation mechanism of Green and Schwarz can be understood in terms of the low energy field theory that emerges from the superstring, which is a slightly modified form of d=10 supergravity. One finds that the dangerous Lorentz and gauge anomalies cancel, if and only if, the gauge group is SO_{32} or $E_8 \times E_8$. The ordinary superstring theory cannot incorporate $E_8 \times E_8$. The apparent correspondence between the low energy limit of anomaly free superstring theories and anomaly free supergravity theories provided the motivation that led to the discovery of a new string theory, by J. Harvey, E. Martinec, R. Rohm and myself, whose low energy limit contained an $E_8 \times E_8$ gauge group--the heterotic string. The heterotic string is a closed string theory that produced by a stringy generalization of the Kaluza-Klein mechanism of compactification, gauge interactions. These are determined by consistency to be $E_8 \times E_8$ or Spin $32/Z_2$. It is of more than academic interest to construct this theory since its phenomenological prospects are much brighter.

4. THE FREE HETEROTIC STRING

Free string theories are constructed by the first quantization of an action which is given by the invariant area of the world sheet swept out by the string or by its supersymmetric generalization. The fermionic coordinates of superstrings can be described either as ten-dimensional spinors or by ten-dimensional fermionic vector fields. Similarly the sixteen, left-moving coordinates of the heterotic string can be described by 32 real fermionic coordinates, by the local coordinates on the group manifolds of $E_8 \times E_8$ or spin $(32)/Z_2$, or by sixteen bosonic coordinates. Here we shall take the right-moving fermionic coordinates to be described by a ten-dimensional spinor and the left-moving internal coordinates to be sixteen bosonic fields. This will have the advantage of making the ten-dimensional supersymmetry manifest and of yielding a rather physical picture of the left-moving internal space. The price we pay is that this formulism is only tractable in light cone gauge, so that we must relinquish manifest Lorentz invariance.

The manifestly supersymmetric action for the heterotic string is given by the Green-Schwarz action for the right-movers plus the Nambu action for the left-movers. The superstring action is that of nonlinear sigma model on superspace (the super-translation group manifold) with a Wess-Zumino term. Consider an element of the supertranslation group, $h = \exp i(X \cdot P + \theta \cdot Q)$, where X_μ ($\mu=0,1,\ldots,a$) and θ^a (a ten-dimensional Majorana-Weyl spinor) are the superspace coordinates, P and Q are the generators of translations and supersymmetry translations. Then

$$\Pi_\alpha = h^{-1}(x) \partial_\alpha h(x) = (\partial_\alpha X^\mu - i\bar\theta\gamma^\mu \partial_\alpha \theta) P_\mu + \partial_\alpha \theta \cdot Q, \tag{1}$$

and defining $tr(P_\mu P_\nu) = \eta_{\mu\nu}$, $tr(Q_a Q_b) = 0$, $tr(Q_a Q_b P^\mu) = (\gamma^\mu C^{-1})_{ab}$, we have

$$S_R = -T/2 \{ \int d^2\xi\, eg^{\alpha\beta}\, tr\, \Pi_\alpha \Pi_\beta + \int_M d^3\xi\, \epsilon^{\alpha\beta\gamma}\, tr\, \Pi_\alpha \Pi_\beta \Pi_\gamma$$

$$+ \int d^2\xi e\, \lambda^{++} (e_+^\alpha \Pi_\alpha)^2 \}. \tag{2}$$

Here $\xi^\pm = \tau \pm \sigma$ are two dimensional light cone coordinates, e^α_a is the two-bein ($e^\alpha_a e^{\beta a} = g^{\alpha\beta}$). The second term is the Wess-Zumino term, where M is a three-dimensional manifold whose boundary coincides with the world sheet. The last term enforces the constraint that the coordinates are only right moving.

The left movers are given by a similar action

$$S_L = -T/2 \int d^2\xi\, e[\frac{1}{2} g^{\alpha\beta} \partial_\alpha X^A \partial_\beta X^A + \lambda^{--} (e_-^\alpha \partial_\alpha X^A)^2], \tag{3}$$

where $A=0,\ldots 25$.

When expanded, the full action is recognizable as the Green-Schwarz action for right movers alone plus the Nambu action for 26 left movers. It is therefore invariant under ten-dimensional Poincare transformations, $N=1$ supersymmetry, which acts on the right movers alone, and 16-dimensional Poincare transformations of X^I ($I=A-9=1,2,\ldots 16$). In addition it is invariant under local two-dimensional reparametrizations and the local fermionic transformation of Green and Schwatrz. These local symmetries enable us to choose a gauge--"light cone gauge"--where $g^{\alpha\beta} \sim \eta^{\alpha\beta}$, $X^+(\sigma,\tau) = \pi/T\, P^+\tau + x^+$, and $\gamma^+\theta = 0$. The resulting dynamics is given by the light cone action

$$S_{HeT} = -\frac{1}{2\pi} \int d^2\xi\, [(\partial_\alpha X^i)^2 + \frac{i}{2} S(\partial_\tau + \partial_\sigma) S + (\partial_\alpha X^I)^2 + \lambda[(\partial_\tau - \partial_\sigma) X^I]^2], \tag{4}$$

where we have chosen units in which the string tension $T = \frac{1}{2\pi\alpha'} = \frac{1}{\pi}$. The physical degrees of freedom are now manifest--eight transverse coordinates X^i (i=1...8), a Majorana-Weyl-light cone right moving spinor S^a ($(1+\gamma^{11})S^a = \gamma^+ S^a = 0$), and sixteen left-moving coordinates X^I (I=1,...,16). The equations of motions are then

$$\partial^2 X^i = 0; \qquad (\partial_\tau + \partial_\sigma) S^a = 0; \qquad (\partial_\tau - \partial_\sigma) X^I = 0; \qquad (5)$$

to which we must append the constraints that follow from the gauge fixing.

Canonical quantization of the transverse and fermonic coordinates is straightforward. Since we are dealing with closed strings, the fields are periodic functions of $0 \leq \sigma \leq \pi$ and can be expanded as

$$X^i(\tau-\sigma) = \frac{1}{2} x^i + \frac{1}{2} p^i (\tau-\sigma) + \frac{i}{2} \sum_{n \neq 0} \frac{\alpha_n^i}{n} e^{-2in(\tau-\sigma)}$$

$$X^i(\tau+\sigma) = \frac{1}{2} x^i + \frac{1}{2} p^i (\tau+\sigma) + \frac{i}{2} \sum_{n \neq 0} \frac{\tilde{\alpha}_n^i}{n} e^{-2in(\tau+\sigma)} \qquad (6)$$

$$S^a(\tau-\sigma) = \sum_{n=-\infty}^{+\infty} S_n^a e^{-2in(\tau-\sigma)}, \qquad (7)$$

where

$$[x^i, p^j] = i\delta^{ij}, \quad [\alpha_n^i, \alpha_m^j] = [\tilde{\alpha}_n^i, \tilde{\alpha}_m^j] = n\delta_{n+m,0} \delta^{ij}$$

$$[\alpha_i^n, \tilde{\alpha}_j^m] = 0$$

$$\{S_m^a, S_n^b\} = (\gamma^+ h)^{ab} \delta_{n+m,0}. \qquad (8)$$

The quantization of the left moving coordinates, X^I, is more subtle. We must impose the second class constraint (which is actually the equation of motion) that

$$\phi(\sigma,\tau) = (\partial_\tau - \partial_\sigma) X^I = 0. \qquad (9)$$

But this is not consistent with the cannonical commutator of X^I and its conjugate momentum $P^I = \frac{1}{\pi} \partial_\tau X^I$

$$[X^I(\sigma,\tau), P^J(\sigma',\tau)] = i\delta^{IJ} \delta(\sigma-\sigma'). \qquad (10)$$

We therefore modify the commutation relations, ala Dirac, defining

$$[A,B]_{DIRAC} = [A,B] - \sum_{\phi_i,\phi_j} [A,\phi_i] C_{ij} [\phi_j B],$$

where the sum runs over all constraints ϕ_i whose commutator is $[\phi_i,\phi_j] = (C^{-1})_{ij}$. The Dirac bracket is then consistent with the constraint, $[A,\phi_i] = 0$, which can be imposed as an operator identity. In our case we must modify Eq. (10) to read

$$[X^I(\sigma,\tau), P^J(\sigma',\tau)] = \frac{i}{2} \delta^{IJ} \delta(\sigma-\sigma').$$

$$[X^I(\sigma,\tau), X^J(\sigma',\tau)] = -\frac{i}{4} \delta^{IJ} \text{Sgn}(\sigma-\sigma'). \tag{11}$$

Therefore $X^I = X^I(\tau+\sigma)$ has the expansion

$$X^I(\tau+\sigma) = x^I + P^I(\tau+\sigma) + \frac{i}{2} \sum_{n \neq 0} \frac{\tilde{\alpha}_n^I}{n} e^{-2in(\tau+\sigma)}, \tag{12}$$

where upon quantization

$$[\tilde{\alpha}_n^I, \tilde{\alpha}_m^J] = n\delta_{n+m,0} \delta^{IJ}$$

$$[x^I, P^I] = \frac{i}{2} \delta^{IJ}. \tag{13}$$

Note the factor of 1/2 in the commutator of X^I and P^I, which implies that $2P^I$ is the generator of translations in the internal space. The allowed values of P^I must be restricted since X^I must be a periodic function of σ. They will be determined by the structure of the internal sixteen dimensional space.

In light-cone gauge $X^+(\tau,\sigma) = x^+ + p^+\tau$ and X^- is determined by solving the constraints that result from choosing a conformally flat metric on the world sheet. If we expand $X^-(\tau,\sigma)$ as

$$X^-(\tau,\sigma) = x^- + p^-\tau + \frac{i}{2} \sum_{n \neq 0} \frac{1}{n} (\alpha_n^- e^{-2in(\tau-\sigma)} + \tilde{\alpha}_n^- e^{-2in(\tau+\sigma)}) \tag{14}$$

then α_n^- is given, as in the fermionic string, by ($n \neq 0$)

$$\alpha_n^- = \frac{1}{p^+} \sum_n \alpha_m^i \alpha_{n-m}^i + \frac{1}{2p^+} \sum_m (m - \frac{n}{2}) \bar{S}_{n-m} \gamma^- S_m \tag{15}$$

and $\tilde{\alpha}_n^-$ is constructed, as in the bosonic string, as ($n \neq 0$)

$$\tilde{\alpha}_n^- = \frac{1}{p^+} \sum_m (\tilde{\alpha}_m^i \tilde{\alpha}_{n-m}^i + \tilde{\alpha}_m^I \tilde{\alpha}_{n-m}^I). \tag{16}$$

(Note, in these formulas $\alpha_0^i = \tilde{\alpha}_0^i = \frac{1}{2} p^i$, $\tilde{\alpha}_0^I = p^I$.)
These same constraints determine p^-, the generator of τ translations conjugate to X^+, and thereby the mass operator $m^2 = 2p^+p^- - (p^i)^2$ of the string

$$\tfrac{1}{4}(mass)^2 = N + (\tilde{N}-1) + \tfrac{1}{2} \sum_{I=1}^{16} (p^I)^2, \tag{17}$$

where $N(\tilde{N})$ are the normal ordered number operators for the right (left) movers

$$N = \sum_{n=1}^{\infty} (\alpha_{-n}^i \alpha_n^i + \tfrac{1}{2} n\, S_{-n} \gamma^- S_n)$$

$$\tilde{N} = \sum_{n=1}^{\infty} (\tilde{\alpha}_{-n}^i \tilde{\alpha}_n^i + \tilde{\alpha}_{-n}^I \tilde{\alpha}_n^I). \tag{18}$$

The subtraction of -1 in (2.17) is due to the normal ordering of $\tilde{N}$, which is unnecessary in the case of the right movers due to fermion-boson cancellations. This subtraction can also be seen to be necessary to ensure ten dimensional Lorentz invariance. Finally the factor of $(p^I)^2/2$ comes from the internal momentum and winding number (see below) of the left movers.

In addition there is a further constraint that requires

$$N = \tilde{N} - 1 + \tfrac{1}{2} \sum_{I=1}^{16} (p^I)^2. \tag{19}$$

This constraint has a simple physical explanation. Since there is no distinguished point on a closed string, we are free to shift the origin of the σ coordinate by an arbitrary amount Δ. This is achieved by the unitary operator

$$U(\Delta) \equiv e^{2i\Delta(N - \tilde{N} + 1 - 1/2 \sum_I (p^I)^2)}, \tag{20}$$

which (recall that $[X^I, P^I] = \tfrac{i}{2} \delta^{IJ}$) satisfies $U(\Delta) F(\tau,\sigma) U^+(\Delta) = F(\tau,\sigma+\Delta)$, where F can be x^i, x^I or S^a. The operator $U(\Delta)$ must therefore equal the identity operator on the space of physical states, which must therefore satisfy eq. (19). Again the subtraction constant, -1, is due to the normal ordering of $\tilde{N}$ and is necessary for Lorentz invariance.

Now let us consider the internal left-moving coordinates X^I whose properties are responsible for the new features of the heterotic string. These coordinates can be thought of as parametrizing an internal space T. It is unlikely that T could be curved without producing an inconsistent theory, since the resulting two-dimensional field theory of X^I would be an interacting nonlinear σ model with conformal anomalies. Therefore we consider only flat internal manifolds, taking T to be a sixteen-dimensional torus.

Since closed strings contain gravity in their low energy limit one expects that a compactified closed string theory will contain massless vectors associated with the isometries of the compact space. In the case of a flat 16-dimensional torus this would yield the gauge bosons of $[U(1)]^{16}$. A remarkable feature of closed string theories is that for special choices of the compact space there will exist additional massless vector mesons. These are in fact massless solitons of the closed string theory. They combine with the Kaluza-Klein gauge bosons to fill out the adjoint representation of a simple Lie group whose rank equals the dimension of T. In the case of the heterotic string the structure of T is so severely limited that only two choices are consistent. These produce the gauge mesons of $Spin(32)/Z_2$ or $E_8 \times E_8$!

To see this take T to be the most general torus, which may be thought of as R^{16} modulo a lattice Γ generated by 16 basis vectors e_i^I (i=1...16). We identify the center of mass coordinate X^I with its translation by $\pi\, e_i^I$ [$\sqrt{e_i^I \cdot e_i^I}$ is the diameter of the torus in the ith direction]

$$X^I \equiv X^I + \pi \sum_{i=1}^{16} n_i\, e_i^I \equiv X^I + \pi\, L^I \qquad (21)$$

with n_i integers. On such a torus the allowed center of mass momenta P^I lie on the dual lattice, Γ^*, generated by e_i^{*I} (i=1...16), defined by

$$\sum_{I=1}^{16} e_i^I \cdot e_j^{*I} = \delta_{ij}. \qquad (22)$$

In other words $\exp(2\pi i\, P^I \cdot L^I)$, which represents a translation around T by the element L^I (recall that $2P^I$ generates translations), must equal one, so that

$$P^I = \sum_{i=1}^{16} m_i\, e_i^{*I} \quad (m_i = \text{integer}). \qquad (23)$$

For the heterotic string, which is a periodic function of σ on T,

$$X^I = x^I + p^I\tau + L^I\sigma + \ldots = X^I(\tau+\sigma) = x^I + p^I(\tau+\sigma) + \text{oscillators} \quad (24)$$

represents a string configuration which winds around the torus n_i times in the ith direction, with momenta P^I which equals the winding number L^I. This is in fact a soliton--a classical solution of the equations of motion which is characterized by topological quantum numbers (n_i) that classify the maps of the one sphere ($0 \leq \sigma \leq \pi$) onto T ($\pi_1(T) = Z^{16}$). Such solitons must be included in the spectrum of the interacting closed string, since a string with $L^I = 0$ can split into one with $+L^I$ and $-L^I$.

Next we recall the constraint of Eq. (19), which requires that the allowed values of $(P^I = L^I)^2$ are <u>even</u> integers. Therefore, since all $(L^I + L^J)$ are allowed windings, $L^I \cdot L^J$ must be an integer (since $(L^I + L^J)^2$, $(L^I)^2$ and $(L^J)^2$ must all be even numbers). The winding numbers L^I must therefore lie on an <u>integer</u>, <u>even</u> lattice, so that the "metric" $g_{ij} = \sum_{I=1}^{16} e_i^I e_j^I$ is integer valued, and g_{ii} = even. The momenta P^I lie on the dual lattice Γ^*, which for integer Γ, contains Γ. [Since every vector in Γ, $V^I = \sum_{i=1}^{16} n_i e_i^I$, can be written as a vector in Γ^*, $V_i = \Sigma m_i e_i^{*I}$, with $m_i = \sum_{j=1}^{16} n_j g_{ji}$ = integer.] In general Γ^* contains more points than Γ, including some that are not of even length squared. For example if Γ = hypercubic lattice with spacing $\sqrt{2}$, Γ^* = hypercubic lattice with spacing $1/\sqrt{2}$. From a geometrical point of view we would, however, expect that <u>all</u> vectors in Γ^* are allowed momenta, in which case Γ^* must also be integer and even. But then $\Gamma \supset \Gamma^*$ and since $\Gamma^* \supset \Gamma$ the lattice must be self dual, $\Gamma = \Gamma^*$.

In that case

$$L^I = p^I + \sum_{i=1}^{16} n_i e_i^I \quad (n_i = \text{integer}), \quad (25)$$

and the diameters of the torus are all equal to $\sqrt{2} = \sqrt{\frac{2}{\pi T}}$.

Actually, for the free string we need not take Γ^* to be integer and even. But if we do so, the non geometric nature of the momenta would produce trouble at the interacting level where we would find two-dimensional (global) diffeomorphism anomalies that render string loops nonunitary or sick.

Even self dual lattices are extremely rare. They only exist in 8n dimensions. In sixteen dimensions there are two such lattices, $\Gamma_8 \times \Gamma_8$ and Γ_{16}. The first is the direct product of two Γ_8's, where Γ_8 is the root lattice of the exceptional Lie algebra E_8; the second is the weight lattice of Spin(32)/Z_2, generated by the weights of the adjoint of SO(32) plus one of the spinor representations. Let us consider $\Gamma_8 \times \Gamma_8$, generated by e_i^I, the roots of $E_8 \times E_8$. In this case the torus T is the "maximal torus" of $E_8 \times E_8$, generated by $\exp iH_i$, where H_i constitute the Cartan subalgebra of $E_8 \times E_8$.

The appearance of these special tori and the consequent emergence of the gauge group $G(G=E_8 \times E_8$ or $\text{Spin}(32)/Z)$ might seem mysterious. To those familiar with the theory of affine Lie (Kac-Moody) algebras the mathematical framework is familiar; nonetheless let us take a physicist's approach. Consider the massless states of the theory. These must satisfy $N = \tilde{N}-1 + 1/2 \Sigma(P_I)^2 = 0$. The bosonic states with $\tilde{N}=1$, $P^I=0$, consist of gravitons ($\tilde{\alpha}^i_{-1}|0\rangle$) and 16 corresponding gauge bosons ($\tilde{\alpha}^I_{-1}|0\rangle$). These massless vectors are expected--they arise from the $U(1)^{16}$ isometry of the torus. The corresponding conserved charges are simply the components of the internal (left-handed) momenta P^I. However there are additional massless vectors, solitons of the string theory, with winding numbers $L^I=P^I$ such that $(P^I)^2=2$. For the allowed self-dual lattices there are precisely 480 such vectors. The solitons have nonvanishing $U(1)^{16}$ charges, which can be identified with the non-zero weights of the adjoint representation of G, and they are massless vector mesons. It is well known that the only consistent theory of massless vector bosons with nonvanishing charges is a local gauge theory. But why couldn't the gauge group be $U(1)^{496}$? The reason is that the interactions will allow a soliton of charge, which equals winding number, P^I to break up into two solitons of charge P^I+K^I and $-K^I$ respectively. Thus each gauge boson couples to every other one, and the corresponding gauge group must be G. To show this explicitly we should construct the generators of G in the Fock space of the string oscillators. This we shall do below when we discuss interactions, since these generators are simply the vertex operators for a string to emit a massless gauge boson of vanishing momenta.

Let us now consider the full spectrum of the heterotic string. The physical states are simply direct products of the Fock spaces $|\rangle_R \times |\rangle_L$ of the right moving fermionic string and the left moving bosonic string, subject to the constraint, which ensures that the masses are non negative,

$$(\text{mass})^2 = 8N \geq 0$$
$$N = \tilde{N} + \frac{1}{2}(P^I)^2 - 1. \qquad (26)$$

The right-handed ground state is annihilated by α_n^i and S_n^i ($n>0$) and N. It consists of 8 bosonic states $|i\rangle_R$ and 8 fermionic states $|a\rangle_R$, which form a massless vector and spinor supermultiplet. The left-moving ground state consists of $\tilde{\alpha}_{-1}^i|0\rangle_L$ ($\tilde{N}=1$, $P^I=0$) and $\tilde{\alpha}_{-1}^I|0\rangle_L$ ($\tilde{N}=1$, $P^I=0$) and $|P^I, (P^I)^2=2\rangle_L$. The most general state

$$\prod_i \alpha_{-n_i}^i \prod_j S_{-m_j}^{a_j} \prod_k \tilde{\alpha}_{-n_k}^k |\text{Ground State}\rangle,$$

can be decomposed into irreducible representations of D=10, N=1 supersymmetry and of the group G.

The demonstration that these states are indeed Lorentz invariant and supersymmetric is straightforward. It is a consequence of the fact that the generators of Lorentz and supersymmetry transformations act separately on the right and left movers, and that the left movers are Lorentz invariant in 26 dimensions and the right movers Lorentz invariant and N=1 supersymmetric in 10 dimensions.

The ground state is a direct product of $|i\rangle_R + |a\rangle_R$ with $\tilde{\alpha}_{-1}^i|0\rangle_L$ (which yields the N=1 D=10 supergravity multiplet), and with $\tilde{\alpha}_{-1}^I|0\rangle_L$ and $|(P^I)^2=2\rangle_L$ (which yields the N=1 Yang Mills supermultiplet in the adjoint of $E_8 \times E_8$). The higher mass states are easily assembled into SO(9)×G multiplets. For example the first massive level, with (mass)² = 8, has N=1 and ($\tilde{N}$, $(P^I)^2$) = (2,0), (1,2) or (0,4). We separately assemble the right and left moves into SO(9) multiplets and take direct products. The right movers contain $\alpha_{-1}^i|j\rangle_R$ and S_-^a (b)$_R$ which fill out the SO(9) representations $\underline{44}$ = ▢ and $\underline{84}$ = ☰ , as well as the fermion states $\alpha_{-1}^i|a\rangle_R$ and $S_{-1}^a|i\rangle_R$ which form the 128 of SO(a). In the left-moving sector we have the ($E_8 \times E_8$, SO(9)) representations: ((1,1), $\underline{44}$) which contain $\tilde{\alpha}_{-1}^i \tilde{\alpha}_{-1}^j|0\rangle_L$, ((248,1) + (1,248), $\underline{9}$) which contains $\tilde{\alpha}_{-1}^I|(P^I)^2=2\rangle_L$, and ((3875,1) + (1,2875) + (248,248) + 2(1,1), $\underline{1}$) which contains $|(P^I)^2=4\rangle$. Altogether, at this level, we have 18,883,584 physical degrees of freedom!

At higher mass levels the number of states will increase rapidly, not only will we have the usual proliferation of higher spin states but also we will get even larger representations of $E_8 \times E_8$. The number of states of mass M in the heterotic string increases as $d(M) \underset{M\to\infty}{\sim} \exp\beta(\beta_H M)$, with β_H given by $(2+\sqrt{2}) \pi \sqrt{\alpha'}$, the mean of corresponding factors for the fermionic superstring and the 26 dimensional bosonic string.

The heterotic string theory has, by now, been developed to the same stage as other superstring theories. Interactions have been introduced and shown to preserve the symmetries and consistency of the theory, radiative corrections calculated and shown to be finite.

5. STRING PHENOMENOLOGY

In order to make contact beween the string theories and the real world one is faced with a formidable task. These theories are formulated in ten flat space-time dimensions, have no candidates for fermionic matter multiplets, are supersymmetric and contain an unbroken large gauge group--say $E_8 \times E_8$. These are not characteristic features of the physics we observe at energies below a Tev. If the theory is to describe the real world one must understand how six of the spatial dimensions compactify to a small manifold leaving four flat dimensions, how the gauge group is broken down to $Su_3 \times SU_2 \times U_1$, how supersymmetry is broken, how families of light quarks and leptons emerge, etc. Much of the recent excitement concerning string theories has been generated by the discovery of a host of mechanisms, due to the work of Witten and of Candelas, Horowitz and Strominger, and of Dine, Kaplonovsky, Nappi, Seiberg, Rohm, Breit, Ovrut, Segre, and others, which indicate how all of this could occur. The resulting phenomenology, in the case of the $E_8 \times E_8$ heterotic string theory is quite promising.

The first issue that must be addressed is that of the compactification of six of the dimensions of space. The heterotic string, as described above, was formulated in ten dimensional flat spacetime. This however is not neccessary. Since the theory contains gravity within it the issue of which spacetime the string can be embedded in is one of the string dynamics. That the theory can consistently be constructed in perturbation theory about flat space is equivalent to the statement that ten dimensional Minkowski spacetime is a solution of the classical string equations of motion. Such a solution yields the background expectation values of the quantum degrees of freedom. We can then ask are there other solutions of the string equations of motion that describe the string embedded in, say, four dimensional Minkowski spacetime times a small compact six dimensional manifold?

At the moment we do not possess the full string functional equations of motion, however one can attack this problem in an indirect fashion. One method is to deduce from the scattering amplitudes that describe the string fluctuations in ten dimensional Minkowski space an effective Lagrangian for local fields that describe the string modes. Restricting one's attention to the masseless modes, the resulting Lagrangian yields equations which reduce to

Einstein's equations at low energies, and can be explored for compactified solutions. Another method is to proceed directly to construct the first quantized string about a trial vacuum in which the metric n_{ab} (as well as other string modes) have assumed background values. In this approach one starts with the action of Equation (2), or its supersymmetric generalization, but allows $n_{ab}(x)$ to be the metric of a curved manifold. A consistent string theory can be developed as long as the two dimensional field theory of the coordinates $\chi^\alpha(\sigma,\tau)$ is conformally invariant. This is a nontrivial requirement, since the theory described by (2) is an interacting nonlinear σ-model. The condition that the two dimensional theory be conformal invariant is equivalent to demanding that the string equations of motion are satisfied. Thus one can search for alternative vacuum states by looking for σ-models (actually supersymmetric σ-models), for which the relevant β functions (which are local functions of the metric $n_{ab}(x)$ and its derivatives) vanish. In addition one must check that the anomaly in the commutators of the stress energy tensor is not modified. Given such a theory one can construct a consistent string theory and if $n_{ab}(x)$ describes a curved manifold the string will effectively be embedded in this manifold.

Remarkably there do exist a very large class of conformally invariant supersymmetric σ-models, that yield solutions of the string classical equations of motion to all orders and describe the compactification of ten dimensions to a product of four dimensional Minkowski space times a compact internal six dimenional manifold. These compact manifolds are rather exotic mathematical constructs (they are Kahler and admits a Ricci flat metric--i.e. they have SU_3 holonomy) and are called "Calabi-Yau" manifolds. In general they have many free parameters (moduli) which, among the rest, determine their size. Once again, this is an indication of the enormous vacuum degeneracy of the string theory, at least when treated perturbatively, and leads to many (at the present stage of our understanding) free parameters. This abundance of riches should not displease us, at the moment we would like to know whether there are any solutions of the theory which resemble the real world, later we can try to understand why the dynamics picks out a particular solution.

In the case of the heterotic string it is not sufficient to simply embed the string in a Calabi-Yau manifold. One must also turn on an SU_3 subgroup of $E_8 \times E_8$ gauge group of the string. This is because the internal degree of freedom of the heterotic string consist of right-moving fermions, which feel the curvature of space-time, and left-moving coordinates which know nothing of the space-time curvature but are sensitive to background gauge fields. Unless there is a relation between the curvature of space and the curvature (field strength) of the gauge group there is a right left mismatch which gives rise to anomalies. Therefore one must identify the space-time curvature with the gauge curvature (embed the spin connection in the gauge group). One does this by turning on background gauge fields in an SU_3 subgroup of one of the E_8's, thereby breaking it down to E_6 (or possibly O_{10} or SU_5).

These Calabi-Yau compactifications, produce for each manifold K, a consistent string vacuum, for which the gauge group is no larger than $E_6 \times E_8$ and N=1 supersymmetry is preserved. Furthermore there now exist massless fermions which naturally form families of quarks and leptons. Recall that after Kaluza-Klein compactification the spectrum of massless chiral fermions is determined by the zero modes of the Dirac operator on the internal space. Since, for heterotic string, the gauge and spin connections are forced to be equal one can count the number of chiral fermions by geometrical arguments. The massless fermions fall into __27__'s of E_6. This is good, E_6 is an attractive grand unified model and each __27__ can incorporate one generation of quarks and leptons. The number of generations is equal to half the Euler character of the manifold (which counts the number of "handles" it has), and is normally quite large. If there exists a discrete symmetry group, Z, which acts freely on K, one can consider the smaller manifold K/Z, whose Euler character is reduced by the dimension of Z. By this trick, and after some searching, manifolds have been constructed with 1,2,3,4,... generations. It seems that to be realistic we must restrict attention to manifolds with three, or perhaps four, generations.

The compactification scheme also produces a natural mechanism for the breaking of E_6 down to the observed low energy gauge group. If K/Z is multiply connected one can allow flux of the unbroken E_6 (or of the E_8, for that matter) or to run through it, with no change in the vacuum energy. The net effect is that when we go around a hole in the manifold through which some flux runs we must perform a nontrivial gauge transformation on the charged degrees of freedom. These noncontractible Wilson loops act like Higgs bosons, breaking E_6 down to the largest subgroup that commutes with all of them. By this mechanism one can, without generating a cosmological constant, find vacua whose unbroken low energy gauge group is, say $Su_3 \times SU_2 \times U_1 \times$ (typically, an extra U_1 or two). Moreover there exists a natural reason for the existence of massless Higgs bosons which are weak isospin doublets (and could be responsible for the electroweak breaking at a Tev), without accompanying color triplets.

Many of the successful features of grand unified models, such as the prediction of the weak mixing angle, carry over, and many of the unsuccessful predictions, such as quark lepton mass ratios, do not.

Of course it is also necessary to break the remaining N=1 supersymmetry. For this purpose the extra E_8 gauge group might be useful. Below the compactification scale it yields a strong, confining gauge theory like QCD, but without light matter fields. In general this sector would be totally unobservable to us, consisting of very heavy glueballs, which would only interact with our sector with gravitational strength at low energies. However there could very well exist in this sector a gluino condensate which can serve as source for supersymmetry breaking.

Thus the heterotic string theory appears to contain, in a rather natural context, many of the ingredients necessary to produce the observed low energy physics. I do not mean to suggest that there are not many problems and unexplained mysteries. There exists the danger (common to many grand unified models, especially supersymmetric ones) of too rapid proton decay, there is no deep understanding of why the cosmological constant, so far zero, remains zero to all orders, and when supersymmetry is broken, at least by the mechanism discussed above, the theory tends to relax back to ten dimensional flat space. Nonetheless, the early successes are very reassuring and they give one the feeling that there are no insuperable obstacles to deriving all of low energy physics from the $E_8 \times E_8$ heterotic string theory.

6. OUTLOOK

I do not want to leave the impression that string theory has brought us close to the end of particle physics. Quite the opposite is the case. Not only are there many unsolved problems and deep mysteries that need to be understood before one can claim success, in addition we have only begun to probe the structure of these new theories. I prefer, therefore, to conclude with a list of open problems.

6.1 What is String Theory?

We do not fully understand the deep symmetry principles and symmetries that underly string theories. To date these theories have been constructed in a somewhat adhoc fashion and often the formulism has produced, for reasons that are not totally understood, structures that appear miraculous.

6.2 How Many String Theories Are There?

Do there exist more consistent theories than the known five? Do there exist fewer, in the sense that some of the ones we know already are perhaps different manifestations (different vacua?) of the same theory?

6.3 String Technology

This is not a question but a program of development of the techniques for performing calculations within string theory, including control of multiloop perturbation theory and the construction of manifestly covariant and supersymmetric methods of calculation. In addition one needs to develop, in a manifestly covariant approach, a useful second quantized formulation of the theory--string field theory.

6.4 What is the Nature of String Perturbation Theory

Does the perturbative expansion of the string theory converge? If not, when does it give a reliable asymptotic expansion? How can one go beyond perturbation theory?

6.5 String Phenomenology

Here there are many issues that remain to be resolved. They can all be included in the question--can one construct a totally realistic model which agrees with observation and why is it picked out?

6.6 What Picks the Correct Vacuum?

This is one of the greatest mysteries of the theory, which seems to have an enormous number of acceptable vacuum states. Why then don't we live in ten dimensional flat space? How does the value of the dilaton field get fixed and thereby the dilaton acquire a mass? Does the vanishing of the cosmological constant survive the physical mechanism that lifts the vacuum degeneracy?

6.7 What Is the Nature of High Energy Physics?

By this I mean what does physics look like at energies well above the Planck mass scale? This is a question that is addressable, in principle, for the first time and might be of more than academic interest for cosmology. Does the string undergo a transition to a new phase at high temperatures and densities? Can one avoid in string theory the ubiquitous singularities that plague ordinary general relativity?

6.8 Is There a Measurable, Qualitatively Distinctive, Prediction of String Theory

String theories can make many "postdictions" (such as the calculation of mass ratios of quarks and leptons, Higgs masses, gauge couplings, etc.). They can also make many new predictions (such as the masses of the various supersymmetric partners). These would be sufficient to establish the validity of the theory, however one could imagine conventional field theories coming up with similar pre or post dictions. It would be nice to predict a phenomenon which might be accessible at observable energies and is uniquely characteristic of string theory.

REFERENCES

I have made no attempt to give detailed references to the papers in this rapidly growing field. An incomplete set of references to recent work is given below.

Reviews:
Unified String Theories (Proceedings of the String Workshop at Santa Barbara, World Scientific, 1986).
Superstrings (Reprint Volume, edited by J. Schwarz, World Scientific, 1985).

Superstring Anomalies
M.B. Green and J.H. Schwarz, Phys. Lett. 149B (1984) 117;
L. Alvarez-Gaume and E. Witten, Nucl. Phys. B234 (1983) 269.

Heterotic String
D.J. Gross, J.A. Harvey, E. Martinec and R. Rohm, Phys. Rev. Lett. 54 (1985) 502, Nucl. Phys. B256 (1985) 253, and to be published.

String Phenomenology
P. Candelas, G. Horowitz, A. Strominger and E. Witten, Nucl. Phys. B258 (1985) 46; E. Witten, Nucl. Phys. B258 (1985) 75.

LOW ENERGY SUPERSTRING THEORY

Burt A. OVRUT[+]

Department of Physics, The Rockefeller University,
New York, New York 10021

The spontaneous breakdown of the E_6 gauge group of low energy superstring theory is discussed, and a method given for determining the residual gauge group. We show that certain chiral supermultiplets can be naturally light and that, as a consequence, the residual gauge group $SU(3)_C \times SU(2)_L \times SU(2)_R \times U(1)_Y \times U(1)$ is singled out.

1. INTRODUCTION

The recent proof [1] that superstring theory is anomaly free for gauge groups $E_8 \times E_8$ and $SO(32)$ has led to enormous interest in that subject [2]. In particular, it was shown in Ref. [3] that the superstring vacuum state can compactify in such a way as to produce a possibly realistic theory of particle physics in four dimensions at low energy. In this talk we will discuss one aspect of this low energy theory, the spontaneous breakdown of the E_6 gauge group [4]. We begin by reviewing some aspects of the work in Ref. [3]. These authors showed that the zero mass modes of the superstring form a modified version of the Chapline-Manton N=1 superstring theory in ten space-time dimensions. Furthermore, the vacuum state of this theory can compactify into $M_4 \times K$, where M_4 is Minkowski space and K is a six dimensional compact manifold. It was shown that an unbroken, local N=1 supersymmetry remains in M_4 provided that K is Ricci flat and has $SU(3)$ holonomy group. A large number of such manifolds, called Calabi-Yau manifolds, exist. When K is simply connected the gauge group is reduced from $E_8 \times E_8$ to $E_6 \times E_8$ (the group $SO(32)$ does not lead to chiral fermions in four dimensions so we will not consider it). The E_8 factor group corresponds to a "shadow" world involving an adjoint <u>248</u> vector superfield of gauge fields and gauginos. We will

---

[+] On leave of absence from the University of Pennsylvania, Philadelphia, PA 19104, USA

Report Number DOE/ER/40033B-99 RU85/B/133.
Work supported in part by the Department of Energy under contract number DE-AC02-81ER40033B.000.

ignore this shadow world, and E_8, in this talk. The four dimensional superfields (zero modes of K) transforming non-trivially under E_6 are A) an adjoint $\underline{78}$ of gauge fields and gauginos, and B) $(n_L X)\underline{27}$, $(n_R X)\overline{\underline{27}}$ of left chiral fermions and scalar partners.

Defining $N=n_L-n_R$, then N is the number of light families. It can be shown that $N=(1/2)\chi(K)$ where χ is the Euler characteristic of K. For the simply connected Calabi-Yau manifolds $N\geq 36$. For example, there is such a manifold K_0 with $\chi=200$ and, hence, $N=100$ families. This apparent disaster can be averted by considering multiply connected spaces. Let K be a simply connected Calabi-Yau manifold and H be a finite group which acts freely (no fixed points) on K. Then $K'=K/H$ is a multiply connected Calabi-Yau manifold with order $\pi_1(K')$= order H, $\chi(K')=\chi(K)/$order H, and $N'=N/$order H. Hence, we expect multiply connected spaces to give rise to fewer light families. As an example consider $K=K_0$. It can be shown that $H=Z_5 \times Z_5$ (order H=25) can act freely on K_0. It follows that $K'=K_0/Z_5 \times Z_5$ leads to $N'=4$ families. At the time of writing this space is phenomenologically the most successful and, henceforth, we discuss only this manifold. The zero mode supermultiplets of $K_0/Z_5 \times Z_5$ are as above with $n_L=5$ and $n_R=1$. Although this method of reducing the number of families may appear ad hoc there is an important ramification. This is that E_6 may be spontaneously broken on $K_0/Z_5 \times Z_5$ [5]. Vacuum state E_6 gauge fields, A_m, on any space K satisfy the tree level equation $F_{mn}=0$. If K is simply connected then A_m can be made to vanish everywhere on K by a gauge transformation. Since vacuum state $A_m=0$ then E_6 remains unbroken. However, if K is multiply connected then A_m cannot be made to vanish everywhere on K by a gauge transformation (there is a topological obstruction). Hence, $A_m \neq 0$ everywhere and E_6 can be spontaneously broken since A_m transforms as a $\underline{78}$ of E_6. This can be restated mathematically by considering the path ordered Wilson loop

$$U=e^{-iT^a \int_\gamma A_m^a dx^m} \qquad (1)$$

where T^a are the generators of E_6, and γ is some path on K. Note that U is an element of E_6. If K is simply connected then $U=1$. However, if K is multiply connected then $U \neq 1$ (necessarily) and, as we will see, E_6 breaking can occur. Consider $K'=K/H$ and let $\mathcal{H}=\{U\}$. Then, as an abstract group $\mathcal{H}\subseteq H$. Furthermore, $\mathcal{H} \subset E_6$. As an example let $H=Z_5 \times Z_5$ and $K'=K_0/Z_5 \times Z_5$. Then, even though K' is multiply connected A_m may vanish. In this case $\mathcal{H}=\{1\}$. However, since K' is multiply connected A_m need not vanish and $\mathcal{H}=Z_5$ (either of them) or $Z_5 \times Z_5$. Note that there are many possible embeddings of Z_5 and $Z_5 \times Z_5$ into E_6. Now fix some vacuum gauge field $A_m{}^a$ and let $\mathcal{H}$ be the corresponding discrete subgroup. If $\mathcal{K}$ is the subgroup of E_6 that commutes with $\mathcal{H}$ then E_6 is spontaneously broken to $\mathcal{K}$. This is so important that we repeat it. If

$$[\mathcal{K}, \mathcal{H}] = 0 \qquad (2)$$

then

$$E_6 \to \mathcal{K} \qquad (3)$$

We turn now to the work discussed in Ref. [4]. Note that there are no Higgs scalars involved in the spontaneous breakdown of E_6. There is, however, an element of the adjoint $\underline{78}$ of E_6, namely

$$T^a \int_\gamma A_m^a \, dx^m \qquad (4)$$

which can be considered as an effective "Higgs" VEV. We want to derive all possible E_6 symmetry breaking patterns. First, consider some general results.

1) The two Wilson loop generators of $Z_5 \times Z_5$ commute. Therefore, the associated effective VEV's can be simultaneously diagonalized. It follows that these VEV's can be extended to a complete basis of the E_6 Cartan subalgebra. Since all elements of this basis commute then

$$\text{rank } \mathcal{K} = \text{rank } E_6 = 6 \qquad (5)$$

2) Let $\widetilde{\mathcal{K}} = SU(3)_C \times SU(2)_L \times U(1)_Y$ be the gauge group of the standard model. Then for correct phenomenology $\widetilde{\mathcal{K}} \subset \mathcal{K}$. Combining 1) and 2) implies

$$SU(3)_C \times SU(2)_L \times U(1)_Y \times U(1)^2 \subseteq \mathcal{K} \qquad (6)$$

3) If $U \in \mathcal{H}$ then $U^5 = 1$ since $(\gamma)^5$ is a contractible loop.
4) If $\mathcal{H} = \{1\}$ then E_6 is unbroken.

There are two methods for determining the possible symmetry breaking patterns of E_6. We will discuss only one of these methods in this talk, referring the reader to the first paper of Ref. [4] for the second method. Decomposition into Maximal Subgroups:

There are three maximal subgroups of E_6 with rank six. They are $SU(3)_C \times SU(3)_L \times SU(3)$, $SU(2) \times SU(6)$, and $SO(10) \times U(1)$. Assume $\mathcal{H} = Z_5 \times Z_5$ and consider maximal subgroup $SU(3)_C \times SU(3)_L \times SU(3)$. The first possible form for the

U's is

$$\begin{pmatrix} 1 & & \\ & 1 & \\ & & 1 \end{pmatrix} \times \begin{pmatrix} \alpha^j & & \\ & \alpha^j & \\ \hline & & \alpha^{-2j} \end{pmatrix} \times \begin{pmatrix} \beta^k & & \\ & \beta^k & \\ \hline & & \beta^{-2k} \end{pmatrix} \qquad (7)$$

where $j,k = 0,4$ and $\alpha^5 = \beta^5 = 1$. Note there are twenty-five distinct U's and each one satisfies $U^5 = 1$. It is clear from these matrices that E_6 is spontaneously broken to

$$\mathcal{H} = SU(3)_C \times SU(2)_L \times U(1)_Y \times SU(2) \times U(1) \qquad (8)$$

The second, and last, possible form for the U's is

$$\begin{pmatrix} 1 & & \\ & 1 & \\ & & 1 \end{pmatrix} \times \begin{pmatrix} \alpha^j & & \\ & \alpha^j & \\ \hline & & \alpha^{-2j} \end{pmatrix} \times \begin{pmatrix} \beta^k & & \\ & \beta^{4k} & \\ \hline & & 1 \end{pmatrix} \qquad (9)$$

where $j,k = 0,4$ and $\alpha^5 = \beta^5 = 1$. Again there are twenty-five distinct U's and $U^5 = 1$. This time, however, E_6 is spontaneously broken to

$$\mathcal{H} = SU(3)_C \times SU(2)_L \times U(1)_Y \times U(1) \times U(1) \qquad (10)$$

The $\mathcal{H}$'s given in (7) and (9) are the only embeddings of $Z_5 \times Z_5$ into $SU(3)_C \times SU(3)_L \times SU(3)$ which preserve $\tilde{\mathcal{H}}$. Now consider maximal subgroup $SU(2) \times SU(6)$. One possible form for the two generators (A) and (B) of $Z_5 \times Z_5$ is

$$(A): \quad \begin{pmatrix} e^{i\alpha} & \\ \hline & e^{-i\alpha} \end{pmatrix} \times \begin{pmatrix} 1 & & & & & \\ & 1 & & & & \\ & & 1 & & & \\ & & & 1 & & \\ & & & & 1 & \\ & & & & & 1 \end{pmatrix} \qquad (11)$$

$$ \quad U(1) SU(6)$$

$$(B): \quad \begin{pmatrix} 1 & \\ & 1 \end{pmatrix} \times \begin{pmatrix} \epsilon^{i\beta} & & & & & \\ & \epsilon^{-i\beta} & & & & \\ & & 1 & & & \\ & & & 1 & & \\ & & & & 1 & \\ & & & & & 1 \end{pmatrix} \qquad (12)$$

$$ \quad SU(2) SU(4) \times (U(1))^2$$

where $\alpha^5 = \beta^5 = 1$, and the subgroup of $SU(2) \times SU(6)$ that commutes with each matrix is listed underneath. Note that the second matrix in both (A) and (B) is in a $\underline{6}$ (not a $\underline{\bar{6}}$) of $SU(6)$. Since $\mathcal{H}$ must commute with both generators (and, hence, all elements) of $Z_5 \times Z_5$, it follows that E_6 is spontaneously broken to

$$\mathcal{H} = SU(4) \times (U(1))^3 \qquad (13)$$

Let us check this result by explicitly constructing the 27x27 dimensional matrices for (A) and (B). We have an important ulterior motive for doing this. The construction is carried out by using the fact that under $SU(2) \times SU(6)$ a $\underline{27}$ of E_6 decomposes as

$$\underline{27} = (\underline{2}, \underline{\bar{6}}) + (\underline{1}, \underline{15}) \qquad (14)$$

where $\underline{15} = (\underline{6} \times \underline{6})_A$. We find that

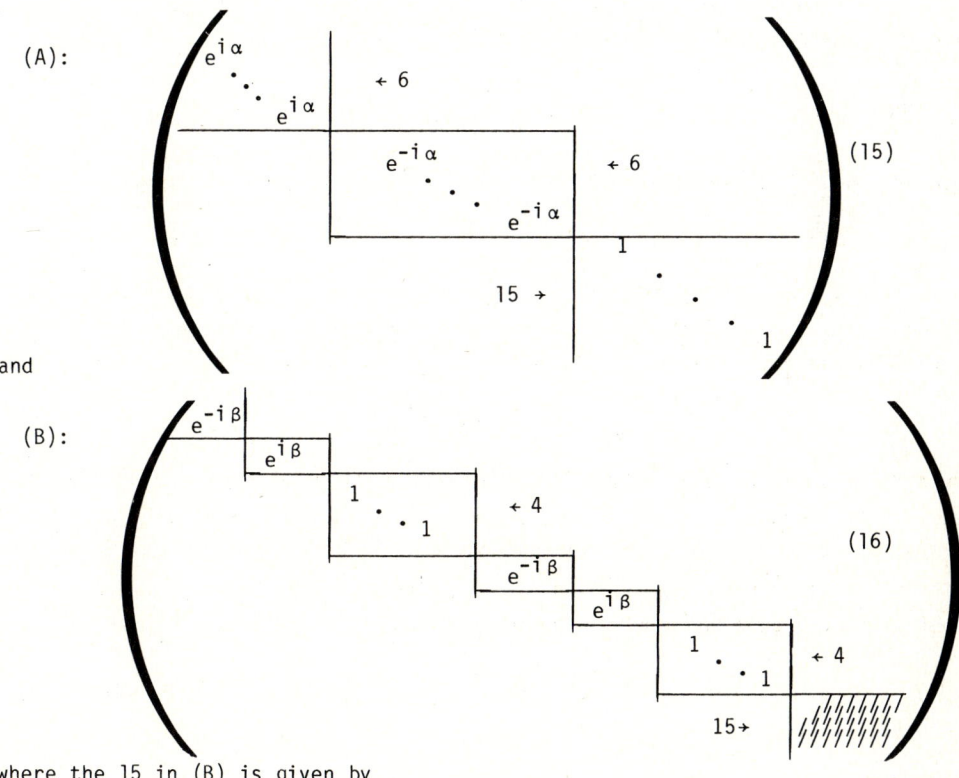

where the 15 in (B) is given by

$$15 = \begin{pmatrix} e^{i\beta} & & & & & & \\ & \ddots & & & & & \\ & & e^{i\beta} & \leftarrow 4 & & & \\ & & & e^{-i\beta} & & & \\ & & & & \ddots & & \\ & & & & & e^{-i\beta} & \leftarrow 4 \\ & & & 7 \rightarrow & & & 1 \\ & & & & & & & \ddots \\ & & & & & & & & 1 \end{pmatrix} \quad (17)$$

Since $\mathcal{H}$ must commute with both (A) and (B) it follows that $\mathcal{H}$ must commute with an element of the $\underline{27}$ of E_6 of the form

$$\underline{27} = 1 + 1 + 4 + 1 + 1 + 4 + 4 + 4 + 7 \quad (18)$$

Decomposing a $\underline{27}$ of E_6 under $SU(2) \times SU(6)$ using (14) and then further decomposing under $SU(4) \subset SU(6)$ we find that

$$\underline{27} = \underline{1} + \underline{1} + \underline{\bar{4}} + \underline{1} + \underline{1} + \underline{\bar{4}} + \underline{4} + \underline{4} + (\underline{6} + \underline{1}) \quad (19)$$

Comparing eqn. (19) with eqn. (18) we conclude that indeed

$$E_6 \rightarrow SU(4) \times (U(1))^3 \quad (20)$$

Why have we done all this? Note that the elements of both (A) and (B) (and, hence, all elements of $Z_5 \times Z_5$) associated with the $SU(4)$ representation $\underline{6+1}$ are unity. Hence, the elements of the effective Higgs VEV, $<\underline{78}>$, associated with the $\underline{6+1}$ representation vanish! Now the fifth $\underline{27}$ and $\underline{\overline{27}}$ zero mode chiral superfields get an effective mass given by

$$\underline{27} \; <\underline{78}> \; \underline{\overline{27}} \quad (21)$$

In this case all components of $\underline{27}$ and $\underline{\overline{27}}$ are massive except for

$$\begin{aligned} \underline{6} + \underline{1} \; \varepsilon \; \underline{27} \\ \underline{6} + \underline{1} \; \varepsilon \; \underline{\overline{27}} \end{aligned} \quad (22)$$

which are naturally light (zero mass prior to spontaneously breaking supersymmetry). We reiterate that the $\underline{6} + \underline{1}$ multiplets are light due to the zeros in $<\underline{78}>$, which in turn are due to the 1's in the U's. We emphasize that these 1's arise naturally from the "quantization" condition $U^5=1$ and are not due to any fine tuning. Phenomenologically this breaking pattern is unacceptable. We would like to find a breaking pattern containing $SU(2)_L$ in which both H and $\overline{H}$ doublets in the fifth 27 are naturally light and their associated $SU(3)_C$ triplets are naturally heavy. In this case, and only this case, can the gauge hierarchy be naturally implemented with the appropriate suppression of nucleon decay. Is there such a breaking pattern? A complete classification of all breaking patterns of E_6 yields the following results. There exists a unique breaking pattern of E_6 for which a pair of H and $\overline{H}$ doublet Higgs superfields in the fifth 27 are naturally light and for which the associated color triplet Higgs superfields are naturally heavy. This breaking pattern is

$$E_6 \rightarrow SU(3)_C \times SU(2)_L \times SU(2)_R \times U(1)_Y \times U(1) \tag{23}$$

We want to emphasize that the imposition of N=1 local supersymmetry in four dimensions is not sufficient to solve the gauge hierarchy problem. Once N=1 supersymmetry is imposed it is still necessary to have H and $\overline{H}$ doublet Higgs superfields which are naturally (no fine tuning) zero mass prior to supersymmetry breaking. As we have seen this requirement puts a severe limitation on the allowed E_6 breaking patterns. It is not impossible, therefore, that $SU(2)_R$ physics will be observed experimentally at energies not far above the electroweak scale.

REFERENCES

1. M. Green and J. Schwartz, Phys. Lett. 149B (1984) 117; Phys. Lett. 151B (1985) 21.

2. D. Gross, J. Harvey, E. Martinec, and R. Rohm, Phys. Rev. Lett. 54 (1985) 502; Heterotic string theory I and II, Princeton Preprints (1985); E. Witten, Phys. Lett. 149B (1984) 351; Phys. Lett. 155B (1985) 151; D. Frampton, H. Van Dam, and K. Yamamoto, Phys. Rev. Lett. 54 (1985) 114; M. Dine, R. Rohm, N. Seiberg, and E. Witten, Phys. Lett. 156B (1985) 55; M. Dine, V. Kaplunovski, M. Mangano, C. Nappi, and N. Seiberg, Superstring Model Building, Princeton Preprint (1985); W. Lang, J. Louis, and B. Ovrut, 16/16 Supergravity: The Low Energy Limit of the Superstring, Univ. of Penn. Preprint (1985); J. Breit, B. Ovrut, and G. Segre, One Loop Effective Lagrangian of the Superstring, univ. of Penn. Preprint (1985); J. Derendinger, L. Ibanez, and H. Nilles, Phys. Lett. 155B (1985) 65; E. Cohn, J. Ellis, C. Gomez, and D. Nanopoulos, CERN Preprint, TH 4159 (1985); P. Binetruy and M. Gaillard, LBL Preprint (1985); T. Hubsch, H. Nishino, and J. Pati, Do Superstring lead to Quarks or to Preons?, Trieste Preprint IC 85/66 (1985).

3. P. Candelas, G. Horowitz, A. Strominger, and E. Witten, Vacuum Configurations for Superstring, Santa Barbara Preprint (1984).

4. J. Breit, B. Ovrut, and G. Segre, E_6 Symmetry Breaking in the Superstring Theory, Univ. of Penn. Preprint (1985); E. Witten, Symmetry Breaking Patterns in Superstring Models, Princeton Preprint (1985); A. Sen, Fermilab Preprint (1985).

5. Y. Hosotani, Phys. Lett. <u>126B</u> (1983) 309; Phys. Lett. <u>129B</u> (1983) 193.

RENORMALIZATION GROUP ANALYSIS OF SUPERSTRING MODELS

Chiara R. NAPPI[*]

Joseph Henry Laboratories, Princeton University, Princeton, NJ 08544, USA

I review the results of the one loop renormalization group (R-G) analysis of the various models which can come from string compactification. Simple extensions of the standard model like $SU(3)_c \times SU(2)_L \times U(1)_Y \times (U(1)^2$ or $SU(2))$, and some other model as well, might give an acceptable phenomenology both for three and four generations.

I. Introduction

Recently, anomaly free d = 10 superstring theories [1] have been discovered with gauge groups $SO(32)$ and $E_8 \times E_8$ [2]. Moreover a starting point for phenomenology has been proposed in [3]. Phenomenology requires that the string theory in d = 10 dimensions be compactified with a vacuum state of the form $M_4 \times K$, where M_4 is a four dimensional Minkowski space and K is some compact six dimensional manifold. The request of unbroken N = 1 supersymmetry in four dimensions (a desirable feature if one hopes that supersymmetry can solve the gauge hierarchy problem) can be satisfied by imposing K to be Ricci flat and Kähler, a so-called Calabi-Yau space [3]. This compactification scheme has several desirable features. First, it breaks $E_8 \times E_8$ down to $E_6 \times E_8$. E_6 is a natural candidate for grandunification [4]. The unbroken E_8, instead, has been postulated to be the "hidden sector" introduced in some low energy supergravity models. Secondly, the fermions come automatically in a chiral structure since they belong to the 27 of E_6, a representation which contains the $10 + \bar{5}$ of $SU(5)$ (or the 16 of $O(10)$). Finally, the number of generations (defined as the number of families (or 27) minus the number of antifamilies (or $\overline{27}$) of opposite chirality) is predicted in terms of the Euler characteristic of the compact manifold K. In most of the simply connected Calabi-Yau spaces built so far, the number of generations turns out to be very

---

[*]Supported in part by NSF grant PHY-80-19754

large. However, if K has a discrete symmetry group $\bar{G}$ which acts freely, then one could consider the manifold $K/\bar{G}$. On this manifold the number of generations is greatly reduced, since the Euler characteristic of $K/\bar{G}$ is the ratio between the Euler characteristic of K and the number of elements of $\bar{G}$. Moreover, on $K/\bar{G}$ the unbroken gauge group is not E_6, but the subgroup M of it whose generators commute with G, a discrete subgroup of E_6 which is the image a homomorphism of $\bar{G}$ into E_6. This way, by going from K to $\tilde{K} = K/\bar{G}$ one gets a reasonably low number of generations [3, 6] and an acceptable phenomenological low energy gauge group [5, 7, 8]. Of course one expects that eventually it will be understood how to pick uniquely from string theory the proper compact manifold $\tilde{K}$ and the homomorphism of its fundamental group into E_6, so to have no ambiguities left in the number of generations, gauge group and Yukawa cou plings of the theory (which also can be determined topologically [6]). In the meantime the approach has been a pragmatic one: try to construct a Calabi-Yau space which would give rise to acceptable phenomenology. Examples have been given with 1, 2, 3 and 4 generations. One needs at least three generations to accomodate all known particles, and too many families have phenomenological problems, as we will discuss shortly. I will describe work done in collaboration with M. Dine, V. Kaplunovsky, M. Mangano and N. Seiberg. In [9] we asked the obvious question whether the various models either with three or four generations could possibly work. We imposed phenomenological constraints coming from proton life-time, neutrino masses, $\sin^2\theta_w$ and the requirement of perturbative unification. Here I will basically concentrate on the R-G analysis. As pointed out in [5], the Georgi-Quinn-Weinberg calculation is valid here since the gauge couplings do obey the standard E_6 relations after E_6 breaking. This is a non trivial statement since the Yukawa couplings, instead, do not need to be E_6 invariant, but only invariant under the unbroken subgroup M [5] (again, useful information in model building). Finally, the last input we need for our analysis: it can happen that some of the particles of a $\overline{27}$ (and necessarily the corresponding particles in a 27, since the number of generations is fixed) stay light after compactification, so that at the end one is left with a given number of generations plus extra particles [5, 7, 8]. Actually, for the reasons explained in [5] a simple case where one can analyze the situation in detail is that in which there is one antifamily only. As in [9], I will restrict myself to this case, since I will need to know exactly the particle content of the theory in order to perform the renormalization group analysis.

II. Gauge Groups and Particle Content.

As already said, if we know the correct vacuum state $M_4 \times \tilde{K}$ and the homomorphism of the fundamental group of $\tilde{K}$ into E_6, the theory would be fully determined, without any adjustable parameter left. Since however we do not know them, our approach will be to investigate all possible groups M which can be derived from E_6 by the method described above and that contain the standard model. There are 27 of them, and they are completely listed in [9]. A model with a gauge group $M \subset E_6$ can exist only if there is a discrete subgroup G of E_6 which commutes with the generators of M but not with any other generator of E_6. So for each of these 27 groups one needs to show that there is a proper subgroup of E_6 which does the job. The first thing to notice, as pointed out in [5], is that one cannot break E_6 directly down to the standard model $SU(3)_c \times SU(2)_L \times U(1)_Y$. The mechanism that breaks E_6 will always force on us at least an extra $U(1)$ as part of the unbroken subgroup. So there are extra weak currents that might survive down to low energy.

The 27 subgroups of E_6 obtained as described above can be classified in various classes:

1) Minimal extensions of the standard model, i.e. groups of the type

$$SU(3)_c \times SU(3)_L \times U(1)_Y \times (U(1) \text{ or } U(1)^2 \text{ or } SU(2)). \qquad (1)$$

2) Models with extended color, i.e. with color group larger than $SU(3)$. For instance:

$$SU(4)_c \times SU(2)_L \times U(1). \qquad (2)$$

3) Models with extended flavor, i.e. where the flavor group is $SU(3)_L$.
4) Early unified models, i.e. models like $SU(5)$ or $O(10)$, where color and flavor are unified in the same group.

Coming now to the particle content of the theory, we have said that each family comes in a 27 representation of E_6. If one decompose E_6 under its maximal subgroup $SU(3)_c \times SU(3)_L \times SU(3)_R$, one gets

$$27 \to (3, 1, \bar{3}) + (\bar{3}, 3, 1) + (1, \bar{3}, 3).$$

The known and unknown particles can be fit in as follows [4]

$$\begin{pmatrix} u \\ d \\ g \end{pmatrix} \quad \begin{pmatrix} \bar{u} & \bar{d} & \bar{g} \end{pmatrix} \quad \begin{pmatrix} H_1^u & H_1^d & e^- \\ H_2^u & H_2^d & \nu \\ e^+ & \bar{\nu} & S_2 \end{pmatrix}$$

One notices that there are exactly two singlets under $SU(3)_C \times SU(2)_L \times U(1)_Y$, one with the proper quantum numbers to represent the right-handed neutrino $\bar{\nu}$ (from now on we will call it S_1), and the other particle S_2 which is singlet under $O(10)$. The other particles above which are not contained in the $10 + \bar{5}$ of $SU(5)$ (and therefore eventually should disappear from the game since they are not observed) are the additional charge $-\frac{1}{3}$ g quarks, and the particles $H_{1,2}^u$ $H_{1,2}^d$ which have the quantum numbers of Weinberg-Salam Higgses. In fact, eventually we would like to use them as Weinberg-Salam Higgses to break $SU(2)_L \times U(1)_Y$ down to $U(1)_{EM}$, and give masses to up and down quarks, and charged leptons.

III. Renormalization Group Analysis

The one loop renormalization group equation is

$$\frac{1}{\alpha_N(\mu)} = \frac{1}{\alpha_{GUT}} + \frac{1}{6\pi} b_N \ln \frac{M_{GUT}}{\mu} \tag{3}$$

where

$$\frac{1}{\alpha_N(\mu)} = \frac{4\pi}{g_N^2(\mu)}$$

and g_N is the gauge coupling constant associated with the group $SU(N)$. The beta function b_N of $SU(N)$ is given by

$$b_N = 9(F-N) \tag{4}$$

where F is the number of families. (In this preliminary analysis we are neglecting any extra light particles.) This has to be compared with the beta function [10]

$$\tilde{b}_N = 6F - 9N \tag{5}$$

that one would get if the particle content of the theory were that of $10 + \bar{5}$ of $SU(5)$ (or 16 of $O(10)$). From (4) it follows that in the case of four families $SU(3)_C$ would not be asymptotically free. But, even worse, one can easily check that both $SU(2)$ and $SU(3)$ have Landau poles. I.e., the couplings $\alpha_{2,3}$ diverge well below any reasonable unification mass. In fact if we start at $\mu = M_W$ with the values of the couplings usually adopted in model building, namely

$$\frac{1}{\alpha_{EM}(M_W)} = 128 \qquad \frac{1}{\alpha_{QCD}(M_W)} = 9 \qquad \frac{1}{\alpha_2(M_W)} = 26$$

and we assume F = 4, we get Landau poles for SU(3) and SU(2) at 10^{13} Gev and 10^{15} Gev respectively. There would be no problem, instead, if the number of generations were three or if the generation content was the $10 + \overline{5}$ of SU(5) or the 16 of O(10). In the latter case we would not have Landau poles even with four generations. However, if the family content is that of a 27 and we have four generations, the only way to avoid Landau poles is to introduce an intermediate scale M_I [11]. As M_I some or all of the unwanted particles of the 27 become heavy, and disappear from the game, so that the beta function flattens down. This would allow us to start with the physical low energy values of the gauge couplings and still get perturbative values at the unification scale. For this to happen, we need $M_I \geq 10^{10}$ Gev. Higher is M_I, less space to run is left to the "bad" beta function (4), so the coupling constant at M_{GUT} does not have the chance to get too big.

To give masses at M_I to the extra particles like g quarks and unwanted H's, we need Higgs scalars whose VEVs do not break the standard model. Such are the two neutral singlets S_1 and S_2. If they get a VEV at the intermediate scale, their VEV will break the starting gauge group M down to a subgroup containing the standard model, hopefully to the standard model itself.

The question that arises is therefore whether this intermediate mass scale can be generated, namely whether we can manage to give expectation values to S_1 or S_2 at a scale $M_I \simeq 10^{10} \div 10^{11}$ Gev. In [9] we argue that this can happen quite naturally. The assumptions are that the "effective" scale of supersymmetry breaking is around M_W, and that $S_{1,2}$ get negative mass^2 of order M_W^2. Moreover one needs to assume that there is no quartic term in the potential (in this case the only scale we could generate is $\langle S_{1,2}\rangle \sim M_W$), but the next order term is instead $O(M_{GUT}^{-2}) S_{1,2}^6$. Then one does generate a scale $M_I \sim \sqrt{M_W M_{GUT}} \sim 10^{10}$ Gev. In order for this scheme to work one needs however D-flat directions of the potential, namely

$$\langle S_{1,2}\rangle = \langle \overline{S}_{1,2}\rangle^\dagger \qquad (6)$$

that is, one needs light Higgses $\overline{S}_{1,2}$ from the antifamilies. As already said, in the case of Calabi-Yau spaces with 1 antifamily only one can easily analyze which elements of the antifamily stay light, and therefore decide whether $S_{1,2}$ are really available to generate the intermediate scale M_I. This question can be answered for each of the groups M in section II, once one knows the discrete subgroup of E_6 which generates it. For instance, it turns out that in the case of groups M of rank five no neutral singlets $S_{1,2}$ can stay light, so no intermediate scale can be generated. Actually, even if neutral singlets from the antifamilies are available to generate M_I, one can show that it is impossible to break any of the groups M in section II to a subgroup of rank less than five. Hence it is not possible to get rid of all extra weak currents at the intermediate scale, but some will survive down to the weak scale. However, it appears that extra weak currents mediated by gauge bosons with masses of a few Tev do not pose any phenomenological problem. A word of warning is due here: this prediction of extra weak currents applies only to the case of Calabi-Yau with one antifamily only. If there is more than one antifamily, one might get more $S_{1,2}$ Higgses, and these could break all extra interactions, leaving us with the pure standard model. But in the case of one antifamily only, we will end up either with $SU(3)_C \times SU(2)_L \times U(1) \times U(1)$ or $SU(3)_C \times SU(2)_L \times SU(2)_R \times U(1)$, at best. For most of the early unified models the VEV's of $S_{1,2}$ can only manage to break the group down to $SU(5)$. For obvious reasons these models need to be discarded. In [9] we also discarded some other models M from the list because they had problems connected with proton decay and neutrino masses. I will not repeat any of these arguments here, but just report the results of the one loop R-G analysis of the surviving models.

IV. Results

First of all, in the case of F = 3, we do not need an intermediate mass scale, since we have no problems with Landau poles. We can therefore contemplate a Grand Desert scenario. The minimal extensions of the standard model in (1) are the only models that work. They give exactly the same result as supersymmetric SU(5) [10] (the calculation in fact is exactly the same), namely $\sin^2\theta_W = 0.206$, $M_{GUT} = 2.10^{17}$ Gev. Moreover $\alpha_{GUT} = 0.11$. Instead left-right symmetric models do not work, since they give $\sin^2\theta > 0.26$ and $M_{GUT} > M_{Planck}$. Similarly, also the extended flavor model have to be excluded because they turn out to have $M_{GUT} > M_{Plack}$ and $\sin^2\theta_W < 0.15$. It is interesting to notice that the extended color model (2) almost works, renormalization-groupwise, both for F = 3 and F = 4, without intermediate scale. In fact it predicts $\sin^2\theta_W = 0.25$ and $M_{GUT} \simeq 6.10^{14}$ Gev, which could be considered borderline. Even for F = 4 α_{GUT} is reasonably small (~ 0.13) because the beta function in (4) is zero for N = 4. However this model seems to have problems with proton decay [9], and probably should be discarded.

Moving now to the case with F = 4, let's assume that we have an intermediate scale and that $S_{1,2}$ couple to g quarks and to the H's to give them a mass $\sim M_I$. In general, we assume that only one pair of H doublets does not couple to $S_{1,2}$ and stays light to actually play the role of W-S Higgses. Once we know the particle content of the theory above and below M_I, and the low energy subgroup, we can write the renormalization group equations for M_{GUT}, $\sin^2\theta_W$ and α_{GUT}. The renormalization group equations for M_{GUT} and $\sin^2\theta_W$ are independent of the number of families, since they contain only differences of beta functions. So the results for M_{GUT} and $\sin^2\theta_W$ are the same for three and four families. Instead, α_{GUT} is small in the case of three families (generically $\alpha_{GUT} \sim 0.05$) and tends to be higher for four families, the exact value depending on the value of M_I. We will consider a model acceptable if the following bounds are satisfied.

$$0 < \alpha_{GUT} < \frac{1}{4} \quad 0.19 < \sin^2\theta_W < 0.24 \quad 10^{15} \text{Gev} < M_{GUT} < 10^{19} \text{Gev}.$$

For each model we play around with different M_I or with more than one pair of Higgs doublets (in case the most economical solution in terms of Higgses does not work) to see if we can satisfy the above bounds.

Here is a list of models which appear to work with an intermediate scale:
1) In the class of minimal extension of the standard model we have

$$SU(3)_C \times SU(2)_L \times U(1)_Y \times U(1) \times U(1)$$
$$SU(3)_C \times SU(2)_L \times U(1)_Y \times U(1) \times SU(2). \tag{7}$$

Both models, if we assume an intermediate mass scale $M_I \sim 10^{10}$ Gev, have $M_{GUT} \sim 10^{17}$ Gev and $\sin^2\theta_W \sim 0.22$. The coupling at M_{GUT} is $\alpha_{GUT} \sim 0.17$. There is also a left-right symmetric model

$$SU(3)_C \times SU(2)_L \times SU(2)_R \times U(1)^2 \tag{8}$$

which looks good. It gives $M_{GUT} \sim 10^{18}$ Gev, $\sin^2\theta_W \sim 0.23$ and $\alpha_{GUT} \sim 0.25$, a little bit on the high side (we have taken $M_I \sim 10^{11}$ Gev).

2) In the class of enlarged flavor models, a good one appears to be the following

$$SU(3)_C \times SU(3)_L \times U(1) \times U(1) \qquad (9)$$

which gives (for $M_I \sim 10^{10}$ Gev and two pairs of Higgs doublets H) $M_{GUT} \sim 10^{17}$ Gev, $\sin^2\theta_W \sim 0.22$ and $\alpha_{GUT} \sim 0.18$.

3) and finally, from the extended color group,

$$SU(4)_C \times SU(2)_L \times U(1) \times U(1) \qquad (10)$$

which has $M_{GUT} \sim 10^{16}$ Gev, $\sin^2\theta_W \sim 0.24$ and $\alpha_{GUT} \sim 0.14$.

All five models have low energy gauge group $SU(3)_C \times SU(2)_L \times U(1) \times U(1)$. The values of M_{GUT} and $\sin^2\theta_W$ are the same for F = 3, 4, but the quoted values of α_{GUT} are those for F = 4. For F = 3, $\alpha_{GUT} \sim 0.05$, as already said.

Actually there is a last remark I want to make. It seems that M_{GUT} and M_{Planck} must be approximately equal to each other in any unified string theory with realistic couplings [12]. Therefore models with extended colors, which tend to have a low M_{GUT}, are definitely excluded. Minimal extensions of the standard model (1) look marginal since $M_{GUT} \sim 10^{17} \div 10^{18}$ Gev, but are probably still okay. On the basis of this criterion the favorite model turns out to be (8) which needs an intermediate mass scale in order to work. But if an intermediate mass scale could not be generated, the models (1), i.e. the minimal extensions of the standard model, are the only surviving candidate, and one would need to have three generations only.

V. Conclusions

I have reported here on the results of the renormalization group analysis of the models obtained by compactifying string theory. I have restricted myself to superstring theories compactified on Calabi-Yau manifolds, the only ones so far which appear to satisfy the equations of motion of superstring theory. Our analysis in [9], considered also the issues of proton life-time and neutrino masses, but was still far from complete. For instance, the question of how to generate the Weinberg-Salam scale (some work in this direction has been done in [13]) and possible phenomenological problems connected with flavor changing neutral currents should be addressed.

Moreover, E_6 is not the only 4-dimensional gauge group that one can get from string theory. An interestig candidate is O(10) [3]. In this case supersymmetry is broken at the compactification scale. Some of the advantages of O(10) over E_6 are discussed in [5]. More phenomenological issues in the context of an O(10) gauge group are currently under investigation [14].

Acknowledgements

I have described work done in collaboration with M. Dine, V. Kaplunovsky, M. Mangano and N. Seiberg in Ref. 9.

References

[1] M. B. Green and J. H. Schwarz, Phys. Lett. 149B (1984), 117.

[2] D. J. Gross, J. A. Harvey, E. Martinec and R. Rohm, Phys. Rev. Lett. 55 (1985), 502, and Princeton preprints (1985).

[3] P. Candelas, G. T. Horowitz, A. Strominger and E. Witten, Nuclear Physics B, in press.

[4] F. Gürsey and P. Sikivie, Phys. Rev. Lett. 36 (1976), 775; P. Ramond, Nucl. Phys. B 110 (1976), 224.

[5] E. Witten, "Symmetry breaking patterns in superstring models", Nucl. Phys. B, in press.

[6] A. Strominger and E. Witten, Comm. in Math. Phys., in press; A. Strominger, to appear.

[7] J. Breit, B. Ovrut and G. Segré, Phys. Lett. B 158 B (1985), 33.

[8] A. Sen, Phys. Rev. Lett. 55 (1985), 33.

[9] M. Dine, V. Kaplunovsky, M. Mangano, C. Nappi, N. Seiberg, Nucl. Phys. B, in press.

[10] S. Dimopoulos, S. Raby and F. Wilczek, Phys. Rev. D 24 (1981) 1681.

[11] See also S. Cecotti, J.-P. Derendinger, S. Ferrara, L. Girardello and M. Roncadelli, CERN preprint TH 4103 (1985).

[12] V. S. Kaplunovsky, Princeton University preprint 1985; M. Dine and N. Seiberg IAS preprint 1985.

[13] M. Mangano, Zeitschrift fur Physik C., to appear.

[14] C. R. Nappi, in preparation.

SUPERSTRINGS AND PREONS

Jogesh C. PATI
Department of Physics and Astronomy, University of Maryland, College Park, MD 20742, USA

1. WHY SUPERSTRINGS AND PREONS?

The superstring theories generate the exciting new prospect that one may finally have a consistent unified quantum theory of all forces including gravity.[1,2] In spite of many attractive features, however, these theories including the heterotic $E_8 \times E_8$ theory appear so far to be beset with serious difficulties in describing the real world. Although one obtains chiral fermions belonging to a desirable representation[3] (27) of E_6, one seems to face either the problem of rapid proton-decay,[4,5] or the problem of inconsistency of renormalization group-analysis[6] or both. Other possible difficulties pertain to the questions of the origin of supersymmetry-breaking and that of the lower mass-scales like m_W. Even if these difficulties are somehow resolved, the greatest stumbling block, it seems to me, would be the problem of the fermion mass-hierarchy and fermion-mixings.[F1]

The purpose of this talk is to suggest[7] that the advantages of the superstring theories as regards (a) uniqueness, (b) parameterlessness and (c) good quantum gravity may be retained and yet the difficulties listed above may be circumvented if the fundamental four dimensional fields are identified with preons[8] rather than with quarks and leptons. Such an identification would also enhance the prospect of an understanding of the fermion mass-hierarchy because of new dynamics and new symmetries, which naturally arise within preonic theories, especially those with supersymmetry.

2. GENERAL CHARACTERISTICS OF A VIABLE PREON-MODEL

While apriori there can be many alternative preon-models, a few general characteristics seem to be rather crucial[9,10] for realizing a viable and an economical preon model. They are the following:

[F1] The only handle one has to address to this problem is topology and one naturally wonders whether topology alone can account for the intricate multistep mass-hierarchy of quarks and leptons spanning over more than five orders of magnitudes, and their mixing angles, as well.

(i) There must exist effectively at least <u>two preonic scales</u> -- i.e. a light scale $\Lambda_H \sim 1$ TeV as well as a superheavy scale $\Lambda_M \sim 10^{14}$-10^{15} GeV -- associated with a "<u>hypercolor</u>" and a "<u>metacolor</u>" gauge force respectively.[10]

(ii) There must also exist supersymmetry preferably in its local form at the underlying preonic level.[11]

The need for the superheavy scale (Λ_M) arises on rather general grounds by demanding consistency of the model with cosmological issues such as the generation of baryon-excess.[10,12] Furthermore, for a large class of preon models, the limits on the strengths of rare processes like $K^0 \leftrightarrow \bar{K}^0$ and $K_L \to \bar{\mu} e$ require that the inverse sizes of the e and the μ-families should exceed at least (30 to 100) TeV.[9,13] That there must also exist the light scale $\Lambda_H \sim 1$ TeV follows by demanding that $SU(2)_L \times U(1)$-breaking and quark-lepton masses arise simultaneously through preonic condensates rather than through the VEV of an elementary Higgs field. These considerations in fact lead one to the conclusion that the inverse size of the heaviest τ and/or a fourth τ' family can not exceed about 1 TeV ($(1/r_0)_{\tau,\tau'} \lesssim 1$ TeV) .[9]

The need for supersymmetry is strongly suggested because it opens the door for evading[14] certain no-go theorems pertaining to (a) chiral symmetry[15] breaking and (b) vectorial symmetry-preservation[16] which apply to ordinary QCD-type theories, but which need to be evaded for preonic theories. Local supersymmetry also opens the door for evading the no-go theorem[17] regarding the formation of massless composite gauge particles, which are needed within certain preon models. Furthermore, a supersymmetric theory provides a natural basis for a fermion-boson or the so-called flavon-chromon type preon-models[18] which have many attractive features. For example, (a) they help satisfy the 't Hooft's anomaly-matching condition in a very simple manner[19] without needing proliferation, and (b) they naturally conserve quantum numbers like B and L at a basic level, which are violated spontaneously. Thus they do not have an intrinsic proton-decay problem unlike some other models.

3. A MINIMAL MODEL WITH FAMILY-REPLICATION

A very economical model incorporating these two features -- with N=1 <u>local</u> supersymmetry -- has been proposed recently.[10] I discuss at the end how such a model can be obtained for example from the heterotic superstring theory.

The model, which we refer to as the "minimal" SUSY model, introduces just four left plus four right-handed chiral superfields, i.e. $\Phi_+^{a,i} = (f_L|C_I)^{a,i}$ and $\Phi_-^{a,i} = (f_R|C_{II})^{a,i}$, with a=1 to 4, each transforming as a fundamental representation "i" of the metacolor gauge symmetry G_M, having a scale-parameter $\Lambda_M \sim 10^{14}$ GeV. It is assumed (a) that supersymmetry breaks dynamically through the formation of the metacolor gaugino-condensate:[20] $\langle \lambda_M \cdot \lambda_M \rangle \sim \Lambda_M^3$; thereby the gauginos become superheavy and <u>decoupled</u>. This induces soft SUSY breaking mass-terms of order $(\Lambda_M^3/M_{p\ell}^2) \sim$ 300 GeV. At this stage, the eight spin-1/2 components

$$f_{L,R}^{a,i} = (u,d|c,s)_{L,R}^i \equiv (f^e|f^\mu)_{L,R}^i \qquad (1a)$$

and the eight spin-0 components

$$c^i \equiv (C_I|C_{II})^i = (r,y,b,\ell|r',y',b',\ell')^i \qquad (1b)$$

of the superfields define independent commuting flavor $SU(4)_L \times SU(4)_R$ and $SU(8)$-color symmetries respectively.[21] The symmetry $SU(8)$-color contains $SU(4)^C$ and $SU(4)^{HC}$ acting on the unprimed and the primed chromons -- i.e. $\{C_I\}$ and $\{C_{II}\}$ -- respectively.

It is assumed that the approximate global flavor-color symmetry. $G = SU(4)_L \times SU(4)_R \times SU(8)^C$ breaks dynamically at the metacolor scale Λ_M into the anomaly-free subgroup $G_0 \equiv SU(2)_L \times U(1)_Y \times SU(3)^C \times SU(4)^{HC}$, which emerges as an effective low energy gauge symmetry, for $E < \Lambda_M$, through the formation of composite gauge bosons of sizes $\sim \Lambda_M^{-1}$. The descent of G into G_0 may take place through the formation of metacolor singlet preonic condensates of the type shown below:

$$
\begin{aligned}
&\text{(i)} \quad \sigma_{L,R}^i \sim (15_{L,R}, 1^C)^{i=1,2} \sim \begin{cases} (\bar{f}_{L,R} \gamma_\mu \vec{\lambda} f_{L,R})(\bar{f}\gamma_\mu \mathbb{1} f) \\ \text{and/or } (\bar{f}_L^a f_R^b)(\bar{f}_R^c f_L^d) \end{cases} \\
&\text{(ii)} \quad \zeta_{\alpha\beta} \sim (1_{L,R}, 63^C) \sim \text{e.g. } (\bar{f}_L \gamma_\mu \mathbb{1} f_L + \bar{f}_R \gamma_\mu \mathbb{1} f_R)(C_\alpha^* \vec{\delta}_\mu C_\beta) \\
&\text{(iii)} \quad \Delta_R^j \sim (1_L \cdot 3_R, \overline{10}^C)^j \sim (f_R f_R C_I^* C_I^*)^j \, .
\end{aligned}
\qquad (2)
$$

The transformation-properties of $\sigma_{L,R}^i$ and ζ are with respect to $SU(4)_{L,R} \times SU(8)^C$, while that of Δ_R is with respect to the familiar subgroup $SU(2)_L \times SU(2)_R \times SU(4)^C$, where $SU(2)_{L,R}$ act on the doublets $(u,d)_{L,R}^i$ and $(c,s)_{L,R}^i$ and $SU(4)^C$ on the quartet (r,y,b,ℓ). The matrices $\vec{\lambda}$ and $\mathbb{1}$ denote the famililar λ-matrices and unit matrix respectively for $SU(4)$. The index j signifies three different family-index combinations -- i.e. Δ_R^{ee}, $\Delta_R^{e\mu}$ and $\Delta_R^{\mu\mu}$ -- corresponding to the entries $f_R^e f_R^e$, $f_R^e f_R^\mu$ and $f_R^\mu f_R^\mu$ in (1). Two distinct σ_L's (i = 1 and 2) are needed with a pattern of VEV's as shown below to break[22] $SU(4)_L$ to the GIM subgroup $SU(2)_L^{e+\mu}$:

$$\sigma_L^1 = \begin{pmatrix} 1 & & & \\ & 1 & & \\ & & -1 & \\ & & & -1 \end{pmatrix} v_1, \quad \sigma_L^2 = \begin{pmatrix} 1 & & & \\ & -1 & & \\ & & -1 & \\ & & & 1 \end{pmatrix} v_2 \ . \tag{3}$$

Likewise for $SU(4)_R$. The action of the condensates shown in (2) is shown below:

$$\begin{array}{ccccc}
SU(4)_L & \times & SU(4)_R & \times & SU(8)^C \\
\downarrow \sigma_L^i & & \downarrow \sigma_R^i & & \downarrow \zeta \\
SU(2)_L^{e+\mu} & \times & SU(2)_R^{e+\mu} & \times & [SU(4)^C \times U(1)_K \times SU(4)^{HC}] \\
& & & \downarrow \Delta_R & \\
& & G_o = SU(2)_L^{e+\mu} & \times & U(1)_Y \times SU(3)^C \times SU(4)^{HC}
\end{array} \tag{4}$$

All four condensates are governed by one and the same scale parameter Λ_M. The symmetry $U(1)_K$ corresponds to the generator Y_K belonging to $SU(8)^C$ which is $(+1|-1)$ in the space of $(C_I|C_{II})$. Note that Δ_R breaks one linear combination of flavon and chromon numbers; it also breaks $SU(2)_R$ and B-L $\subset SU(4)^C$. In fact only the component of Δ_R^{ee} having the composition $u_R u_R \ell^* \ell^*$, and likewise the components $u_R c_R \ell^* \ell^*$ and $c_R c_R \ell^* \ell^*$ of $\Delta_R^{e\mu}$ and $\Delta_R^{\mu\mu}$ respectively, are electrically neutral. Thus only these components having the quantum numbers of dineutrino $(\nu_R \nu_R)$ acquire VEV. They break lepton number L associated with the ℓ-chromon and give a heavy Majorana mass to the composite ν_R's. In the context of a Higgs-mechanism for left-right symmetric theories[23] with elementary Higgs-fields, it has been shown that Δ_R acquire a large VEV ($\gg m_{W_L}$) while $\langle \Delta_L \rangle$ remains essentially zero. Following this result as a guidance, one assumes that an analogous breaking pattern holds for the case of dynamical symmetry breaking as well.

As mentioned above, Δ_R breaks lepton number L but not baryon number B associated with (r,y,b) chromons. We assume that additional metacolor-singlet condensates form which break B and other surviving global quantum numbers -- e.g. the condensate $\Sigma \sim f_R^a f_R^b f_R^c d_R^* c_I^* c_I^* c_I^*$ breaks B and L satisfying $\Delta B = \Delta L = -1$. These, together with Δ_R's, help generate baryon-excess[12] and induce proton-decay. The superheaviness of the metacolor scale $\gtrsim 10^{14}$ GeV turns out to be crucial for the generation of baryon-excess. Note furthermore that the spontaneous breaking of the global quantum numbers of the model like flavon and chromon-numbers will generate Goldstone bosons which will, however, be weakly coupled and thus "invisible" owing to the metacolor scale Λ_M being superheavy. These are some of the reasons -- based on cosmological considerations -- why the metacolor scale is chosen to be superheavy in the first place.

One can also derive a value for the metacolor scale from <u>independent considerations</u> based on renormalization group-equations. Since $SU(3)^C$ and $SU(4)^{HC}$ are unified within $SU(8)^C$ at a momentum scale $\sim \Lambda_M$, corresponding to the scale of the condensates ζ and Δ_R which break $SU(8)^C$, one obtains:

$$(\Lambda_M/\Lambda_H)^{b_4} = (\Lambda_M/\Lambda_{QCD})^{b_3} \tag{5}$$

The exponents b_3 and b_4 determine the β-functions for $SU(3)^C$ and $SU(4)^{HC}$ respectively. With only gauge boson-contributions, we have $(b_3/b_4) = 3/4$, and thus $\Lambda_M \approx (\Lambda_H^4/\Lambda_{QCD}^3)$. Substituting $\Lambda_{QCD} \approx$ (100 to 200) MeV and $\Lambda_H \approx$ 1 TeV (deduced from the scale of $SU(2)_L \times U(1)$-breaking, see later), we thus obtain:

$$\Lambda_M \approx 10^{15}\text{-}10^{14} \text{ GeV}. \tag{6}$$

It is remarkable that this scale, derived on the basis of $SU(8)$-renormalization group-analysis, happens to agree with the constraints arising from cosmology mentioned above and that it also nearly coincides with the typical grand unification-scale.[F2]

<u>Quantum Metacolor Dynamics and Massless Composites</u>: To determine the spectrum of massless composites made by the metacolor force, we make the following dynamical assumptions:

(i) Chiral $SU(2)_L$ and $U(1)_Y$ gauge symmetries are not broken dynamically at the metacolor scale. In turn, these symmetries protect the masses of all the flavons and also of the spin-1/2 composite fermions -- like quarks and leptons -- if they form, although the full global chiral symmetry $SU(4)_L \times SU(4)_R$ breaks partially as shown in (3).

(ii) The spin-0 chromons C_I and C_{II} and also the lowest configuration metacolor singlet spin-1/2 and spin-0 composites, i.e. fC^*V and CC^* remain

---

[F2] Indirectly, we infer that the hypercolor gauge symmetry would have to be $SU(4)$. If it were any "smaller", its scale-parameter could not be much higher than that of $SU(3)^C$, assuming that it is still unified with $SU(3)^C$. It can not be "bigger" than $SU(4)$, otherwise Λ_M would be too low. For example, if

$$G_{HC} = SU(5), \text{ we would have } \Lambda_M \approx (\Lambda_H/\Lambda_{QCD})^{3/2} \Lambda_H \approx 100 \text{ TeV}.$$

massless in the scale of Λ_M (V denotes metacolor gluon).[F3]

(iii) There is a saturation of binding at the level of the <u>lowest configuration</u> metacolor singlet composites like fC^*V and CC^* (see below).

The assumptions (i) and (ii) leading to masslessness of the spin-1/2 composites require that we satisfy 't Hooft's anomaly matching condition. Since the vectorial symmetries $U(1)_f$ and $U(1)_C$ are broken dynamically at Λ_M, the only releveant symmetry for anomaly-matching might have been the full global flavor symmetry $SU(4)_L \times SU(4)_R$. The corresponding anomalies of the preons and of the fC^*V-composites would match if the metacolor gauge symmetry $G_M = SU(N_M)$ with N_M = number of chromons = 8 (for the present model). On the other hand, since $SU(4)_L \times SU(2)_R$ breaks dynamically to $SU(2)_L \times SU(2)_R$ at Λ_M (see (3)), the anomaly of the preons match trivially that of the composites since both vanish, without any restriction on G_M.

Anomaly-matching is, however, only a necessary condition, but not sufficient, to yield massless composite fermions. One may naturally ask: What does protect the $SU(2)_L$ symmetry which in turn forces the flavons and the composite quarks and leptons to remain massless? If quantum metacolor dynamics (QMD) was exactly like QCD, one would have expected the $\langle \bar{f}^a f^a \rangle$-condensate to form, like the $\langle \bar{q}q \rangle$-condensate, and this would have broken even $SU(2)_L$-symmetry. It is good to note, however, that none of the standard proofs showing that chiral symmetry must break[15] in QCD and furthermore that vectorial global symmetries including "baryon number" and "isospin" can not break dynamically in vector-like QCD type theories,[16] apply to supersymmetric QCD-type theories.[14] One can in fact argue that for locally supersymmetric QCD theories a solution permitting preservation of chiral symmetry but a breakdown of supersymmetry through the formation of the gaugino-condensate is at least a consistent one. One is thus led to conjecture that the underlying reason for the protection of some chiral symmetry like $SU(2)_L$ and the simultaneous breaking of vectorial symmetries like B, L and $SU(4)_{L+R}$ including "isospin" through condensates like σ^i and Δ_R is local supersymmetry. Of course, this conjecture

---

[F3]The identifications $q_L = f_L C^*$ and $q_R = f_R C^*$ pose the following dilemma. Since the vertex involving the transition $q_L \to f_L C^*$ should be of the form $[f_L \gamma_\mu (p_f - p_q)_\mu q_L]/\Lambda_M$, the residue of the amplitude for $f_L C^* \to f_L C^*$ scattering at the composite q_L-pole would be damped by $[(m_f - m_q)/\Lambda_M]^2 \to 0$. No such dilemma arises if we identify $q_{L,R}$ with metacolor singlet $(f_{L,R} C^* V)$ or $(f_{L,R} C^* VV)$-composites in appropriate gauge invariant forms where V represents the metagluon.

still awaits a proof.

To sustain our assumption (iii) of masslessness of the spin-0 chromons and the spin-0 CC*-composites, we need to invoke once again supersymmetry and chiral symmetry. If chiral symmetry protects the masses of the fermions, supersymmetry would protect the masses of the bosonic partners. In our case, N=1 local SUSY breaks through the formation of the gaugino-condensate $\langle \bar{\lambda} \cdot \lambda \rangle \sim \Lambda_M^3$. It has been shown that in this case the mass m_C of the chromons is protected by the Planck mass: $m_C \sim (\Lambda_M^3/M_{P\ell}^2) \sim$ few hundred GeV for $\Lambda_M \sim 10^{13.5}$ GeV, as desired. With these to provide some rationale for sustaining our dynamical assumptions (i) and (ii), we turn to the spectrum of the massless composites.

Metacolor Binding (size $\sim \Lambda_M^{-1}$): Following the saturation-assumption, we list the set of lowest configuration metacolor singlet composites of the (f,c,V)-system which form under the influence of the metacolor force. For simplicity of writing, we shall consistently suppress the gluonic component V in the composites fC*V. The composites are classified under $SU(4)^C \times SU(4)^{HC}$; their chiral flavor-transformation-properties are not exhibited, but should be apparent:

$$\psi_{L,R}^a = (f_{L,R}^a \, C_I^*)_{4_C^*,1_H} \equiv \psi_{L,R}^{(o)e,\mu}$$

$$\xi_{L,R}^a = (f_{L,R}^a C_{II}^*)_{1_C,4_H^*} \longrightarrow \text{Hyperfermions}$$

$$\mathcal{D}_o = (C_I C_{II}^*)_{4_C,4_H^*} \longrightarrow \text{Colored hyperbosons}$$

$$\mathcal{E} = (C_{II} C_{II}^*)_{1_C,(1+15)_H} \; ; \; \mathcal{H} = (C_I C_I^*)_{1M,(15+1)_C,1_H} \, . \quad (7)$$

Consistent with our assumptions (i) and (ii)[F4] these composites are massless in the scale of Λ_M. They all have sizes $\sim \Lambda_M^{-1} \sim (10^{14} \text{ GeV})^{-1}$. The

---

[F4] Note that we have not listed the spin-0 composites $\phi = ff$ in (6). This is because in the limit of supersymmetry such composites correspond to the auxiliary component of the composite superfield $\Phi \Phi^*$ and thus should effectively have infinite mass. In the presence of SUSY-breaking at a scale M_S, we expect ϕ to have a mass of order $\Lambda_M^2/M_S \sim 10^{15}$-$10^{17}$ GeV at Λ_M. However, the effective mass of ϕ can be much lower at Λ_H due to renormalization through Yukawa coupling. See remarks later on the possible role of ϕ.

composites ψ carrying flavor and color yield two quark-lepton-famililes which we identify with the fermions of the "bare" e and μ-families.

Hypercolor Binding (size $\sim \Lambda_H^{-1} \sim 1$ TeV^{-1}): The hyperfermions ξ and the colored hyperbosons $\mathcal{O}_0^*$ which are almost point-like composites with sizes $\sim \Lambda_M^{-1}$, bind through the hypercolor force to yield two new families of much bigger sizes $\sim \Lambda_H^{-1} \sim (1 \text{ TeV})^{-1} \ggg \Lambda_M^{-1}$ having precisely the same quantum numbers as the ψ-composites:

$$\chi_{L,R}^a = (\xi_{L,R}^a \mathcal{O}_0^*)_{4_C^*, 1_H} \equiv \chi_{L,R}^{(o)\tau,\tau'} . \tag{8}$$

These new composites are naturally identified with the bare $\tau^{(o)}$ and a fourth $\tau'^{(o)}$-families, which are replicas of the bare $e^{(o)}$ and $\mu^{(o)}$-families respectively.

Note that the second-stage fermionic composites $\chi = (\xi \mathcal{O}_0^*) = (fC_{II}^*)(C_I^* C_{II})$ have precisely the same flavor and color-attributes as the first-stage composites ψ in their core. Yet they can not be regarded as ordinary radial, orbital or quantum pair-excitations of ψ because their sizes are very much bigger than those of ψ. For all probes of momenta $\ll \Lambda_M$, the composites χ will appear as two -- rather than four-body composites with constituents ξ and $\mathcal{O}^*$, which are distinct from those of ψ.

By utilizing the existence of the two scales, <u>the model has thus generated a mechanism for two-fold replication of quark-lepton families, which in turn predicts four rather than three families</u>.

The replication-idea appears to be a particularly desirable feature, because it leads to preon-models which are (a) economical, (b) viable and (c) thoroughly testable (see discussions later). Furthermore, it greatly simplifies the structure of the fermion mass-matrix by reducing the 2N-family problem to essentially that of an N-family problem (see discussions below).

At a deeper level, if one can understand why there must be four left-handed plus four right-handed chiral superfields, one would understand why there is the muon accompanying the electron. This would answer Rabi's famous question: "<u>Who ordered that?</u>" Simultaneously, owing to the replication-mechanism, one can explain why there must be the τ and predict a τ'. I believe that the apparent arbitrariness in the choice of the number and the representation-content of the preonic superfields will ultimately be removed by appealing to higher dimensional theories, in particular the superstring theories. I return to this question later.

Fermion-Masses: By assumption, the hypercolor force, makes hyperfermion-condensates $\langle \xi_L^a \xi_R^b \rangle \equiv A_{ab} \Lambda_\xi^3$, which gives masses to the quarks and the leptons and also break $SU(2)_L \times U(1)_Y$ dynamically. Here, A_{ab} denotes a 4×4-matrix in the chiral flavor space with entries of order unity or zero, whose structure needs to be determined non-perturbatively and $\Lambda_\xi \sim \Lambda_H \sim 1$ TeV. Unlike the case of QCD, however, A_{ab} need not be a unit matrix owing to the presence of supersymmetry at scales $\geq \Lambda_M$, which may trigger the breakdown of even vectorial symmetry at the metacolor-scale and in turn at the hypercolor-scale. We return to this question later.

The fermion masses and mixings depend additionally on the strengths of the effective four fermion-transitions: (i) $\psi_L^a + \xi_R^b \to \psi_R^b + \xi_L^a$, (ii) $\psi_L^a + \xi_R^b \to \chi_R^b + \xi_L^a$ and (iii) $\chi_L^a + \xi_R^b \to \chi_R^b + \xi_L^a$. Ordinarily the strengths of the first two processes would be strongly damped by Λ_M^{-2} and thus the masses of the e and the μ-families would only be of order $(\Lambda_H^3/\Lambda_M^2) \sim 10^{-16}$ MeV. But the situation alters drastically due to the presence of the spin-0 $\mathscr{A}_0$-particles with masses $\sim (\Lambda_M^3/M_{p\ell}^2) \sim 300$ GeV. These can bind with the hypergluons V_μ^H to yield light spin-1 composites $\mathscr{A}_\mu$ with the same internal quantum numbers as $\mathscr{A}_0$ and with masses $\sim \Lambda_H$. Now, $\mathscr{A}_\mu$-exchange would contribute to all three processes listed above; this would be the only relevant contribution to the first two processes, while for the third there can be additional contribution of comparable magnitude. Thus, we have an 8×8 mass-matrix of the following form:

$$M_{8 \times 8} \approx \begin{bmatrix} \psi(e^0,\mu^0) & \chi(\tau^0,\tau'^0) \\ \eta^2 \cdot m_{ab} & \eta \cdot m_{ab} \\ \hline \eta \cdot m_{ab} & (1+X)m_{ab} \end{bmatrix}. \quad (9)$$

Here, $m_{ab} \equiv g^2 (\Lambda_\xi^3/m^2) A_{ab}$ and $\eta \equiv (h/g)$, where g and h are the <u>dimensionless</u> effective coupling constants for the vertices $\chi \to \xi + \mathscr{A}_\mu^*$ and $\psi \to \xi + \mathscr{A}_\mu^*$ respectively. Naively, one expects g and h to be comparable and thus η to be of order unity.[F5] The quantity X denotes contributions from mechanisms other than the one involving $\mathscr{A}_\mu$-exchange. Thus $X \sim \mathcal{O}(1)$ and conservatively $1/10 \leq X \leq 10$.

The simplification brought about by the family-replication idea is obvious. The same 4×4-matrix m_{ab} appears in each of the four blocks in (9). As a

---

[F5] Strictly speaking, one must examine whether the fact that ψ and χ have very different sizes could lead to a large suppression for h compared to g.

result, the 8 × 8-matrix M, which normally would require, e.g. 20 parameters, if it is charge-conserving, real and symmetric, requires at best 8 -- i.e. η, X and six more to describe m_{ab}. Even these eight parameters are in principle calculable in terms of the only two parameters[F6] of the model -- i.e. Λ_M and Λ_H.

Regardless of the structure of the condensate matrix m_{ab}, we can diagonalize (9) in the ψ-χ space. This yields (for $\eta \lesssim 1/3$):

$$m^i_{e,\mu} = (\eta^2 \omega) m^i_{\tau,\tau'}, \quad (i = U, D) . \qquad (10)$$

Here $\omega \simeq X/(1+X)^2$ and the superscript $i = U, D$ corresponds to the up and down-flavors in each family. Since $\omega_{max} = 1/4$ and $\omega \simeq 1/10$ for $X \simeq 1/10$ or 10, we see that for a value of η in its natural range -- e.g. $1/5 \lesssim \eta \lesssim 1/10$, which is certainly within reason, we get

$$m^i_{e,\mu} \simeq (10^{-2} \text{ to } 10^{-3}) m^i_{\tau,\tau'} . \qquad (11)$$

In other words, without choosing η to be too small and for a large range of reasonable values for X, we obtain a hierarchy-ratio of order 10^{-2} to 10^{-3}. The <u>model thus provides a very simple explanation for the two major steps of the mass-hierarchy</u>, i.e. why $(m^i_e/m^i_\tau) \ll 1$ and simultaneously why $(m^i_\mu/m^i_{\tau'}) \ll 1$, in accord with the concept of "naturalness" advocated by Dirac. In so doing, it introduces a novel feature. The physical particles, owing to $\bar\psi\chi$-mixing (see (9)), are now <u>mixtures</u> of the small size $(\psi^{(o)})$ and the large-size $(\chi^{(o)})$ composites:

$$\psi_{physical} = \psi^{(o)} \cos \alpha + \chi^{(o)} \sin \alpha$$
$$\chi_{physical} = -\psi^{(o)} \sin \alpha + \chi^{(o)} \cos \alpha \qquad (12)$$

where $\tan \alpha = \eta/(1+X) \sim (1/5 \text{ to } 1/10)/(1+X) \sim 3$ to 10%. Thus the physical quarks and leptons of the electron and the muon families are mostly very small size-composites $(\psi^{(o)})$, but they do have a small admixture of the large size-composites $(\chi^{(o)})$. This is what makes the replication-idea testable.

The last two steps of the mass-hierarchy which need to be explained involve (i) <u>The interfamily (τ-τ') or equivalently the (e-μ) splittings</u> -- i.e. why

$$m_{t'} > m_t , \; m_{b'} > m_b , \; m_c > m_u \text{ and } m_s > m_d , \text{ and}$$

[F6]Note that if the hypercolor gauge force is generated through composite gauge bosons, as in the minimal SUSY model, then even Λ_H is calculable in terms of Λ_M.

(ii) <u>Up-Down splittings</u> within each family -- i.e. why $(m_t/m_b) \simeq 8-9$, $(m_c/m_s) \simeq 9$, etc.

Barring certain anomalies with regard to the lightest e-family, on which we comment later, the mass-ratios are typically as follows:

$$\left.\begin{array}{l}\text{Replication}\\ \text{Splitting}\end{array}\right\} \quad (\bar{m}_\tau/\bar{m}_e) \sim (\bar{m}_{\tau'}/\bar{m}_\mu) \sim 400$$

$$\left.\begin{array}{l}\text{Interfamily or}\\ \text{e-}\mu\text{ splitting}\end{array}\right\} \quad (\bar{m}_{\tau'}/\bar{m}_\tau) \sim (\bar{m}_\mu/\bar{m}_e) \sim 10\text{-}20$$

$$\left.\begin{array}{l}\text{"Isospin"}\\ \text{splitting}\end{array}\right\} \quad (m_U/m_D)_{\tau,\mu,e} \sim 5\text{-}10 \ . \tag{13}$$

Here, the bar indicates an average mass of a family, the subscripts U and D represent up and down-members of a given family. We have purposely chosen the masses of the members of the electron-family to be somewhat higher than their observed values to exhibit a pattern. This may be justified aposteriori. <u>We see from (13) a progressively decreasing hierarchy</u>. Yet note that not even the Up-Down splitting in any given family is really small enough to suggest an identification with a perturbative electroweak effect. In other words, all three steps in (13) suggest the relevance of NEW PHYSICS involving perhaps new dynamics and/or new symmetries.

To explain the last two steps of the hierarchy exhibited in (13), one needs to know the structure of the condensate-matrix A_{ab} for which a first-principle calculation is not yet in hand. In the context of a left-right symmetric gauge theory with elementary Higgs fields (like Δ_L and Δ_R), it has been shown recently that the Up-Down splittings in any given family can be understood simply in terms of spontaneous breaking of parity[24] -- i.e. of $SU(2)_R$ -- through the VEV $\langle\Delta_R\rangle$, which breaks vectorial "isospin".[F7] It was in fact shown that such a breaking can lead to a sizable "isospin" breaking in the effective Yukawa couplings of the Up and Down quarks with the Higgs multiplet ϕ, already at the scale of $\langle\Delta_R\rangle \gg m_{W_L}$, which in turn can induce large mass-splittings at the electroweak scale. This idea has its analog in a left-right symmetric preon-theory of the type presented here.

One can in fact envisage that a similar mechanism, having its origin in the

---

[F7]Note that the idea of spontaneous breaking of vectorial symmetries like "isospin" and in fact flavor-SU(4) as well as B and L is being entertained for the metacolor preon-dynamics, in the first place, because, as mentioned before, SUSY-QCD opens the door for a breakdown of vectorial symmetries unlike ordinary QCD.

non-perturbative breaking of the full vectorial flavor symmetry $SU(4)_f$ through the condensates such as $\sigma^j_{L,R}$ and Δ_R (see (4)), triggers $SU(4)_f$-breaking in the effective Yukawa interactions of the composite scalars with the composite fermions at the metacolor scale.[F8] These in turn generate $SU(4)_f$ breaking at the hypercolor scale leading to a charge conserving but $SU(4)_f$-breaking

---

[F8]M. Cvetic and I are examining this possibility. The composite scalar which we have in mind is $\phi^{ab} = \bar{f}^a_L f^b_R$. As mentioned before, such a composite is expected to be born superheavy at the metacolor scale with a mass exceeding Λ_M because it arises only as an auxiliary composite field in the limit of SUSY. However, one can show that renormalization group-equations can reduce the effective mass of ϕ ($\bar{m}_\phi$) evaluated at the hypercolor-scale dramatically by even 10 to 12 orders of magnitude if the relevant Yukawa couplings ($h_y^2/4\pi$) of ϕ with the composite fermions (ψ,ξ) are of order 1/10, say. Now these effective Yukawa couplings would clearly break full $SU(4)$-flavor, including of course $SU(2)$-isospin, in the presence of the condensates $\sigma^j_{L,R}$ and Δ_R. Such breaking-effects, which are finite, are not damped, e.g. by the heavy meta-color-scale, as observed in Ref. 24 for the case of isospin. Cvetic and I are examining whether this $SU(4)$-breaking effective Yukawa interactions of ϕ superimposed on the dynamics of the hypercolor gauge-force can generate the desired pattern for the condensate-matrix A_{ab} (see text), consistent with the minimization of the energy.

I should add that radiative effects involving ϕ-exchange can induce sizable $SU(4)$-breaking effects on the vertices of the type $\xi + \mathcal{B}_\mu \to \psi$, which are relevant for the mass-matrix (see text). In the presence of such symmetry-breaking effects, the η-parameter ($\equiv h/g$, see text) for the u-type flavors would be different from that for the d-type flavors. Although this will not alter our explanations of the e-τ and the μ-τ'-mass-hierarchies as long as η_u and $\eta_d \lesssim 1/5$, it turns out that the inequality of η_u and η_d is crucial for generating the weak-mixing elements V_{bc} and V_{bu}, which would otherwise vanish in the model. In the text, I have kept $\eta_u = \eta_d$ for the sake of simplicity.

As an additional remark, it is worthnoting that the condensate matrix $\langle \bar{\xi}^a \xi^b \rangle$ together with the Yukawa coupling of ϕ with $\bar{\xi}\xi$ would naturally induce a VEV $\langle \phi^{ab} \rangle \sim h_{ab} \langle \bar{\xi}^a \xi^b \rangle / \bar{m}^2_{\phi_{ab}}$, although $\bar{m}^2_\phi$ is positive. In other words, ϕ_{ab} being almost point-like at the hypercolor-scale, acts like a <u>composite Higgs-field</u>. However, in the preonic model (Ref. 10), the $\langle \bar{\xi}\xi \rangle$-condensate still provides the driving mechanism for symmetry-breaking. The VEV of $\langle \phi \rangle$ can be relevant for the masses of the light e and μ-families provided the effective $\bar{m}_\phi$ at the hypercolor-scale is $\lesssim$ (100-1000) TeV, which is not unlikely. Now $\bar{m}_\phi$ can not be much lighter than about 100 TeV, either, or else it would induce flavor-changing neutral current-processes like $K_L \to \bar{\mu}e$ with too big an amplitude. The joint effects of the composite ϕ and the condensate $\langle \bar{\xi}\xi \rangle$ are under study.

condensate-matrix of the form:[F9]

$$A = \begin{bmatrix} A_d & 0 \\ 0 & A_u \end{bmatrix}, \quad A_u = \begin{bmatrix} 0 & \varepsilon \\ \varepsilon & 1 \end{bmatrix}, \quad A_d = \begin{bmatrix} 0 & \varepsilon' \\ \varepsilon' & 1 \end{bmatrix} \kappa' . \quad (14)$$

With the very few effective dimensionless parameters[F10] ε, ε' and κ' having values in a natural range (1/3 to 1/10), such a pattern would account for τ-τ'-splittings (i.e. $(m_t,/m_t) \sim \mathcal{O}(-\varepsilon^2)$ and $(m_b,/m_b) \sim \mathcal{O}(-\varepsilon'^2)$) and likewise for e-$\mu$ splittings (i.e. ("m_u"/m_c) $\sim \mathcal{O}(-\varepsilon^2)$; ("$m_d$"/$m_s$) $\sim \mathcal{O}(-\varepsilon'^2)$) as well as for Cabibbo-mixing (sin $\theta_c \approx \varepsilon - \varepsilon'$). With such a pattern for A_{ab}, together with the replication-splitting explained before (eq. (11)), we would be able to understand why "m_u" and "m_d" are so small ($\mathcal{O}(10^{-4})$) compared to the masses of the quarks of the heaviest family (τ'). We would attribute this small number (10^{-4}) to $(\eta^2 \omega) \times (\varepsilon^2$ or $\varepsilon'^2)$ being that small without any of the parameters η, ω, ε or ε' being too small. Because "m_u" and "m_d" are that small ($\sim$(10-50 MeV)), however, we would expect induced radiative corrections which we have omitted to be particularly important for them. The quotation-marks on m_u and m_d signify that these values for the masses are derived without induced radiative effects and that they are expected to be modified substantially due to such effects, more so than the masses of the other quarks. The radiative corrections may have signs <u>opposite</u> to those of "m_u" and "m_d" and may help account for the rather puzzling observed values of m_u and m_d satisfying $|m_u| < |m_d|$.[F11] Whether a condensate matrix of the form (14) can be derived by minimizing the energy is under study.[F8]

4. CRUCIAL EXPERIMENTAL TESTS

As we saw, because of the replication-property and the special fermion mass-generation-mechanism of the model, the e- and the μ-families are primarily small-size composites with sizes $\sim \Lambda_M^{-1}$, while the τ and the τ' are primarily large size composites with sizes $\sim \Lambda_H^{-1}$. But fortunately the physical e and the physical μ have an admixture of order $\delta \equiv \eta/(1+X) \approx 3$ to 10% (for $1/5 \lesssim \eta \lesssim 1/10$ and $X \lesssim 5$) of the bare large-size composites τ^o and τ'^o

---

[F9] The zeros in A_u and A_d may in fact correspond to entries of order ε^2 and ε'^2 respectively, with $|\varepsilon|$ and $|\varepsilon'| < 1/3$, say.

[F10] Note that these are also, in-principle, calculable within the model.

[F11] If the radiative corrections to masses are as large as nearly 1 GeV, they may even lead to e <--> μ switching. A. Datta and I have recently considered the consequences of this possibility.

respectively (see eq. (12)). This permits a host of observable consequences of compositeness involving collider experiments at relatively low energies $\sim \Lambda_H \sim 1$ TeV and also rare processes. Some of these are listed below:

(i) First we expect that the e and the μ-families should exhibit electroweak charge form factors $\sim \delta^2(-\Lambda_H^2)/(q^2-\Lambda_H^2)$ while the τ and τ' should exhibit form factors $\sim (1-\delta^2)(-\Lambda_H^2)/(q^2-\Lambda_H^2)$.

(ii) We also expect that the scattering amplitudes for family non-diagonal processes like $(e^-e^+$ or $q_e\bar{q}_e) \to \tau^-\tau^+$, $q_b\bar{q}_b$ or $q_t\bar{q}_t$, at momenta $< \Lambda_H$, should possess a nonelectromagnetic component $\sim \delta^2(g_H^2/\Lambda_H^2)$ with $g_H^2/4\pi \simeq 1$. This new component would compete favorably with and even exceed the standard electromagnetic amplitude $\sim e^2/q^2$, and would thus show itself prominently through total cross-section, forward-backward asymmetry and polarization-measurements for $q^2 \sim (1/\delta^2)(e^2/g_H^2)\Lambda_H^2 \sim ((1 \text{ to } 3)\Lambda_H)^2 \sim (1 \text{ to } 3 \text{ TeV})^2$, for $\delta \sim 1/10$ to $1/30$. The signal should be visible, of course, already at $|q| \sim (1/3 \text{ to } 1)\Lambda_H \simeq (300 \text{ to } 1000)$ GeV, where the non-electroweak amplitude would be of order 30%.

Note that the family diagonal-processes like $e^-e^+ \to e^-e^+$ and $q_e\bar{q}_e \to e^-e^+$ and even $e^-e^+ \to \mu^-\mu^+$ and $q_e\bar{q}_e \to \mu^-\mu^+$ will not, however, show compositeness to this extent at $q^2 \sim \Lambda_H^2$, since the corresponding amplitude would be of order $\delta^4(g_H^2/\Lambda_H^2)$ or smaller still (by $\sin^2\theta_{Cabibbo}$). We, therefore, predict that the accelerators with CM energies in the range of $(1/3 \text{ to } 1)$ TeV should discover (a) compositeness of e, μ and τ, primarily through family, non-diagonal processes[F12] like $(e^-e^+$ or $q_e\bar{q}_e) \to \tau^-\tau^+$, $q_b\bar{q}_b$ or $q_t\bar{q}_t$ and (b) a breakdown of e-μ-τ universality (i.e. $A(e^-e^+ \to \mu^-\mu^+) \neq A(e^-e^+ \to \tau^-\tau^+)$ etc.). It is thus especially important to improve the efficiency for τ-detection.

It is also important to stress that the amplitude for the process $q_t\bar{q}_t \to \tau^-\tau^+$ would have a component of order $q_H^2(1-\delta^2)/\Lambda_H^2$ due to compositeness which would be comparable to the electroweak amplitude at rather low $q^2 \sim (e^2/g_H^2)\Lambda_H^2 \sim \Lambda_H^2/100 \sim (100 \text{ GeV})^2$. Thus, for $m_t \sim (40 \pm 10)$GeV, we expect that a study of toponium decays into $\tau^-\tau^+$ in the process $e^-e^+ \to$ toponium $\to \tau^-\tau^+$ should show compositeness rather prominently.[F13]

Contrast these from the phenomenological considerations of Eichten, Lane and Peskin,[25] which presumes the relevance of the TeV-scale for the family-diagonal processes involving only the e and the μ-families. The lower limits

---

[F12] If there is e-μ switching due to radiative effects (see F11) $(e^-e^+$ or $q_e\bar{q}_e) \to \tau'^-\tau'^+$ would show large departures, instead.

[F13] I thank George A. Snow for drawing my attention to this point.

on the compositeness-scale extending up to two TeV, recently derived from[26] $e^-e^+ \to e^-e^+$, $\mu^-\mu^+$, and somewhat lower limits derived from[27] $q_e\bar{q}_e \to e^-e^+$, etc. are based on this analysis of Ref. 25. Mindful of cosmological considerations and the limits from $K_L \to \bar{\mu}e$ and $K^0-\bar{K}^0$ processes we have, on the other hand, pointed out[9,10] that there are at least very good reasons to believe that the electron and the muon-families have primarily very small sizes ($\lesssim$ (30 to 100) TeV^{-1}), [F14] while the τ and/or a heavier τ'-family must have sizes or order 1 TeV^{-1}. In this case, one would not expect to see any noticeable signal of compositeness at the TeV-scale by studying processes which involve the fermions of only the electron and the muon-families. <u>We, therefore, can not overemphasize the importance of studying in particular the family non-diagonal processes involving the (e–τ) and the (μ–τ)-combinations for discovering compositeness at the 1 TeV-scale.</u>

(iii) One of the crucial predictions of the model which seems to hold for variants of the mass-matrix as well is the occurrence of $q_d + \bar{q}_s \to \bar{\mu}e$ and, therefore $K_L \to \bar{\mu}e$. This occurs through a box-diagram involving double σ_μ-

---

[F14] We have noted elsewhere (Ref. 9) that it is possible to construct at least semi-viable models in which the compositeness scales of the e and the μ-families can be low (~ 1 TeV) without conflicting with the limits from $K_L \to \bar{\mu}e$ and $K^0-\bar{K}^0$. This can happen, e.g., if the SU(4)-color constituents of the e and the μ-families are different. To realize this situation, introduce e.g., four flavons $f_{L,R} = (u,d,|c,s)_{L,R}^i \equiv (f_I|f_{II})_{L,R}^i$ and eight complex spin-0 chromons $C = (r,y,b,\ell|y',y',b',\ell')^i \equiv (C_I|C_{II})^i$, each transforming as the fundamental representation "i" of a metacolor gauge symmetry G_M with a scale parameter Λ_M, as in the family-replication model. But, unlike the replication model, assume that <u>both</u> C_I and C_{II} are quartets of the familiar SU(4)-color. This amounts to saying that only the diagonal sum of the two SU(4)-symmetries, defined by the sets $\{C_I\}$ and $\{C_{II}\}$ respectively, is realized as an approximate low-energy symmetry. The diagonal sum breaks further into an SU(3)-subgroup which is identified with SU(3)-color. In this case, the SU(4)^{MC} of the replication-model is not realized. We can now construct four metacolor-singlet composite quark-lepton families, all of inverse size ~ Λ_M: $\psi_1 = f_I C_I^*$, $\psi_2 = f_{II} C_I^*$, $\psi_3 = f_I C_{II}^*$, $\psi_4 = f_{II} C_{II}^*$, which can be identified with the e, μ, τ, and τ'-families (barring mixings) in a few alternative ways. But with the e and the μ-families having different sets of chromons (and possibly even the flavons), e.g. with $F_e = \psi_1$ and $F_\mu = \psi_3$ or ψ_4, the straightforward arguments leading to the constraints from $K_L \to \bar{\mu}e$ and $K^0-\bar{K}^0$ do not apply. In this case, all three or four families can have inverse sizes ~ Λ_M ~ 1 TeV. The consistency of this type of model with fermion mass-hierarchy, mixings and in particular cosmology has not yet been demonstrated. <u>Yet, because of this type of model, I still keep an open mind regarding the possibility that the e and μ-families are also primarily large-size composites (size~ 1 TeV^{-1}).</u> We clearly need experimental guidance on this important matter. A model of the type mentioned above was suggested in Ref. 9 and some time later in the second paper of Ref. 28 as well.

exchange involving vertices which give masses and Cabibbo-mixing (see discussions on fermion-masses). The corresponding amplitude is $\simeq g_\eta^4 {}^4/(16\pi^2 m_c^2)$. With constraints from mass-matrix and mixing angles, this leads to a branching ratio $B(K_L \to \bar{\mu}e) \simeq 10^{-7}$ to 10^{-10}, which should be observable in the on-going and forthcoming experiments at BNL. This is a crucial test of the model.

(iv) The model naturally predicts the occurrence of $\tau \to e\gamma$ with a branching ratio which is estimated[10] to be $\simeq (1/4)(10^{-5}\text{-}10^{-6})$. The decay $\tau' \to \mu\gamma$ would be even more prominent[F15], because $m_{\xi_s} \gg m_{\xi_d}$. But $\mu \to e\gamma$ is strongly suppressed.

(v) With the simplest mechanism of mass-generation presented here, the model predicts that either $B_d^0\text{-}\bar{B}_d^0$ or $B_s^0\text{-}\bar{B}_s^0$ mixing must be maximal (i.e. $\Delta m \gg \Gamma$).[29] Such mixing will naturally reflect itself in a pronounced like sign dilepton-production.

(vi) (g-2)-measurements of the electron and the muon should show noticeable departures from standard expectations with an improvement by a factor of 5 to 10 in the accuracy of measurements in each case.[30]

(vii) There must exist a host of new particles with spins 1/2, 0 and 1, which are composites of hyperfermions $\xi^{u,d,s,c}$, with masses of the order of (100 to 2000) GeV. One expects to see an increase in $R(e^-e^+)$ in the energy range of (500-1000) GeV due to excitations of (a) hyperfermion-pairs, (b) spin-0 chromons (C_I), and (c) their point-like composites carrying color and hypercolor like $\mathcal{D}_0$. Some of these will resemble technicolor-spectroscopy, but there will still be distinct elements due to the chromon-component. Note that the hyperfermions have charges ±1/2, while C_I's have charges (1/6, 1/6, 1/6, -1/2) while the C_{II}'s are neutral.

As a digression, it is worth noting that the model utilizes the idea of dynamical symmetry breaking like the technicolor models,[31] but it is much more economical than the technicolor-models involving extended technicolor. One can also show[32] that with four flavors, the model possesses GIM-like mechanism at the preon-level and does not lead to excessive $|\Delta S| = 2$ flavor changing neutral current-processes like $K^0\text{-}\bar{K}^0$ unlike familiar technicolor models. Thus the preon-model will have the richness of technicolor-spectroscopy and even more but without the problems of the standard technicolor models.

(viii) There would have to be a fourth family; the expected masses of t', b' and τ' are in the ranges of (250-400), (40-100) and (25-60) GeV respec-

---

[F15] For the case of e <--> μ-switching, one would expect to see $\tau \to \mu\gamma$-decays, instead.

tively.

Thus the model has several rather intriguing and stringent predictions on the basis of which it can live or die. This is contrary to the prevailing notion that composite models are devoid of rather hard predictions. It is worth stressing that the majority of the predictions listed above are crucially tied as much to the replication-idea as to the specific mechanism for the generation of fermion-masses and mixings, which was presented.

5. FROM SUPERSTRINGS TO PREONS

I now discuss how the field content of preonic theories of the type presented here may be obtained from superstring theories. I shall first concentrate on the heterotic superstring theory,[2] based on the gauge symmetry $E_8 \times E_8$ in d = 10. As is well known, it is generally assumed that the ten dimensional theory compactifies into $M^4 \times K$, where M^4 is the four dimensional Minkowski space and K a compact six dimensional Calabai-Yau manifold with SU(3) holonomy.[3-5] The existence of such vacuum solutions at the tree level is known. Such a compactification leaves an unbroken N = 1 localsupersymmetry at the Planck (or compactification) scale. It, furthermore, breaks $E_8 \times E_8$ into $E_8 \times E_6$ if K is simply connected, and into a lower symmetry such as $E_8 \times SU(3) \times SU(2)_L \times [U(1)]^3$, if K is multiply connected. In other words, the topology of K determines the pattern of the primary symmetry breaking. It also determines the massless zero mode matter superfields in d = 4. These turn out to be in the form of several <u>copies</u> of 27's and $\overline{27}$'s of E_6 (even when E_6 is broken). The number of generations N defined by $n_{27} - n_{\overline{27}}$ is given by half the Euler characteristic of K which turns out to be reasonably low if once again K is multiply connected. Models with 1, 2 and 4 generations are known.

Following Ref. 7, I now show that the precise field-content of (four left + four right)-handed superfields of the minimal model sketched above can be derived from the d = 10, $E_8 \times E_8$ superstring theory provided that the compactification to d = 4 leads to two copies of 27's of E_6 (i.e. $N \equiv n_{27} - n_{\overline{27}} = 2$) and that E_6 breaks at the compactification scale to a subgroup $G_0 = SU(4)_M \times \tilde{G}$, where $\tilde{G}$ = either $[U(1)]^3$ or $SU(2) \times [U(1)]^2$, or $[U(1)]^2$ or $SU(2) \times U(1)$.-

The symmetry $SU(4)_M$ is identified with the metacolor gauge symmetry which generates the preon binding force. The symmetry $\tilde{G}$, on the other hand, breaks completely at Λ_M, dynamically, due to preon-condensates. That a two copy-model can arise for several alternative choices of K has been noted in Ref. [4]. A possible breaking of E_6 into $SU(4)_M \times \tilde{G}$ with $\tilde{G} = SU(2) \times [U(1)]^2$ has also been shown to exist in the literature [4,5]. Following similar methods, one can argue that $\tilde{G} = [U(1)]^3$ is also permissible.[7]

We now examine the field content of this configuration and show how it can match the one of the minimal preon-model.

The Field Content:

In the absence of the effective Higgs-field VEV $\langle H^a \rangle$, transforming like an adjoint 78 of E_6, we obtain, through compactification, a set of massless 27's and $\overline{27}$'s of E_6 such that $N = n_{27} - n_{\overline{27}} = 2$. In the presence of $\langle H^a \rangle \neq 0$, owing to the effective coupling of $27 \cdot \overline{27} \cdot H(78)$, each of the $\overline{27}$'s pair off with a 27 and become superheavy $\sim \Lambda_C$, while the excess of 27's over $\overline{27}$'s given by $N = 2$ remains massless. Each of these 27 is a left chiral superfield. To be concrete choose $\tilde{G} = U(1)_L \times U(1)_R \times U(1)_N$, where $U(1)_{L,R}$ denote the diagonal generators of $SU(2)_{L,R} \subset SO(10) \subset E_6$, while $U(1)_N$ commutes with $SO(10)$. (Note that with a preonic identification for the minimal model, none of the components of E_6, not even $U(1)_{L,R}$, are to be identified with the familiar flavor-color symmetries, which arise only through effective local symmetries at the composite level.) Each 27 transforms under $SU(4)_M \times U(1)_L \times U(1)_R \times U(1)_N$ as follows:

$$27 = [16_1 + 10_{-2} + 1_4 \text{ of } SO(10) \times U(1)_N]$$

$$= A(4, \pm 1, 0)_1 + \tilde{B}(4^*, 0, \pm 1)_1$$

$$+ C(6, 1, 1)_{-2} + D(1, \pm 1, \pm 1)_{-2} + E(1, 0, 0)_4 \ . \tag{15}$$

The $U(1)_N$-charge is denoted in each case by a subscript. The entire set is complex owing to this $U(1)_N$-charge. The spin-1/2 and spin-0 components of the superfields are labelled as follows:

$$A^{4,i} = \begin{pmatrix} f_L^{4,i} \\ C^I \end{pmatrix}, \quad \tilde{B}^{4,j} = (\overline{B}^{4^*,j})^\dagger = \begin{pmatrix} f_R^{4,j} \\ C^{II} \end{pmatrix}$$

$$C^I = \begin{pmatrix} \psi_L^6 \\ S^6 \end{pmatrix}^I, \quad D^J = \begin{pmatrix} \psi^D \\ S^D \end{pmatrix}^J, \quad E^I = \begin{pmatrix} \psi^E \\ S^E \end{pmatrix}^I \ . \tag{16}$$

The labels $\underline{4}$ and $\underline{6}$ denote $SU(4)_M$-representations, while i, j, I and J run over allowed values of other quantum numbers (i.e. $U(1)_{L,R}$) and also allow for the presence of two copies of 27 of E_6.

The metacolor force becomes strong at the scale $\Lambda_M \sim 10^{14}$–10^{15} GeV. Following Refs. [10] and [21], we assume first of all that it makes the metacolor gaugino-condensate $\langle \tilde{\lambda}_M \cdot \tilde{\lambda}_M \rangle \sim \Lambda_M^3$, which in turn breaks supersymmetry and induces soft SUSY breaking terms of order $\Lambda_M^3 / M_{P\ell}^2$. This makes the metacolor gauginos superheavy and thereby decoupled.

Second, we assume that owing to the strong metacolor force a bilinear condensate $\langle \psi^6 \psi^6 \rangle$ and appropriate combinations of the quartic condensates $\langle f_R f_R C^* C^* \rangle$ form as well. These, first of all, break $\tilde{G}$, i.e. $U(1)_L \times U(1)_R \times U(1)_N$, completely. Second, the condensate $\langle \psi^6 \psi^6 \rangle$ makes not only the sextets ξ but also the metacolor singlets M and N superheavy. This comes about, because the $SU(4)_M \times \tilde{G}$-symmetry, which is operative between M_{Planck} at Λ_M, leads to a superpotential of the form

$$W = \omega_1 A.\bar{B}.D + \omega_2 A.A.C + \omega_3 \bar{B}.\bar{B}.\xi + \omega_4 D.D.E + \omega_5 C.C.E . \tag{17}$$

In the presence of the ω_5-coupling, the condensate $\langle \psi^6 \psi^6 \rangle$ induces an effective VEV $\langle E \rangle \sim \Lambda_M$. This in turn generates a heavy mass of order Λ_M for E as well as for D (due to the ω_4-coupling), but not for A and $\bar{B}$. Note that E couples directly only to C and D but not to A and $\bar{B}$.

Thus, subject to our assumption about the formation of the condensates which, a priori, is at least feasible, the sextet C as well as the singlets D and E become superheavy and decouple at Λ_M. The only massless fields which remain are the quartets A and $\bar{B}$.

Since $U(1)_L \times U(1)_R \times U(1)_N$ is broken completely, we may now drop the distinctions which arise due to the corresponding charges. Allowing still for two copies of 27, we thus have from (15) altogether four left-handed quartets $A^4, i=1$ to 4 plus four left-handed antiquartets $\bar{B}^{4*}, j=1$ to 4 (or equivalently four right-handed quartets $B^4, j=1$ to 4). This is precisely the field content of the minimal preon model presented before. For the $E_8 \times E_8$ superstring theory, the metacolor gauge symmetry G_M must be identified with $SU(4)_M$.

To summarize, an identification of the fundamental fields in $d = 4$ with those of the "minimal" supersymmetric preon-model (Ref. [10]) is indeed possible. This can preserve the good features of the superstring theories while circumventing the shortcomings associated with a quark-lepton identification of these fields. In particular, the possible serious difficulties of rapid proton-decay and inconsistency of renormalization group-analysis clearly disappear with a preonic identification.

As we saw, the preonic identification permits, of course, spontaneous violations of B, L, C and CP at a temperature scale $\sim \Lambda_M \sim 10^{14}$ GeV, and, thereby, an adequate generation of baryon-excess in the early universe. Because $\Lambda_M \sim 10^{14}$ GeV, we would still expect proton to decay with a lifetime in an observable range $\sim 10^{31}$-10^{34} yrs, as in grand-unification models like SO(10) and SU(16).

It is possible to consider the so-called "maximal" supersymmetric preon models in which flavor and color are exact global (or possibly even local) symmetries even in the limit of SUSY, together with hypercolor and metacolor. The derivation of such variants from $E_8 \times E_8$ superstring theories is being examined by Hübsch and myself.

In another context, the preonic identification retains the good features of dynamical electroweak symmetry breaking associated with the technicolor idea but avoids the proliferations and difficulties of the standard technicolor scenario. The existence of the metacolor force with a high scale (Λ_M) is a peculiarity of preonic theories, which indeed serves multiple desirable purposes including spontaneous breaking of SUSY and the binding of the quarks and the leptons.

One advantage of the preonic idea as illustrated by the model of Ref. [10] is that it naturally brings with it, contrary to prevailing notion, a host of intriguing <u>crucial</u> and testable predictions involving rare decays such as $K_L \to \bar{\mu}e$ and reactions such as $q\bar{q} \to \tau^-\tau^+$, etc. with CM energies of the order of 300-1000 GeV. Thus, if these ideas are even grossly correct, we will, first of all, know about it from experiments in the near future. Most important, that will prevent the idea of a "grand desert" (10^2 to 10^{14} GeV) from being extended to that of a "super-desert" (10^2 to 10^{19} GeV).

Finally, in case I have conveyed the impression that preonic ideas solve all problems, I must state that this is, of course, still far from the truth. One has at best certain plausible scenarios which I believe show promise. But one is still groping in the dark. Much work needs to be done on the dynamics of compactification of superstring theories on the one hand and on the dynamics of preonic theories involving the questions of (a) supersymmetry breaking, (b) chiral symmetry - preservation at the metacolor scale and (c) spontaneous breaking of vectorial symmetries, on the other hand. One is certain of at least one point. Experiments in the very near future can exclude the basic line of construction of the preon-models presented here, if it happens to be wrong.

ACKNOWLEDGEMENT

It is a pleasure to thank George Lazarides and other organizers for providing a very pleasant environment for the symposium and also for their warm hospitality. The research was supported in part by the National Science Foundation.

REFERENCES

1) M. B. Green and J. H. Schwarz, Phys. Lett. $\underline{149B}$ (1984) 117; $\underline{151B}$ (1985), 21, and references therein.

2) D. J. Gross, J. A. Harvey, E. Martinec and R. Rohm, Phys. Rev. Lett. $\underline{54}$ (1985) 502.

3) P. Candelas, G. T. Horowitz, A. Strominger and E. Witten, Princeton University, Preprint NSF-ITP-84-170 (1984).

4) E. Witten, "Symmetry breaking patterns in superstring theories", Princeton University, Preprint (1985); A. Strominger and E. Witten, Princeton University, preprint (1985).

5) J. D. Breit, B. A. Ovrut and G. Segre, University of Pennsylvania, preprint UPR-0279 T (1985).

6) M. Dine, V. Kaplunovsky, M. Mangano, C. Nappi and N. Seiberg, Princeton University, preprint (1985).

7) T. Hübsch H. Nishino and J. C. Pati, IC/85-66, Phys. Lett. to appear.

8) J. C. Pati and Abdus Salam, Phys. Rev. $\underline{D10}$, (1974) 275; For a review and other references see e.g. M. Peskin, Proc. Int. Symp. on Lepton and Photon INter. at High Energies, Bonn (1981) Ed. W. Pfeil, p. 880; L. Lyons, Prog. Part. and Nucl. Phys. $\underline{10}$, (1983) 227; H. Harari, Weizman Institute, preprint (to appear in the Proc. of the 1984 Summer School, St. Andrews).

9) J. C. Pati, Phys. Rev. D Rap. Com. $\underline{30}$, (1984) 1144.

10) J. C. Pati, Phys. Lett. $\underline{144B}$, (1984) 375.

11) J. C. Pati and Abdus Salam, Nuc. Phys. $\underline{B214}$ (1983) 109; ibid $\underline{B234}$ (1984) 223; R. Barbieri, Phys. Lett. $\underline{121B}$ (1983) 43.

12) P. Mohapatra, J. C. Pati and H. Stremnitzer, Md. preprint 85-200.

13) I. Bars, Proc. XVIIth Rencontre de Moriond, Les Arcs, France, (1982) Ed. J. Tran Thanh Van, Vol. 1, p. 54 (1982).

14) J. C. Pati, H. Sharatchandra and M. Cvetic (in preparation); M. Cvetic (unpublished).

15) See e.g. D. Weingarten, Phys. Rev. Lett. $\underline{51}$ (1983) 1830 and S. Nussinov, Phys. Rev. Lett. $\underline{51}$ 2081, E. Witten, Phys. Rev. Lett. 51 (1983) 2351.

16) C. Vafa and E. Witten, Nucl. Phys. $\underline{B234}$ (1984) 173.

17) S. Weinberg and E. Witten, Phys. Lett. $\underline{96B}$, 59 (1980).

18) J. C. Pati and Abdus Salam (Ref. 8).

19) R. Barbieri, R. N. Mohapatra and A. Masiero, Phys. Lett. $\underline{105B}$ (1981) 369.

20) F. Ferrara, L. Girardello and H. Niles, Phys. Lett. $\underline{125B}$ (1983) 457.

21) J. C. Pati and Abdus Salam, (Ref. 11).

22) A. Davidson and J. C. Pati, Nucl. Phys. B175 (1980) 175.

23) J. C. Pati and Abdus Salam, Phys. Rev. D10 (1974) 275; R. N. Mohapatra and J. C. Pati, Phys. Rev. D11 (1975) 566; ibid (1975) 2558; G. Senjanovic and R. N. Mohapatra, Phys. Rev. D12 (1975) 1502; R. N. Mohapatra and G. Senjanovic, Phys. Rev. Lett. 44 (1980) 912.

24) D. Chang, R. N. Mohapatra, P. Pal and J. C. Pati, "Spontaneous Breakdown of Parity as the Origin of Isospin Breaking", Md. Phys. Publication 86-19.

25) E. J. Eichten, K. D. Lane and M. E. Peskin, Phys. Rev. Lett. 50 (1983) 811.

26) PETRA experiments reported at Kyoto Conference, August 1985. To appear in the Proceedings.

27) C. Rubbia, UA1 experiments, Proc. Kyoto Conference, August 1985 (to appear).

28. J.C. Pati (Ref. 9); O. W. Greenberg, R. N. Mohapatra and S. Nussinov, Phys. Lett 148B, (1984) 465.

29. A. Datta and J. C. Pati, ICTP preprint, IC/85/145, Phys. Lett., to be published.

30. R. Godbole, Trieste preprint (to appear).

31. For a review and relevant references, see e.g. E. Farhi and L. Susskind, Phys. Rep. (1982).

32. J. C. Pati, Proc. XXI Intern. Conf. on High Energy Physics (Paris, July 1982) C3, p. 197.

SCALAR SECTOR OF GAUGE THEORIES AND THE QUEST FOR A UNIFIED THEORY

M.A.B. BÉG *
The Rockefeller University, New York, New York 10021

The state of the scalar sector in gauge theories is critically examined. Much of the discussion is predicated on the generally accepted --albeit not rigorously proven--result that self-coupled scalar fields constitute a trivial dynamical system characterized by a unit S-matrix. A process of elimination leads to the conclusion that canonical electroweak theory can find consistency only after grand unification embedding is followed by a further synthesis; if there be but four basic forces, this would mean with gravity. Alternatives considered include a provisional scenario that permits one to live with the standard model over a substantial energy domain, and a bold proposal to dispense with elementary scalars altogether by introducing hypercolor-based dynamical symmetry breaking. Experimental implications of these constructs are reviewed; it is suggested that they are sufficiently interesting to warrant investigation even if the theories that yielded them fail to survive. Prospects of the unification program, with each of its steps evidently needed to solve embarrassing theoretical problems, are briefly discussed in the light of recent attempts to tame gravitation.

1. INTRODUCTION

The subject I have been asked to review, the scalar sector of modern gauge theories, has evolved to a point where the discussion will take us quickly into deeper waters, into questions pertaining to the consistency of physical theories and to the need for unification of the fundamental interactions.

That unification is a desirable end in itself has been long taken for granted. The success of the primal synthesis that led to electromagnetism and the more recent triumphs of electroweak-theory[1] may be cited as vindications of the underlying philosophy. Yet there are many reasons at this time to pause and reflect, to question the rules of the game and the manner in which it is being played. Despite many proposals for a further synthesis, to unify the electroweak and strong forces, no satisfactory model has so far emerged.[1-3] In fact, as we shall see, existing models are

---

Report Number DOE/ER/40033B-95 RU85/B/128

* Work supported in part by the Department of Energy under contract number DE-AC02-81ER40033B.000.

in more serious trouble than is generally realized. Gravitation continues to defy all attempts to describe it, in a consistent way, in the language of quantum field theory; indeed, within the framework of conventional field theory in (1+3)-dimensional space, a unified description of the four fundamental interactions -- the so-called Holy Grail of particle physics -- appears to be beyond reach.[4] It behooves us therefore to ask if the quest is warranted, if the unification-dogma can be put on firmer foundations than are afforded by the subjective criteria of elegance and taste. That such questions are relevant, and in need of an answer, becomes obvious once one realizes that there exist schools of scientific thought, albeit mostly in biology, that see nothing wrong with diversification[5] -- the antithesis of unification.

In this talk, progressive unification will be isolated as the most promising remedy for the shortcomings of existing gauge theories--which stem largely, though not exclusively, from their scalar sectors. This is anticipated in the following statement of a key result: Within the framework of the canonical methodology, it is <u>necessary</u> to go beyond grand unification, and--if there be no other forces--unify with gravity, for a consistent electroweak theory. (Whether this would also be sufficient is not known.) There is a disturbing corollary: Until a consistent gravity-theory is available, electroweak theory can not be formulated in a manner that is both canonical and consistent.

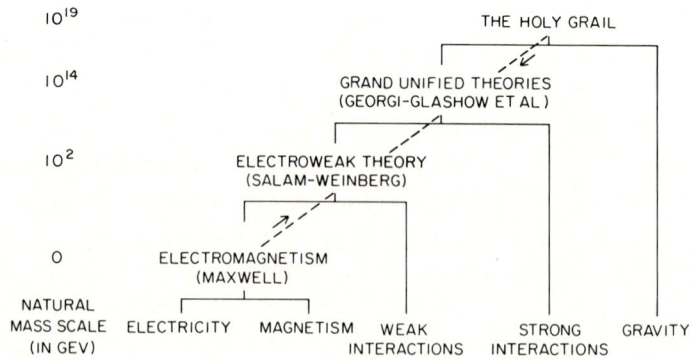

FIGURE 1
Progressive unification, circa 1985 C.E.
(Ascending: conventional theory. Attempting descent: superstring theory).

Figure 1 illustrates the seemingly paradoxical state of affairs. The end of the journey determines whether one can start on the quest for the Grail.

The primary input, that leads to the above result, is the triviality of the pure $\lambda\phi^4$-theory. Tentative theoretical scenarios, which permit one to live with this triviality and impose a manner of consistency on electroweak theory, inevitably lead to constraints on the parameters of the theory. These have been fairly well discussed in the literature[6-8]. However, I shall briefly review one proposed scenario[6], largely because it will give me an opportunity to state some recently-derived results, communicated to me in preprint form but not yet in print. I shall also go over the present status of attempts to cut the Gordian Knot presented by elementary scalars, by abandoning the canonical methodology and using QC'D (hypercolor/technicolor dynamics) to trigger the Higgs mechanism.[9] Finally, after listing some of the more serious difficulties confronting conventional theory, I shall try to give an assessment of the outlook in the light of very recent theoretical developments, mainly in superstring theories, which hold promise of giving us a handle on gravitation.[4]

In the following, the arguments are mostly of a heuristic nature; this is partly out of choice and partly out of necessity; if an alternate, rigorous, derivation of a result is available, the fact will be duly noted.

2. CONSISTENCY CRITERIA

The consistency of a physical theory is certainly questionable if the following conditions are not satisfied:

A. No fundamental lengths or ad-hoc cut-offs are required to define the theory.

B. All observables, defined and determined by the theory, are predicted to be finite and are expressible in terms of a finite number of parameters.

Henceforth, A and B will be deemed to be <u>necessary</u> conditions for consistency. This is not to suggest that theories which violate these criteria can not in some context be regarded as meaningful; they will simply not be consistent, as fundamental theories, in the sense in which we shall use the term. Note that B encompasses, in language that makes no appeal to perturbation theory, both finite and renormalizable theories.

No attempt will be made to enumerate criteria that may be regarded as sufficient for consistency.

3. STATUS OF SOME PHYSICALLY RELEVANT QUANTUM FIELD THEORIES

3.1 Quantum Electrodynamics

This venerable gauge theory, based on the group U(1), has been described[10] as "...the most successful theory...". Agreement between low-order perturbative calculations and experimental data is, however, the sole yardstick by which success is being measured here. Consider a long-standing problem.

Let $\bar{\alpha}(-q^2)$ be the running or renormalization group invariant fine structure constant at space-like momentum q. At the one loop level, we have[1]

$$\bar{\alpha}(-q^2)^{-1} = \alpha(m^2)^{-1} - \frac{1}{3\pi} \ln(-q^2/m^2) \qquad (3.1)$$

where m may be chosen to be the electron mass; with this choice $\alpha(m^2)$ may be identified with the renormalized fine structure constant: $\alpha(m^2) \simeq 1/137$.

Now physics dictates that $\bar{\alpha}$ be positive semi-definite; this implies that

$$\alpha(m^2) \leq \left[\frac{1}{3\pi} \ln(-q^2/m^2) \right]^{-1} \qquad (3.2)$$

for all momenta q in the theory. Thus if the theory is required to be valid even in the limit $|q| \to \infty$, that is to say, no ad-hoc cut-offs are introduced, one is forced to the result that the theory is a free theory:

$$\alpha(m^2) = 0. \qquad (3.3)$$

This paradox, first noted by Landau[11], has been the subject of endless debates. Two schools of thought have emerged. The first considers the paradox to be an artifact of the calculational procedure[12]; implicit in this viewpoint is the suggestion that the β-function has a second zero, a non-trivial fixed point of the renormalization group, that can be perceived when a better calculus is in hand. The second, more in tune with Landau as well as contemporary views about QED, accepts the absence of a fixed point and thereby the reality of the paradox; underlying this view is the recognition that QED as an isolated theory is <u>not</u> consistent, that it must therefore be regarded as an effective theory which describes only low-energy phenomena[13]. To estimate the critical energy or momentum beyond which the theory must breakdown, we may use Eq. (3.2)

$$\frac{2}{3\pi} \ln \frac{|q_{max.}^{QED}|}{m_e} \leq \frac{1}{\alpha} \qquad (3.4)$$

or

$$|q_{max.}^{QED}| \leq m_e \exp\left(\frac{3\pi}{2\alpha}\right)$$

$$\simeq 5 \times 10^{276} \text{GeV} \qquad (3.5)$$

The enormous magnitude of $|q_{max.}^{QED}|$ makes one wonder if the Landau paradox merits serious attention, except at the very fundamental level.

3.2 Quantum Chromodynamics (QCD)

The principal difference between QCD and QED, from our standpoint, is that the former is based on a simple gauge group, namely SU(3). For the running coupling constant, one now has the expression[14]:

$$\bar{g}(-q^2)^{-2} = g(\mu^2)^{-2} + \frac{33-2n_F}{48\pi^2} \ln(-q^2/\mu^2) \qquad (3.6)$$

where μ is some parameter with the dimension of mass and n_F is the number of quark-flavors, presumably less than 17.

An immediate consequence of the positivity of the second term, on the right hand side of Eq.(3.6), is that the theory is paradox-free; there is no obvious impediment to its standing alone as a consistent theory.

3.3 Salam-Weinberg Electroweak Theory

Let us remind ourselves of the traditional interpretation of this theory.[1] In the absence of scalars, there would be four massless gauge fields: $W^\mp$, Z and γ. The scalar sector, viewed as an isolated system consisting of a self-interacting complex doublet, exhibits spontaneous symmetry breaking and thereby yields three massless Goldstone modes plus one massive neutral particle even under CP -- in precise analogy with the σ-model. Coupling between the gauge-field and scalar sectors enables the Goldstone bosons to combine with the gauge quanta, to generate massive $W^\mp$ and Z particles; the massive scalar survives as an artifact of the Higgs mechanism. It is important to note that this interpretation would be untenable if the scalar sector leads to a unit S-matrix; free field theories do not yield Goldstone bosons.

Consider then the scalar sector, decoupled from fermions and gauge-fields. For the quartic self-coupling, we have a result analogous to Eq.(3.1):

$$\bar{\lambda}(-q^2)^{-1} = \lambda(m_W^2)^{-1} - \frac{3}{4\pi^2} \ln(-q^2/m_W^2) \qquad (3.7)$$

where we have defined the renormalized coupling at the W-mass. The value of this coupling constant is determined by the mass of the left-over Higgs boson:

$$m_H = m_W (2\lambda \sin^2\theta_W / \pi\alpha)^{1/2}$$
$$\simeq \sqrt{\lambda} \cdot 355 \text{ GeV} \qquad (3.8)$$

Here α is the fine-structure constant and θ_W is the Salam-Weinberg angle; the numbers in Eq.(3.8) correspond to $\sin^2\theta_W \simeq 0.22$.

Since positivity of λ is required for the stability of the theory, the arguments that led to $\alpha=0$ can be repeated; the statement of the Landau

paradox for this theory is thus seen to be
$$\lambda = 0. \quad (3.9)$$

To avert this catastrophe, we might consider cutting off all momenta at some value $q_{max.}^{EW}$. However, for $m_H \simeq 600$ GeV, a not unreasonable a priori value of Higgs mass, we find that

$$\left| q_{max.}^{EW} \right| \simeq 800 \text{ GeV} \quad (3.10)$$

This momentum is within the reach of existing accelerators, far too low for comfort. Unlike the situation with QED, the consistency question is not merely a point of principle; there seems to be a real physical problem which merits careful scrutiny. Quite obviously the first question that must be answered is whether the Landau argument is correct in this case, if the pure $\lambda\phi^4$-theory is indeed trivial.

4. MORE ABOUT THE $\lambda\phi^4$-THEORY

This theory is known to be trivial[15] if d, the dimensionality of space-time, exceeds 4; for d < 4, it is known to be non-trivial[16]. For d=4, it has been established that the non-relativistic limit[17] of the theory is trivial; this is consistent with, but does not imply, the triviality of the Lorentz invariant theory. Regrettably, there are no other (interesting) rigorous results at this time. Nonetheless, it is widely believed that the theory is trivial; the belief is largely based on experience with lattice calculations, which invariably lead to a zero renormalized coupling in the continuum limit.[18] Other calculations also yield the same result.[19] In the subsequent discussion, we accept the triviality of the pure $\lambda\phi^4$-theory in (1+3) - dimensional space.

5. PROVISIONAL SCENARIO FOR RESCUE OF THE STANDARD MODEL

Theoretical scenarios that permit one to live with the canonically realized $SU(3) \otimes SU(2) \otimes U(1)$ - based standard model, despite the manifest shortcomings of the electroweak sector, have been described in the literature.[6-8] The following brief exegesis of one such proposal[6] serves to underline its provisional nature.

The framework rests on a progressive triad: (a) It is proposed that the coupled Higgs--Gauge-Field sector is non-trivial. (b) To uphold(a), it is required that $y = \bar{\lambda}(t)/g_\perp(t)^2 \lesssim O(1)$ for "large t". [Here $g_\perp$ is the U(1) gauge coupling; $t \equiv (1/2)\ln(-q^2/m_W^2)$.] (c) To implement (b), one demands that y be driven towards an ultra-violet stable fixed point of the renormalization group.

One is thus led to consider the renormalization-group equations for the model:

$$\frac{d\bar{g}_i}{dt} = \beta_i(\bar{g}_j, \bar{\lambda}, \bar{G}) \qquad (5.1)$$

$$\frac{d\bar{\lambda}}{dt} = \beta_\lambda(\bar{g}_j, \bar{\lambda}, \bar{G}) \qquad (5.2)$$

$$\frac{d\bar{G}}{dt} = \beta_G(\bar{g}_j, \bar{\lambda}, \bar{G}) \qquad (5.3)$$

Here g_j, (j = 1,2 or 3), are the three gauge couplings in the model and G denotes Fermion-Higgs Yukawa couplings.

When one-loop expressions for the β-functions are substituted into Eqns.(5.1)-(5.3), and they are then analyzed with the help of a VAX-780, two interesting results emerge. First, unless there exist fermions with (current) masses of the order of or in excess of 80 GeV, the G's do not affect the analysis and may be dropped. Second, y can not go to a fixed point unless its initial value (i.e. at t=0) is bounded from above; this bound translates into one on the mass of the Higgs boson:

$$m_H < 125 \text{ GeV}. \qquad (1 \text{ loop}) \qquad (5.4)$$

That the scenario is self-consistent may be checked through numerical experiments; it was found that if the bound (5.4) is satisfied, y is indeed of O(1) for values of t up to 80; if it is violated, the growth of y with increasing t is well-nigh explosive. That the bound (5.4) is stable under inclusion of two-loop contributions to the β-functions has been established through the work of Babu and Ma[20]; they found

$$m_H < 130 \text{ GeV} \qquad (1 \text{ loop and 2 loops}) \qquad (5.5)$$

Finally the analysis has been extended to the case of N-Higgs doublets by Bovier and Wyler[21], with the result that the mass of the lightest Higgs particle satisfies

$$m_H < \frac{125}{\sqrt{N}} \text{ GeV}. \qquad (5.6)$$

For N=2, Eq. (5.6) agrees with an independent analysis by Babu and Ma.[22] These authors, in an earlier paper[22], also extended the analysis of Ref. 6 to the case of four electroweak generations.

Let us now look at the other side of the coin. There is the problem of the Landau ghost pole at[6]

$$\Lambda_G \simeq 4 \times 10^{41} \text{ GeV} \qquad (5.7)$$

To avoid running into it, one may simply cut off all momenta at, say, 10^{37} GeV; this energy is so large that one may justifiably argue that it can not possibly have any effect on electroweak physics at 100 GeV. However, by introducing a cut-off, we have violated our own definition of consistency for a fundamental theory -- as opposed to an effective theory or high phenomenology. It is in this sense that the logic is flawed and the scenario must be regarded as provisional.

6. THE PROMISE OF GRAND UNIFICATION

The first possibility that logically presents itself, for resolving the difficulties of the standard model, is grand unification[1-3]. If the model is embedded in a gauge theory based on a semi-simple group, then -- subject to known constraints on the particle spectrum, to which nature hopefully conforms -- the running fine structure constant automatically maintains its positivity; the burden of the Landau paradox is removed from QED and it becomes a consistent theory. This is an important achievement; all contexts in which it can be realized deserve to be examined with care.

Consider the scheme based on the group SU(5). In the so-called minimal version, Higgs fields transforming according to the fundamental and adjoint representations (**5** and **24**) are used to obtain the symmetry breaking pattern: $SU(5) \to SU(3) \otimes SU(2) \otimes U(1) \to SU(3) \otimes U(1)$. Fifteen real fields are absorbed in the Higgs mechanism; nineteen are left-over, to bear witness to the canonical symmetry-breaking process.

Various recent analyses[23,24] confirm each other in the finding that all quartic couplings which occur in the Higgs sector grow with increasing energy, that for at least one the growth is explosive. This absence of asymptotic freedom signals the presence of a Landau paradox and must be deemed to be a harbinger of inconsistency.[24] Furthermore, since SU(5) is a simple group, no provisional scenario, of the type discussed for the standard model, can be constructed. These considerations overshadow the much debated aesthetic inadequacy of the theory.[1,2]

It may be argued that minimal SU(5) is already a dead theory[3]; it runs into difficulties with fermion masses and predicts too rapid a rate for proton decay. However, there exist a variety of ways to modify the theory so as to rid it of its problems with low-energy phenomenology, for example through the introduction of a **45**-plet of Higgs fields. These modifications do not alter the argument outlined above in any essential way.

We seem to have come to the end of the trail, with canonical grand unification. The promise it held, for a consistent embedding of electroweak theory, is not fulfilled. Since the problems that defy resolution stem from (elementary) scalar fields, this may be an opportune time to pause and look at the status of attempts to get rid of them.

7. CUTTING THE GORDIAN KNOT: PROPOSALS TO DISPENSE WITH ELEMENTARY SCALARS

7.1 Prologue

Attempts to generate symmetry-breaking in a dynamical way, without benefit of elementary scalars, have a long history. The seminal ideas emerged from the pioneering work of Anderson[25] and Nambu[26]. In the modern gauge-theoretic context, the need for the dynamical option was established, and the interpretation of scalars as phenomenological props was suggested, in a paper by Sirlin and the present speaker.[27] Early proposals[28] have the feature that symmetry breaking is an ultra-violet effect.[29] Physics dictates, however, that the origins of symmetry-breaking lie in the infra-red sector, with symmetry-restoration taking place in the far ultra-violet. It is perhaps a testament to the poverty of human imagination that the only concrete proposal, at this time, that conforms to this requirement is the one that goes under the names of hypercolor and technicolor.[9] (Technicolor, the name used by Susskind, happens to be a registered trade mark in the United States; the switch to hypercolor was made by some to avoid running afoul of the law!)

Note that electroweak theory, by itself, cannot be rescued by getting rid of scalars; problems stemming from the U(1) factor in the gauge group will remain. In the following discussion, the intent to embed in a grand unified theory, based on a semi-simple group, should be understood.

7.2 Hypercolor

This alternative to the canonical methodology postulates the existence of a new confining gauge interaction described by QC'D, a theory similar to QCD except for its much larger mass scale ($\Lambda_{QC'D} \sim 10^3 \Lambda_{QCD}$). Spontaneous breaking of chiral symmetries, in hyperflavor space, provides the Goldstone bosons needed for triggering the Higgs mechanism. The construct has little difficulty in reproducing the canonical results for $W^{\mp}$ and Z masses, and has the added virtue of being experimentally distinguishable from the canonical theory both at low (10-100 GeV) and, more convincingly, at intermediate (~ a few TeV) energies.[9,30] These features of hypercolor are fairly well understood; for extensive discussions I refer you to the standard review articles.[9]

On the debit side, hypercolor has been beset from the beginning by a variety of problems. Some of these have been recently "solved". To appreciate the quote marks, consider three cognate problems: (i) generation of quark current masses, (ii) emergence of flavor changing neutral currents (FCNCs) and $\Delta S = 2$ transitions and (iii) the stability of the $\Delta I_{wk} = 1/2$ rule.

7.2.1 Quark Current Masses

The mechanism for mass-generation[31] is predicated on the observation that in the absence of Higgs couplings, the quark -- gauge-field sector is invariant under the group

$$G_H^{QUARK}: SU(n)_L^{d,u} \otimes SU(n)_R^d \otimes SU(n)_R^u \otimes U(1)$$

where the left-handed group operates in the space of both "up" and "down" quarks, the first right handed group operates only in the space of "down" quarks etc., n being the number of "up" or "down" quarks. Gauging of G_H is mandatory, if one is to avoid unwanted Goldstone bosons. With sufficiently heavy gauge fields, these horizontal gauge interactions need not disrupt the neat agreement between experiment and electroweak theory; in addition one can generate masses which simulate current masses at low energies.

The quintessential features of the problems at hand are best brought out by making some simplifying assumptions. Let us work with just four quark flavors and assume that hypercolor-bearing quarks, the hyperquarks, are devoid of the attribute of ordinary color. The following assignments under the horizontal group then suggest themselves: $(d_w, s_w, d_i')_L^T$ and $(u_w, c_w, u_i')_L^T$ transform as n-plets under $SU(n)_L^{d,u}$, $(d_w, s_w, d_i')_R^T$ and $(u_w, c_w, u_i')_R^T$ under $SU(n)_R^d$ and $SU(n)_R^u$ respectively. Here i (=1,2 ... $n_C' n_F'$) labels the hyperquarks, $n \equiv 6 + n_C' n_F'$, n_C' and n_F' being the number of hypercolors and hyperflavors; the suffix W, omitted on the hyperquarks, indicates that we are dealing with weak, as opposed to mass, eigenfields. The resulting fermion -- gauge-field interaction may be written as:

$$\delta\mathcal{L}_I = g_1 [(\bar{d}_i' \gamma_\mu d_w + \bar{u}_i' \gamma_\mu u_w)_L \, Y_1^{i\mu}$$
$$+ (\bar{d}_i' \gamma_\mu s_w + \bar{u}_i' \gamma_\mu c_w)_L \cdot Y_2^{i\mu}]$$
$$+ g_2 [(\bar{d}_i' \gamma_\mu d_w)_R \cdot Y_3^{i\mu} + (\bar{d}_i' \gamma_\mu s_w)_R \cdot Y_4^{i\mu}]$$
$$+ g_3 [(\bar{u}_i' \gamma_\mu u_w)_R \cdot Y_5^{i\mu} + (\bar{u}_i' \gamma_\mu c_w)_R \cdot Y_6^{i\mu}] \qquad (7.1)$$

+ terms diagonal in quark and hyperquark spaces + H.C.

where the g's are the coupling constants for the three SU(n) groups. With the additional assumption of no hyperflavor mixing, the gauge fields with well-defined mass are related to the fields in Eq.(7.1) via

$$\tilde{Y}_\ell^{i\mu} = a_{\ell m} Y_m^{i\mu} \qquad (7.2)$$

a being a transformation of the group SO(6):

$$a_{\ell m} a_{\ell n} = a_{m\ell} a_{n\ell} = \delta_{mn} \qquad (7.3)$$

The current mass of, say, the s-quark can now be calculated from Fig. 2; using the procedure described in ref. 31, for handling the momentum dependence of the dynamical hyperquark mass, one obtains

$$m_s = \sum_\ell a_{\ell 2} a_{\ell 4} (\delta M_\ell^2/M^2)$$

$$(n_F' n_C' g_1 g_2/4\pi^2) \cdot m_D^3/M^2$$

$$\ell n(M^2/m_D^2) \qquad (7.4)$$

where $\delta M_\ell^2 \equiv M_\ell^2 - M^2$, M being the mean gauge-field mass, and $m_D \equiv m_D(\Lambda_{QC'D}^2)$. Eq. (7.4) reduces to

$$m_s \simeq 12 \text{ GeV} (\sum_\ell a_{\ell 2} a_{\ell 4} \, \delta M_\ell^2/M^2) \qquad (7.5)$$

for rather reasonable values of the various parameters: $m_D \simeq 1$ TeV, $M \simeq 10$ TeV, $n_F' n_C' \simeq 10$; $g_1 \simeq g_2 \simeq 1$.

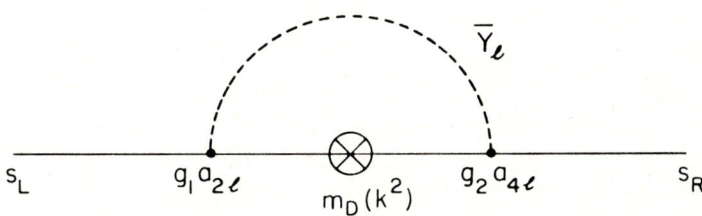

FIGURE 2
Graph depicting contributions to s-quark current mass. k is the momentum of the internal fermion, a hyperquark; other symbols are defined in the text.

The main point, brought out by Eq. (7.5), may now be stated as follows: To get adequate masses for the heavier quarks, it is not necessary to lower M to values that may be unacceptable to weak interaction phenomenology, contrary to the fears of some early explorers.[9] Most widely cited was the possibility of dangerously large couplings involving FCNCs and consequent $\Delta S = 2$ amplitudes of an unacceptable magnitude; to a reexamination of these topics let us now turn.

7.2.2 Flavor Changing Neutral Currents

Consider a quark-space diagonal piece of the Lagrangian

$$\delta \mathcal{L} = g_2 \, [\bar{D}_W(\theta_1) \, \gamma_\mu \, \tfrac{1}{2} \, \tau_i \, D_W(\theta_1)]_R \cdot E_R^{\mu i} \qquad (7.6)$$

where $D_W(\theta_1)^T = (d_W, s_W)$. This interaction arises from the SU(6) sub-group of the $SU(n)_R^d$ factor in the gauge group; the piece displayed corresponds to retaining only the color singlet field in the reduction of SU(6) gauge fields under $SU(3)_C \otimes SU(2)_F$: $35 = (1,3) + (8,1) + (8,3)$. Now if $E^{\mu 1}$ and $E^{\mu 3}$ mix, and if the mixing angle be 2χ, the quarks coupling to mass-eigenfields are $D_W(\theta_1-\chi)$. The effective $\Delta S = 2$ interaction stemming from (7.6) through the exchange of one gauge boson is thus

$$\delta\mathcal{L}_{eff.}^{\Delta S=2} = (g_2^2 / 4M^2)(\delta M^2 / M^2) \cdot \sin^2 2(\theta_1-\chi)$$
$$(\bar{d}\,\gamma_\mu s)_R \cdot (\bar{d}\,\gamma^\mu s)_R + H.C. \tag{7.7}$$

where $\delta M^2 = M(E^\mp)^2 - M(E^3)^2$. Note that the Cabibbo angle is $\theta = \theta_1 - \theta_2$, where θ_2 is the mixing angle in the (u,c) sector; the formula analogous to Eq.(7.7) for $\Delta C = 2$ would involve θ_2.

The strength of FCNC couplings is thus determined by angles such as $\theta_1-\chi$ which can not be inferred from low energy physics; they may be calculable, in principle, but present techniques are not adequate for the task; in the present state of the art, one can regard them as adjustable free parameters.

More interesting than this interim solution of the $\Delta S=2$ problem is the natural suppression of such amplitudes through what has been called[31] Pseudo-GIM. This effect has its origins in the $SU(2)_F$ symmetry encountered earlier; in the exact symmetry limit $m_s = m_d$, $m_c = m_u$, $\delta M^2 = 0$ and the $\Delta S = 2$ amplitude vanishes; the factor $\delta M^2/M^2$ in Eq. (7.7) thus corresponds to pseudo-GIM suppression. The mechanism has been shown to persist up to the one - QFD -loop level[32]; it thus helps keep the double gauge boson exchange contribution to $\Delta S = 2$ at a level no worse than in the canonical theory.

7.2.3 The $\Delta I_{wk} = 1/2$ Rule

In the canonical theory[1], the Higgs sector has an O(4) - symmetry ($\equiv SU(2)_L \otimes SU(2)_R$) which collapses to $SU(2)_{L+R}$ after spontaneous symmetry breaking; this custodial SU(2) guarantees the $\Delta I_{wk} = 1/2$ rule,

$$\rho \equiv (m_W^2 / m_Z^2 \cos^2\theta_W) = 1 \tag{7.8}$$

to all orders in the scalar coupling. Furthermore, $\rho=1$ and $m_u \neq m_d$ are compatible with each other at the tree level. At the one-loop level

$$\Delta\rho \sim \frac{\alpha^2}{m_W^2 \sin^2\theta_W} \cdot \left[\frac{2m_u^2 m_d^2}{m_u^2 - m_d^2} \cdot \ln(m_d^2/m_u^2) \right.$$
$$\left. + m_u^2 + m_d^2 \right] \tag{7.9}$$

The situation in the hypercolor scenario is very different.[33,34] Quark mass generating interactions break custodial SU(2) and imply SU(2) - breaking interactions among the hyperquarks. Corrections to the $\Delta I_{wk} = 1/2$ rule may thus be linear in the mass splittings; indeed a crude estimate yields[34]

$$\Delta\rho \simeq (m_u - m_d) / 8\pi^2 \Lambda_{QC'D} \tag{7.10}$$

While Eq. (7.10) does not give any immediate cause for alarm, better measurements and calculations of $\Delta\rho$ could lead to a decisive confrontation with experiment. This expression of hope assumes, of course, that a reliable calculational procedure will be available in the not too distant future.

7.2.4 Present Status of Hypercolor

The preceding discussion permits us to appraise the present status of the hypercolor-based dynamical alternative to the canonical theory.

(a) The elemental problems that afflicted the scenario in its early days are soluble, but--apart from an occasional glimmer of elegance, such as provided by pseudo-GIM--the solutions can not exactly be described as attractive.

(b) Further progress is impeded by the mathematical intractability of the formalism, a shortcoming common to all schemes for dynamical symmetry breaking. While almost everything is calculable, in principle, nothing can be calculated with present technology. Parameter proliferation can not be checked.

(c) As indicated earlier, hypercolor can play a useful role only in the context of grand unified theories based on semi-simple groups. To date, however, it has not been possible to construct a satisfactory working model--one in which the symmetry-breaking pattern may be deemed to be physically acceptable.

(d) Perhaps the only bright feature of hypercolor is its immediate experimental relevance. Despite the apparent insolubility of the underlying theoretical problems, it is possible to extract some well defined predictions;[9,30] these should certainly be tested. One can then either nail the coffin -- without some experimental support it would be hard to take the subject seriously -- or learn to live with the construct and devise ways and means of molding it into an acceptable theory.

8. SALVATION THROUGH GRAVITY?

On the brink of defeat in our effort to get rid of elementary scalars, by invoking a new ad-hoc force between particles that have not yet manifested themselves, let us consider an alternate strategy for achieving consistency. We retreat to a grand unified theory with elementary scalars, interpret it as an effective theory valid upto some cut-off momentum, then remove this momentum through a further synthesis with gravitation.(Presumably this would "neutralize" the scalars, via a mechanism as yet unknown.) Why gravitation? First, gravity is the only omniscient force, aware of all elementary particles. Second, and this is perhaps a more important consideration, unless a fifth force exists, there is no other possibility for embedding a grand unified theory in a larger context. However, as indicated earlier, there is a difficulty here; within the framework of quantum field theory in (1+3)-space, no formulation of gravity that may be deemed consistent by our definition is available, in all likelihood none exists. This has to do with the very singular nature of gravity.

For Einstein gravity in the absence of matter, a remarkable cancellation of divergences occurs at the one loop level.[35] At the two loop level, however, non-renormalizability again manifests itself.[36] When matter is brought in, through scalar fields, for example, renormalizability is lost at the one loop level.[35]

An interesting perspective emerges if one generates gravity, as part of supergravity, through local supersymmetry. Pure supergravity theories, theories in which the spectrum consists entirely of gravity supermultiplets, exhibit one loop finiteness[37] just as pure gravity does; furthermore, unlike the situation in pure gravity, the finiteness persists (at least) upto the two-loop level[38]; this circumstance fueled intense activity during the late 70's and early 80's, and led many to speculate that a sensible description of the coupled matter-gravity system may be in hand. Unfortunately, however, the matter spectrum suggested by the group $SU(3) \otimes SU(2) \otimes U(1)$ can not be forced into the procrustean bed of gravity supermultiplets; even one-loop renormalizability is therefore not attained in the real world.[39]

It seems therefore that in the presence of gravity, grand unified and electroweak theories make less sense than they did on their own, that the cut-off dependence that mars the canonical formulations of these theories is exacerbated, not removed. However, in my judgment, this is a hasty and rash conclusion. The proper moral to be drawn is that one has to go outside the framework of local field theory in (1+3)-space to describe quantum gravity. Once an ultra-violet convergent gravity-theory is available, one may proceed with some optimism to complete the logic of grand unification. Recent developments in superstring[40] theory give cause for hope.

9. SUMMARY AND OUTLOOK

Accepting the widely believed but hitherto unproven result that the pure $\lambda\phi^4$-theory is trivial, we considered a scenario that permits us to work with the standard model of electroweak interactions; it is a provisional scenario in that the Landau ghost problem is not solved; all momenta are carried to values as large as feasible, but effectively cut off at a value below the ghost mass.

Our definition of consistency, for a fundamental theory, does not allow for pathologies such as cut-off dependence. To remove the cut-off we considered grand-unification, and were led to conclude that canonically-realized grand unified theories, based on semi-simple or simple groups, were themselves in need of a cut-off; here, even the provisional strategy proposed for handling the standard model did not work.

Cognizant of the source of the difficulties so far encountered, we resurrected and reviewed in some detail a proposal to cut the Gordian Knot and dispense with elementary scalars altogether. This hypercolor alternative has many interesting experimental consequences that merit investigation; from a theoretical standpoint, however, we found it wanting.

The next logical step was bringing gravity into the picture, in the hope that the ad-hoc cut-offs will thereby be removed or replaced by the Planck mass in a natural way; within the normal field-theoretical framework, this hope was not fulfilled, because of the very singular nature of gravitation.

Where, then, do we now stand? Whither do we go?

One conceivable viewpoint is that we have built on sand, that the proper course for resolving the problems of asymptotically non-free theories is to define them via calculational techniques that yield a non-unit S-matrix. So long as a watertight proof of triviality remains elusive, one is certainly entitled to think along these lines.

The sensible thing to do, however, is to continue with the unification program. Canonical grand unification, its many shortcomings notwithstanding, did lead to consistency for pure QED; it seems proper, therefore, to regard its structure as logically incomplete rather than flawed in any fundamental way; indeed it is not unnatural to expect that bringing the remaining force, gravity, into the picture will shed light on the symmetry-breaking process, at least enough to help resolve the problems posed by elementary scalars. To take this last but essential step; it may be necessary to extend the concepts embodied in quantum field theory to permit it to accommodate gravitation.

The rapidly developing subject of superstrings may well yield the appropriate extension of local field theory. While an ultra-violet convergent description of gravity appears to be now available, from our standpoint we are not quite over the finish line. The problem now is whether one can safely descend to the real world, from 10^{19}GeV--the natural mass in superstring theories -- to 100 GeV, and make contact with electroweak physics. We are dealing with theories that are almost parameter free and therefore, in principle, with great predictive power; the acid test will of course come when suitable observables, which have the virtue of lending themselves to definitive theoretical evaluations, have been identified; until then we have to rely on aesthetic criteria.

To conclude, let us note that, from the standpoint of this report, each step of the unification process is needed to solve specific physical problems and thereby serves a useful purpose. This is welcome indeed, for it elevates unification from a religious dogma to a sound scientific principle.

REFERENCES AND FOOTNOTES

1) For a review, see: M.A.B. Bég and A. Sirlin, Phys. Reports 88, 1(1982).

2) A very complete review of grand unification, and many other topics of interest to us, may be found in: P. Langacker, Phys. Reports 72, 185 (1981). The SU(5) - based theory is also discussed in Ref. 1; for recent developments, see the papers in Ref. 3.

3) P. Langacker, Comments on Nuc. and Part. Phys., in press (1985); D. Nanopoulos, ibid, in press (1985). J. Lo Secco, ibid, in press (1985); M. Goldhaber and W.J. Marciano, ibid, in press (1985).

4) J.H. Schwarz, Comments on Nucl. and Part. Phys. 13, 103 (1984), and references therein; ibid, in press (1985), and references therein.

5) F.J. Dyson, in: The Aesthetic Dimension of Science, ed. D.W. Curtin (Philosophical Library, New York, 1982).

6) M.A.B. Bég, C. Panagiotakopoulos and A. Sirlin, Phys. Rev. Lett. 52, 883 (1984).

7) R. Dashen and H. Neuberger, Phys. Rev. Lett. 50, 1897 (1983).

8) D.J.E. Callaway, Nucl. Phys. B233, 189 (1984).

9) For reviews, see: M.A.B. Bég and A. Sirlin, Ref. 1; E. Farhi and L. Susskind, Phys. Reports 74, 277 (1981).

10) R. Jost, The General Theory of Quantized Fields, Lectures in Applied Mathematics, ed. M. Kac (Am. Math. Soc., Providence, Rhode Island, 1965). The appraisal of QED in this book remains valid, twenty years after.

11) L.D. Landau, in:Niels Bohr and the Development of Physics (McGraw Hill, New York, 1955).

12) N.N. Bogoliubov and D.V. Shirkov, Introduction to the Theory of Quantized Fields (Interscience, New York, 1959), Sec. 43.2.

13) Cf. W. Thirring, Principles of Quantum Electrodynamics (Academic Press Inc., New York, 1958) p. 199.

14) H.D. Politzer, Phys. Rev. Lett. $\underline{30}$, 1346 (1973); D.J. Gross and F. Wilczek, ibid $\underline{30}$, 1343 (1973).

15) M. Aizenman, Phys. Rev. Lett. $\underline{47}$, 1 (1981); J. Frohlich, Nucl. Phys. $\underline{B200}$ [FS4], 281 (1982).

16) D.C. Brydges, J. Frohlich and A.D. Sokal, Commun. Math. Phys. $\underline{91}$, 191 (1983).

17) M.A.B. Bég and R.C. Furlong, Phys. Rev. $\underline{D31}$, 1370 (1985).

18) K.G. Wilson, Phys. Rev. $\underline{B4}$, 3184 (1971); K.G. Wilson and J. Kogut, Phys. Reports $\underline{126}$, 78 (1974); G.A. Baker Jr. and J. Kincaid, Phys. Rev. Lett. $\underline{42}$, 1431 (1979) and J. of Stat. Phys. $\underline{24}$, 469 (1981); G.A. Baker Jr., L.P. Benofy, F. Cooper and D. Preston, Nucl. Phys. $\underline{B210}$, 273 (1982); C.M. Bender, F. Cooper, G.S. Guralnik, R. Roskies and D.H. Sharp, Phys. Rev. $\underline{D23}$, 2976 (1981), $\underline{24}$, 2772 (E) (1982); B. Freedman, P. Smolensky and D. Weingarten, Phys. Lett. $\underline{113B}$, 481 (1982).

19) W.A. Bardeen and M. Moshe, Phys. Rev. $\underline{D28}$, 1372 (1983), and references therein; N.D. Gent and A. Vladikas, Phys. Lett. $\underline{151B}$, 285 (1985).

20) K.S. Babu and E. Ma, University of Hawaii Preprint (1985).

21) A. Bovier and D. Wyler, Zurich ETH Preprint (1984).

22) K.S. Babu and E. Ma, Phys. Rev. $\underline{D31}$, 2861(1985) and University of Hawaii Preprint (1984).

23) J. Ellis, M.K. Gaillard, A. Peterman and C. Sachrajda, Nucl. Phys. $\underline{B164}$, 253(1980).

24) M.A.B. Bég and R.C. Furlong, unpublished.

25) P.W. Anderson, Phys. Rev. $\underline{112}$, 1980 (1958).

26) Y. Nambu and G. Jona-Lasinio, Phys. Rev. $\underline{122}$, 345 (1961).

27) M.A.B. Bég and A. Sirlin, Annu. Rev. of Nuclear Science $\underline{24}$, 379(1974).

28) R. Jackiw and K. Johnson, Phys. Rev. $\underline{D8}$, 2386 (1973); J.M. Cornwall and R.E. Norton, Phys. Rev. $\underline{D8}$, 3338 (1973).

29) L. Dolan and R. Jackiw, Phys. Rev. $\underline{D9}$, 3320 (1974).

30) M.A.B. Bég, in: Recent Developments in High Energy Physics, Proc. Orbis Scientiae, 1980, Coral Gables, eds. B. Kursunoglu, A. Perlmutter and L.F. Scott (Plenum, New York, 1980) p. 23.

31) M.A.B. Bég, Phys. Lett. $\underline{124B}$, 403 (1983).

32) M.A.B. Bég, Phys. Lett. $\underline{129B}$, 113 (1983).

33) A. Zepeda, Phys. Lett. $\underline{132B}$, 407 (1983).

34) T. Appelquist, M.J. Bowick, E. Cohler and A.I. Hauser, Phys. Rev. D31, 1676 (1985).

35) G. 't Hooft and M. Veltman, Ann. Inst. Henri Poincaré 20, 69 (1974).

36) M.H. Goroff and A. Sagnotti, Cal.-Tech preprint (1985).

37) M.T. Grisaru, P. van Nieuwenhuizen and J.A.M. Vermaseren, Phys. Rev. Lett. 37, 1662 (1976).

38) M. Grisaru, Phys. Lett. 66B, 75 (1977).

39) P. van Nieuwenhuizen and J.A.M. Vermaseren, Phys. Lett. 65B, 263 (1976).

40) For a list of recent papers, see the citations in the second paper of Schwarz in Ref. 4.

COMPOSITE HIGGS AND COMPOSITE FERMIONS

Howard M. GEORGI

Lyman Laboratory of Physics, Harvard University, Cambridge, MA 02138 U.S.A.

I will talk today about models in which the Higgs boson and the quarks and leptons are composite states. I am sure that all of you have heard enough talks about composite models to be suspicious of the subject. But I ask you to withhold the assumption of guilt by association. Just because you have heard so many dumb talks doesn't mean that the idea is dumb. I will try to avoid the usual pitfalls by discussing a very simple class of models and carefully spelling out the calculational tools that I use.

The work I will discuss today has its roots in an idea that I and my students (notably David Kaplan, Mike Dugan, David Kosower, Ann Nelson, and Lisa Randall) have been studying for the last two years. The idea is that the Higgs boson of the simplest SU(2) x U(1) model may exist but be a composite pseudo-Goldstone boson, like the pion, built out of strongly interacting fermions at a scale larger than the SU(2) x U(1) breaking scale, $v \simeq 250$ GeV.[1] This idea can be realized under the following conditions:

1.) There are interactions (which we call ultracolor) that get strong at a scale $\Lambda_{uc} \gg v$.

2.) These interactions have a global chiral symmetry G that is broken spontaneously down to a subgroup H, producing Goldstone bosons.

3.) The unbroken subgroup H contains an SU(2) x U(1) factor. If this were not the case, the electroweak SU(2) x U(1) would necessarily be broken at the scale Λ_{uc} and there would be no Higgs boson at all.

4.) The Goldstone boson manifold, G/H, contains an SU(2) x U(1) doublet. If the unbroken subgroup H is identified with the electroweak SU(2) x U(1), this doublet has the quantum numbers of the Higgs boson. Still, this is not enough to allow us to identify it with the Higgs doublet, because a true Goldstone boson has only derivative interactions. G must be explicitly broken by interactions that are weak at the ultracolor scale.

5.) The effects of weaker interactions break G explicitly in such a way as to almost align the electroweak SU(2) x U(1) gauge symmetry with the unbroken SU(2) x U(1) subgroup of H. A small misalignment can then break the SU(2) x U(1) gauge symmetry. This process can be equivalently described by saying that the G breaking interactions produce a potential for the

doublet pseudoGoldstone boson that gives it a small VEV.
We know how to construct explicit models based on QCD-like ultracolor interactions in which the required symmetry breaking is done entirely by weak gauge interactions (SU(2) x U(1) itself and some additional gauge interaction, such as an axial U(1)).[1] In these models, the self interactions of the Higgs, that in the standard model are arbitrary non-gauge interactions, are actually produced by the symmetry breaking effects of the weak gauge interactions. The Higgs mass is calculable in terms of the parameters in the effective chiral theory that describes the Goldstone bosons and is of order gv where g is a typical weak gauge coupling.[2] In fact, because the physics that produces the Higgs mass in these theories is the same as that which produces the $\pi^+ - \pi^0$ mass difference in QCD, the relevant parameters can be read off from the wallet cards, with the result that the Higgs mass is about 1.7 $M_W/N^{1/2}$ for SU(N) ultracolor.

Unfortunately, models based on QCD-like gauge ultracolor groups do not seem to give an attractive picture of the other important non-gauge interactions of the standard model, the Yukawa couplings of the Higgs to quarks and leptons. It is possible that these couplings are generated by the symmetry breaking effects of four-fermion operators associated with physics above the ultracolor scale. This is analogous to the extended technicolor idea in technicolor models. As in the technicolor case, this possibility leads to very cumbersome models that really only put off the puzzle of flavor to a higher energy scale. But there is another possibility that arises if the quarks and leptons, as well as the Higgs, are composite. If the composite quarks and leptons transform chirally under the unbroken symmetry group H of the ultracolor interactions, then in the absence of H symmetry breaking, they can have no Yukawa couplings. But then the weak gauge interactions can break the the symmetry and generate Yukawa couplings proportional to powers of the gauge coupling g^2.

Like the composite Higgs potential, the Yukawa couplings of the composite Higgs to composite fermions can be analyzed using an effective chiral theory as a calculational tool. Given a particular pattern of spontaneous symmetry breaking, G/H, by the strong (ultracolor) theory and given the transformation properties of the massless chiral fermions under H (not G!), the physics at energies small compared to the ultracolor scale can be determined in terms of a few numbers that parameterize the effects of the ultracolor interactions. In reference 3, I show explicitly how this works in a simple example in which

G = SU(3) x SU(3) x U(1);   H = SU(3) x SU(2) x U(1);

and the chiral fermions transform under the unbroken SU(3) x SU(2) x U(1) just as the quarks and leptons transform under color and electroweak gauge interactions. In this model, if color SU(3), electroweak SU(2) x U(1) and one additional U(1) is gauged, the model can exhibit the composite Higgs mechanism. Yukawa couplings to the chiral fermions are generated in order $g^2/16\pi^2$, where g is a typical gauge coupling. This model is not very interesting because flavor does not appear explicitly. It is hidden in the details of the strong interactions that we know nothing about. But one fact is striking. The fermions are lighter than the W and Z and the Higgs. Their masses are at most of order $g^2 v$. This encouraged me to look for models incorporating flavor in more interesting ways.

In reference 3 the only calculational tool used was that of the effective chiral theory.[4] Given a particular pattern of spontaneous symmetry breaking by the strong (ultracolor) theory and the transformation properties of the massless chiral fermions under the unbroken subgroup of the original global symmetry, the physics at energies very small compared to the confinement scale Λ can be determined in terms of a few numbers that parameterize the effects of the ultracolor interactions. This is fine as far as it goes, but we cannot be confident that the symmetry breaking pattern and fermions we have assumed can actually be realized in any ultracolor theory unless we have some understanding of the ultracolor dynamics that produces light chiral fermions.

In this talk, I suggest a dynamical picture for a class of ultracolor theories. This dynamical picture determines the symmetry breaking pattern and the structure of massless fermions of the ultracolor theory in an algorithmic way. I will first motivate this dynamical picture in a particularly simple subclass of theories. Then I will show that we can use this dynamical picture to identify a more interesting subclass of models that clearly contains structures that look like quarks with color and flavor.

The only strong gauge interaction that we understand in any detail is QCD. I propose to use our insight into the nature of the physics of QCD by considering the class of theories in which several QCD-like theories are put together in interesting ways. I will argue that this class of models is, in fact, very rich, despite the simplicity of the components. I will try to identify the rules according to which the simple LEGO-like building blocks can be put together into structures with sufficient complexity to describe the low energy world.

Usually, I will be considering only fermions that transform under the simplest N dimensional representations of an SU(N) group. The interesting features will arise because there are several such groups under which the fermions transform simultaneously. In the present state of our ignorance of strongly coupled quantum field theories, I think that we have a much better chance of understanding such models than models involving large representations of peculiar groups. To discuss such LEGO-like models, it is useful to use a notation such as the following.* The gauge groups are simply labeled descriptively;

    SU(N)   an SU(N) gauge group.

An underlined group name labels a global symmetry group;

    <u>SU(N)</u>   an SU(N) global group.

The left-handed (LH) fermions are indicated by directed lines (a line with an arrow) or graphs. If a line goes into (out of) an SU(N), the corresponding fermion transforms as an $N(\bar{N})$ under that SU(N). Thus

    SU(N)—←—~N of SU(N)

    SU(N)—→—~$\bar{N}$ of SU(N).

For example, the color SU(3) theory with three flavors of massless quarks would look like this:

    <u>SU(3)</u>$_R$—→—←—SU(3)$_C$—←—→—<u>SU(3)</u>$_L$

It is clear that a notation of this sort is useful because it has been invented several times independently by people studying extended technicolor models. To build an ETC model in which all the breaking of ETC is dynamical requires very complicated structures involving many different gauge groups. The resulting models in this notation tend to resemble a Moose's antlers. Thus Savas Dimopoulos calls this Moose notation. Sometimes only one arrow on a fermion line is enough. For example, the theory described by the Moose

    <u>SU(M)</u>—→—SU(N)—→—<u>SU(M)</u>

just describes a relabeling of a QCD-like SU(N) gauge theory with an SU(M) x SU(M) global symmetry under which the LH fermions are $(N,\bar{M},1) + (\bar{N},1,M)$. We expect that if M is not too big, the chiral global symmetry will break spontaneously down to a diagonal SU(M) (generated by the sum of the generators of the two original SU(M)'s in some basis). In the process, all the fermions acquire dynamical masses of the order of Λ_{UC}.

Later, we will also consider theories in which some of the gauge groups are SO(N) groups under which the fermions transform like N-vectors. I will represent a LH fermion transforming as an N of SO(N) as

    SO(N)———.

No arrow is required on the fermion line because the representation is real. A simple Moose involving an SO(N) gauge group is the "orthogonal Moose":

    <u>SU(N)</u>—←—SO(M)

describing N SO(M) vector LH fermions with an SU(N) global symmetry. presumably, if N is not too large, this global chiral symmetry breaks spontaneously down to an SO(N) subgroup and all of the fermions acquire dynamical mass.

---

*This is a change (for typographical convenience) in the notation introduced in my talk and in reference 5, but the idea is the same.

Much of the work on theories of massless composite fermions has been based solely on 't Hooft's anomaly consistency condition and simple arguments about decoupling. As a consequence, much of it has been a waste of time. The work of Witten and Vafa[6] shows clearly that consistency is not enough! In vector-like theories, chiral symmetries tend to break spontaneously, rather than surviving to protect massless fermions. The moral is that unless we have some dynamical argument, in a given theory, that leads us to expect the existence of unbroken chiral symmetries and massless composite fermions, they probably aren't there. Nevertheless, there are theories in which massless fermions almost surely exist and can reasonably be called "composite".

$$SU(N) \longrightarrow SU(M) \longrightarrow SU(N) \longrightarrow SU(M).^5$$

This Moose describes a theory with an $SU(M) \times SU(N)_{global} \times SU(M) \times SU(N)_{gauge}$ symmetry under which the LH fermions transform as:

$$(1,\bar{N},M,1) + (1,1,\bar{M},N) + (M,1,1,\bar{N}).$$

Note that the gauge symmetry is free of anomalies. It is easy to see the anomaly cancellation in the Moose notation. Each gauge group must have the same number of fermions leading in and out. If we assume that the $SU(M) \times SU(N)$ chiral symmetry survives in this model (which I call an even linear Moose), there is a very simple way of saturating the anomaly constraints that has an immediate interpretation in a confining picture. Just take one fermion of each type and form a LH gauge singlet fermion state. Under the global $SU(M) \times SU(N)$, these composite fermions transform as an $(M,\bar{N})$. But why should we believe this? One answer is that we can analyze the theory in two other ways and arrive at similar conclusions. The linear Moose theory depends on two parameters, one dimensional and one dimensionless. We can take the dimensional parameter to be the mass scale, M_{UC} at which one or both of the gauge interactions become strong and confining. The dimensionless parameter measures the relative strength of the two gauge interactions. For example, we could take it to be $R_{M/N} = \ln(g_M(M_{UC})/(g_N(M_{UC}))$ where g_M and g_N are the gauge couplings. We are primarily interested in theories in which $R_{M/N} \simeq 0$, so that both gauge interactions get strong and confining at the same scale, but the point is that when $R_{M/N}$ is very large or very small, we can reliably analyze the low energy physics. The reason is that the gauge group in this model is not very chiral. If one gauge group gets strong at a much higher scale than the other, the model looks like QCD with additional weak gauge interactions and some spectator fermions.

For example, suppose that $R_{M/N}$ is very large. Then at energies of order M_{UC}, g_N is very small and the weak $SU(N)$ gauge interactions should have no effect on the confining QCD-like dynamics of the $SU(M)$ gauge group. The strongly interacting fermions have an approximate $SU(N) \times SU(N)$ chiral symmetry and we know from our experience with QCD that this breaks spontaneously down to SU(N). In this case, however, there is no vacuum alignment question. Because one of the SU(N)'s is gauged, all of the Goldstone bosons produced by the spontaneous symmetry breaking are eaten by the Higgs mechanism. No bound states of the strongly interacting fermions remain in the theory as massless states. But the $(1,\bar{N},M,1)$ of fermions in the original Moose do remain in the low energy theory. They no longer carry any gauge interactions at low energies because the SU(N) gauge symmetry has been spontaneously broken. However, the spontaneous symmetry breaking leaves unbroken the global "diagonal" SU(N) symmetry generated by the sum of the original gauge and global SU(N) generators. Thus the global symmetry of the low energy theory is an SU(M) x SU(N) under which these fermions transform as an $(M,\bar{N})$. This symmetry forbids any mass terms for these fermions, so they remain as massless states in the low energy theory.

Obviously, if $R_{M/N}$ is very negative, so that g_M is very small at M_{UC}, something analogous happens. Again there is an SU(M) x SU(N) global symmetry in the theory below M_{UC}, and again there are massless fermions transforming as an $(M,\bar{N})$ of fermions in the original Moose.

Thus, if we look at the low energy physics of this Moose, at energies well below M_{UC}, as a function of $R_{M/N}$, the physics looks the same for very large $R_{M/N}$ and for very small $R_{M/N}$. It is reasonable to expect that the physics will look the same in between. We can approach $R_{M/N} \simeq 0$ from either side and the physics is just what we expect from the simple confining picture in the center. As long as we encounter no phase transition for any value of $R_{M/N}$, we can describe the massless fermions in this theory in any of the three languages. But I think that for $R_{M/N} \simeq 0$, it is reasonable to call these composite fermions.

The linear Mooses above can be immediately generalized to include more strong gauge groups. For example, the Moose

SU(N)$\rightarrow$ SU(M)$\rightarrow$ SU(N)$\rightarrow$ SU(M)$\rightarrow$ SU(N)$\rightarrow$ SU(M)

behaves just like the Moose with two gauge groups. Independent of the order in which the four gauge groups become strong, there is always some global SU(M) x SU(N) symmetry that remains in the low energy theory and some set of massless fermions transforming as $(M,\bar{N})$ saturating the anomaly conditions, just as a confining picture suggests. Again, if all the gauge groups get strong at the same scale, it is reasonable to call these composite fermions.

Likewise, an odd linear Moose such as

SU(M)$\rightarrow$ SU(N)$\rightarrow$ SU(M)$\rightarrow$ SU(N)$\rightarrow$ SU(M)

behaves like QCD. Independent of the order in which the gauge interactions get strong, the SU(M) x SU(M) global symmetry of the model is spontaneously broken down to a diagonal SU(M) and all the fermions get a dynamical mass. Presumably, this behavior persists when all the gauge groups get strong at once.

Thus the linear Mooses fall into two sets. The odd linear Mooses, that is, those with an odd number of gauge groups, produce spontaneous global symmetry breaking but not massless chiral fermions. The even linear Mooses leave the global chiral symmetry unbroken and produce massless chiral composite fermions that saturate the anomaly conditions.

I hope that you are convinced that the odd linear Mooses have massless composite fermions. But while it is convincing in these examples, the process of looking at the theory in all possible limits of the dimensionless parameters is still, at best, only a kind of consistency check on the nature of the physics when all the parts get strong at once. In more elaborate theories, it does not always lead to a unique answer to the question of whether there are massless chiral fermions. Without further motivation, let me suggest my dynamical picture.

The central idea is easy to state and derives from our experience with QCD. I assume that every gauge group that gets strong in a theory of this kind produces a dynamical mass for all the fermions that transform under it. At first hearing, this sounds trivial. It is not because each of these dynamical masses can break other gauge symmetries and *I assume that the dynamical masses form anyway, even if the gauge symmetry is itself broken by other dynamical masses.* Before discussing this assumption further, let me illustrate how it works on the even linear Moose. The dynamical masses produced by the two gauge groups break the gauge symmetries completely down to an SU(M) x SU(N) global symmetry. The dynamical mass produced by the SU(N) gauge interaction (or chiral symmetry breaking condensate, whichever you prefer) breaks the $SU(M)_{global}$ x $SU(M)_{gauge}$ symmetry down to a diagonal global SU(M), while the SU(M) breaks the $SU(N)_{global}$ x $SU(N)_{gauge}$ down to a diagonal global SU(N) x SU(M). Notice that in this process, all the Goldstone bosons associated with the chiral symmetry breaking effects of both sets of dynamical masses are removed by the Higgs mechanism. For related reasons, there is no vacuum alignment problem.

If we label the fermions as follows

$$SU(N) \xrightarrow{A} SU(M) \xrightarrow{B} SU(N) \xrightarrow{C} SU(M)$$

we can schematically represent the Majorana dynamical mass matrix (in a Majorana basis) as

$$(A^T, B^T, C^T)\gamma^0 M \begin{pmatrix} A \\ B \\ C \end{pmatrix}$$

where

$$M = \begin{vmatrix} 0 & \Lambda_M & 0 \\ \Lambda_M & 0 & \Lambda_N \\ 0 & \Lambda_N & 0 \end{vmatrix}.$$

Here Λ_M (Λ_N) is the dynamical mass produced by the strong SU(M) (SU(N)) gauge group. In fact, this mass matrix describes 3MN fermions. Each of the blocks is an MN × MN matrix. But by a choice of gauge we can take each of them to be proportional to the identity matrix. This matrix has MN eigenvectors with zero eigenvalves. These are the massless fermions in the Moose Calculus picture. They transform as an $(M, \bar{N})$ under the global SU(M) × SU(N) that is left unbroken by the dynamical mass matrices.

In this picture, we have an interpretation of the massless fermions that interpolates continuously from $R_{M/N} \gg 1$ (where $\Lambda_M \gg \Lambda_N$ and the massless fermions are essentially entirely the C-type fermions) to $R_{M/N} \ll -1$ (where $\Lambda_N \gg \Lambda_M$ and the massless fermions are A-type fermions). Furthermore, it is not just a consistency condition, but a real dynamical picture. I like to think of it as a "Higgs phase" interpretation of the massless composite fermions.

When an odd linear Moose is analyzed in the same picture, one finds that again the separate symmetries are broken down to a diagonal SU(M) and SU(N). But here the SU(N) is an unbroken gauge symmetry. The global SU(M) × SU(M) symmetry is effectively broken down to the diagonal SU(M). The dynamical mass matrix of the fermion has no zero eigenvalues. There are no massless fermions.

Linear extensions of the orthogonal Moose work in a similar way. The SUN) global symmetry is always broken down to an SO(N) subgroup, there is a remaining unbroken SO(M) gauge symmetry, and all the fermions have dynamical mass.

As for the even linear Mooses, there is no alignment problem associated with any of these Mooses. The gauge and global symmetries are large enough to allow us to put the condensates into any convenient canonical form (for example, to choose them all proportional to the identity matrix in the appropriate space).

An embarrassing (to me) illustration of the power of the Moose Calculus can be seen in the analysis of the triangle Moose discussed by myself and John Preskill in reference 5. On the basis of consistency arguments (stronger than 't Hooft's) we concluded that these models might contain interesting massless chiral fermions. The Moose Calculus, on the other hand, predicts that the global symmetries in the triangle Moose are spontaneously broken and that no massless fermions survive. I now believe the Moose Calculus result.

The even linear Mooses, by themselves, are not useful for building interesting models because the massless composite fermions they yield carry global

chiral U(1) symmetries that forbid Yukawa couplings and masses even if their
global chiral nonabelian symmetries are broken. The charge that generates
this chiral U(1) is +1 on the A and C fermions and -1 on the B fermions. The
massless chiral fermions have charge +1. In a general even linear Moose, the
U(1) charge just alternates, ±1, along the line. Because this U(1) commutes
with all the global symmetries of the theory (which in this case are all linearly realized), it cannot be broken by weak gauge symmetries except through
instanton effects. Such effects can produce masses and Yukawa couplings for
the massless composite fermions, but the models are very cumbersome and inef-
ficient. It is far more interesting to consider models in which the chiral
U(1) symmetries can be broken perturbatively. Consider the following Moose:

$$SU(N) \longrightarrow SU(K) \longrightarrow SU(N) \longrightarrow SU(K+M)$$
$$\uparrow$$
$$SO(M)$$

This is constructed by combining the even linear Moose with the Moose

$$SO(M) \longrightarrow SU(N) \longrightarrow SU(M)$$

which (like the orthogonal Moose) breaks its global SU(M) symmetry down to
SO(M), producing dynamical Majorana masses for all its fermions. The basic
strategy for combining the linear Mooses is to identify one of the gauge groups,
in this case the SU(N), so that the two Mooses share a common strong gauge
interaction. The arrows that determine the fermion representation are chosen
so that the shared global unitary symmetry (in this case SU(K+M)) is larger
than the corresponding global symmetries of the separate Mooses (SU(K) and
SU(M)). This will allow us to produce fermion masses and Yukawa couplings by
weakly gauging subgroups of the SU(K+M).

In the Moose Calculus analysis, the dynamical masses produced by the
SU(K) and SO(M) gauge groups act just as they do in the separate linear
Mooses. The SU(K) condensate breaks the $SU(N)_{global} \times SU(N)_{gauge}$ symmetry
down to a diagonal global SU(N). The SO(M) condensate breaks the $SU(N)_{gauge}$
down to an SO(N). But the SU(N) condensate splits the fermions transforming
as (K+M) under the global SU(K+M) into two subsets. K of them condense with
the fermions that carry the gauge SU(K) while the remaining M condense with
the fermions that carry the gauge SO(M). In the process the SU(N) x SU(K+M)
global symmetry is broken down to SO(N) x SU(K) x SO(M). Associated with the
even linear Moose part are massless fermions transforming as (N,K,1) under the
unbroken symmetry.

There are two global U(1) symmetries of this Moose. If the charges of the
fermions are denoted as

$$\underline{SU(N)} \xrightarrow{q_1} SU(K) \xrightarrow{q_2} \underline{SU(N)} \xrightarrow{q_3} \underline{SU(K+M)}$$
$$\uparrow q_4$$
$$SO(M)$$

the two U(1)'s correspond to the following values:

Q_1: $q_1=1$, $q_2=-1$, $q_3 = K/(K+M)$, $q_4 = 0$;

Q_2: $q_1=q_2=0$, $q_3= M/(K+M)$, $q_4 =-1$.

The symmetry associated with Q_2 is completely broken by the Majorana dynamical masses of the SO(M) fermions. However, the symmetry associated with Q_1 combines with the U(1) subgroup of SU(K+M) that commutes with SU(K) x SO(M) to leave an unbroken U(1) symmetry under which the massless composite fermions transform nontrivially (with charge 1). Thus mass terms for the massless fermions are forbidden both by the unbroken nonabelian SU(K) global symmetry and the unbroken U(1) global symmetry.

This unbroken U(1) does not prevent us from generating masses or Yukawa couplings. Because it contains a piece of the SU(K+M), it can be explicitly broken by weak gauge interactions. For example, suppose that we weakly gauge an SO(K+M) subgroup of the SU(K+M) global symmetry of the Moose. This explicitly breaks both the U(1) and the SU(K) symmetries. Thus no global symmetry remains to forbid mass terms and we might expect masses to develop. To see this explicitly, we must discuss the alignment of the weakly gauged SO(K+M) with respect to the unbroken SU(K) and SO(M). This can be analyzed quantitatively using the formalism of Coleman et al.[4] I will not reproduce this analysis here. Qualitatively, however, we might expect on the basis of experience with QCD that the vacuum alignment will preserve as much of the gauge symmetry as possible. In this case, that would be an SO(K) x SO(M) subgroup of the SO(K+M).

If the vacuum aligns in this way, the symmetry of the low energy theory is a global SO(N) and a weakly gauged SO(K) and SO(M). The light fermions are an (N,K,1) under this symmetry. But these fermions develop a mass of the order of the square of the SO(K+M) gauge coupling times Λ_{UC}. This can be seen in the language of Coleman et al., as in reference 3. But in the Moose Calculus picture, it can also be seen graphically, as follows.

The (N,K,1) fermions are linear combinations of the fermions labeled by q_1 and q_3 above. Those labeled by q_3 are in an SO(K+M) multiplet with fermions that transform as (N,1,M) and have a dynamical Majorana mass as a result of mixing with the fermions labeled by q_4. Thus the light fermions can emit a massive SO(K+M) gauge boson and an (N,1,M) fermion, which in turn can change chirality because of its Majorana mass and reabsorb the gauge boson to become a light fermion with the opposite chirality. The result is a Majorana mass term for the light fermion of order $g^2 \Lambda_{UC}$, where g is the SO(K+M) gauge coupling at the ultracolor scale

The point of this kind of connection between linear Mooses is the following: The Moose Calculus tells us that when they are combined in this way, the fermion mass matrix splits up into pieces, one for each linear Moose. Thus the massless fermions in the simple even linear Mooses are still present in these combinations. But now, weakly gauged subgroups of the larger global symmetries can cause transitions between the massless fermions and the dynamically massive fermions that result in masses or Yukawa couplings to composite Higgs.

As a final simple example of the Moose Calculus, consider the following Moose:

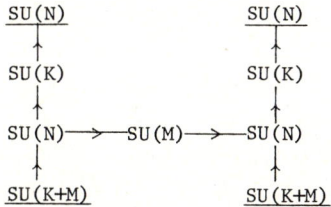

The reader will recognize this as a combination (in the sense discussed above) of two even linear Mooses (the vertical sections) and an odd linear Moose (the horizontal bridge between the even linear Mooses). In this case, the Moose Calculus tells us that the SU(N) x SU(K+M) x SU(K+M) x SU(N) breaks down to SU(N) x SU(K) x SU(M) x SU(K) where the SU(N) and SU(M) are diagonal subgroups. There are two sets of massless fermions, one for each even linear Moose, transforming as $(N,\bar{K},1,1)$ and $(\bar{N},1,1,K)$ under the unbroken symmetries.

In this model, there is one unbroken U(1) symmetry that does not involve any of the SU(K+M) charges. But it is just a baryon number under which all the N's (under the unbroken diagonal SU(N) that is generated by the sum of charges of each of the SU(N)'s) carry charge 1 and all the $\bar{N}$'s carry charge -1. Thus it does not get in the way of the light fermion masses.

The unbroken SU(N) symmetry in this theory behaves like color. If this SU(N) is weakly gauged (by weakly gauging the diagonal SU(N) subgroup of the SU(N) x SU(N) of the original theory--presumably the vacuum aligns so that the gauge SU(N) symmetry is unbroken), the massless composite fermions behave like colored quarks and antiquarks.

The two unbroken SU(K) symmetries, on the other hand, act as independent flavor symmetries that keep the quarks massless. When the flavor symmetries are broken by weak gauge interactions, the quarks will develop masses. If the flavor symmetry is badly broken, a hierarchical structure of masses can develop.

For example, suppose that we gauge a flavor SU(2) subgroup of the two

SU(K+M)'s under which the K+M of each group transforms as the K+M dimensional representation of SU(2). Now there is a serious alignment problem. There are many ways in which the SU(2) generators may align with respect to the global SU(K)'s and SU(M). Which alignment is picked out depends on the signs and magnitudes of the various symmetry breaking terms induced by the SU(2) gauge couplings. An alignment that can occur (if the signs are right) is one in which the M smallest T_3 values of the SU(2) ((M-K-1)/2 to -(K+M-1)/2) are in the subspace on which the SU(M) acts while the K largest T_3 values ((K+M-1)/2 to (M-K+1)/2) are in the subspace on which the SU(K)'s act.[*] In this case, one flavor gauge boson exchange gives mass only to the SU(K) quark with the smallest SU(2) T_3 value (M-K+1)/2, because only this state can be reached from the dynamically massive states by the flavor raising operators. Thus only one of the quarks gets mass in order g^2. Exchange of two flavor gauge bosons can produce mass for the quark with next highest T_3 value in order g^4. And so on. Thus the quark masses in this model can have a hierarchical structure.[†]

Although I will not discuss examples here, it is possible to incorporate weak SU(2) x U(1) and the composite Higgs mechanism in models of this kind. The composite Higgs can arise quite naturally in the sector of the theory that produces dynamically massive fermions carrying the SU(M) symmetry. Although I have no examples, as yet, with a nontrivial KM matrix, it seems possible that the large global flavor symmetries in models of this kind may eventually prove useful in supressing flavor changing neutral current effects. This, in turn, could allow us to avoid pushing Λ_{UC} up to a very large mass scale by fine tuning. Obviously, the issues of the KM structure and CP violation are among the most interesting questions that must be addressed in models of this kind.

What I like best about the tool kit for model building that I have set forth in this paper is the richness of the structures that can be built with these simple components. It remains to be seen whether Nature makes use of these tools, but I am confident that model building of this kind will deepen our insight. We can begin to see, in strongly interacting gauge theories, the same kind of surprising richness that we have already explored in systems with spontaneously broken weak gauge interactions.

---

[*] In any such alignment (in which the projection operator onto the SU(M) subspace commutes with T_3 of flavor), the flavor T_3 gauge symmetry will remain unbroken.

[†] The analytic (in g^2) contributions to the light quark mass involve independent strong interaction parameters in such order in g^2. Thus the calculation of these terms requires a detailed understanding of the ultracolor interactions. However, in this model, the leading contribution to the masses of the lighter quarks are nonanalytic in g^2 ($g^2 \ln g^2$, for example) and are calculable in terms of the mass of the heaviest quark and the flavor coupling g.

REFERENCES

1) See, for example, M. Dugan, et al., Nucl. Phys. B254 (1985) 299.

2) H. Georgi, D. Kaplan, and P. Galison, Phys. Lett. 143B (1984) 152.

3) H. Georgi, Phys. Lett. 151B (1985) 57.

4) S. Coleman, J. Wess, and B. Zumino, Phys. Rev. 177 (1969) 2239; C. Callan et al., Phys. Rev. 177 (1969) 2247.

5) H. Georgi and J. Preskill, Phys. Lett. B156 (1985) 369.

6) E. Witten, Phys. Rev. Lett. 51 (1983) 2351.

PHENOMENOLOGY OF COLOUR EXOTIC FERMIONS

Dieter LÜST[*]

Sektion Physik, Ludwig-Maximilians-Universität, Munich, F.R.G.

and

George ZOUPANOS[+]

Max-Planck-Institut für Physik und Astrophysik, Munich, F.R.G.

We discuss the phenomenological consequences of a dynamical scenario according to which the electroweak symmetry breaking and generation of fermion masses is due to fermions that transform under high colour representations. Particular emphasis is given to the predictions for rare processes and to the spectrum of high colour boundstates.

1. INTRODUCTION

One of the most important unsolved problems of the standard electroweak model[1] is the generation of the weak boson and fermion masses. A first attempt to replace the elementary Higgs scalar by a fermion-antifermion composite was made in the technicolour scenario[2]. There the $SU(2)_L \times U(1)_Y$ gauge symmetry is broken dynamically by the condensates of the techni-fermions. The generation of quark and lepton masses necessitated the enlargement of the technicolour gauge group to the extended technicolour (ETC). However once the ETC gauge boson masses are fixed by fitting the observed fermion spectrum, large flavour changing neutral currents (FCNC) are induced in tree order[3].

Here we shall follow another suggestion for dynamical mass generation[4-12], which is much more attractive than the technicolour scenario, since the dynamical breaking of the electroweak symmetry requires only ordinary Q.C.D. with additional fermions in higher than triplet representations of the $SU(3)_C$ gauge group. Chiral symmetry breaking in the exotic quark sector may occur at much larger mass scales than the ordinary triplet chiral symmetry breaking and therefore could produce the right values of the W and Z boson masses. Such a dynamical scheme is supported by recent lattice Monte Carlo calculations[13].

---

[*]Address after October 1, 1985: California Institute of Technology, Pasadena, U.S.A.

[+]Alexander von Humboldt fellow. On leave from Physics Department, National Technical University, Athens, Greece.

To generate ordinary fermion masses, one has to embed the $SU(3)_c$-colour group into a larger gauge group G_s and subsequently to break it down to $SU(3)_c$. Then the massive $G_s/SU(3)_c$ gauge bosons mediate transitions between ordinary fermions and fermions with high colour (echo-fermions) if they all belong to a common irreducible representation of G_s [5,6].

As we shall discuss later it is very appealing that these models contrary to ETC do not suffer from FCNC due to tree order heavy gauge boson exchange. FCNC processes involving the heavy gauge bosons are induced by high order diagrams but nevertheless they are safe. However the real difficulty for this scenario arises from the existence of massless Goldstone bosons, a defect which also appears in the technicolour scheme.

2. DYNAMICAL SYMMETRY BREAKING FROM FERMIONS IN HIGH $SU(3)_c$-REPRESENTATIONS

Let us consider a QCD theory with N_r massless new quarks Q_r in $SU(3)$ representations $r > 3$. Then the QCD Lagrangian possesses in each fermion sector r a separate global chiral symmetry (neglecting the effects of the axial anomaly):

$$G_r = U(N_r)_L \times U(N_r)_R \tag{2.1}$$

At some value of the effective strength of the colour forces, it is assumed that condensates of the high colour fermions are formed:

$$\langle \bar{Q}_r Q_r \rangle = \mu_r^3 \tag{2.2}$$

These condensates break the global symmetry G_r to the vector symmetry $U(N_r)_{L+R}$. Therefore N_r^2 Goldstone bosons, which are high colour fermion boundstates arise with decay constant F_{π_r}. In analogy to the decay constant of the ordinary pion we set:

$$\mu_r \approx 2.5 \, F_{\pi_r} \tag{2.3}$$

Switching on the $SU(2)_L \times U(1)_Y$ gauge symmetry the $W^\pm$ and Z° bosons acquire masses absorbing three of the Goldstone bosons

$$M_{W^\pm} = \frac{1}{2} g_2 \sum_i F_{\pi^\pm_{r_i}}$$

$$M_{Z^\circ} = \frac{1}{4}(g_1^2 + g_2^2) \sum_i F_{\pi^\circ_{r_i}} \tag{2.4}$$

where g_2, g_1 are the coupling constants of $SU(2)$ and $U(1)$. Since isospin is conserved the $F_{\pi^\pm_r}$ for every r are the same. Therefore the successful mass relation $M_W = M_Z \cos\theta_W$ is preserved. To reproduce the known values of the weak gauge boson masses one has to require:

$$\sum_i F_{\pi_{r_i}} = 250 \text{ GeV} <=> \sum_i \mu_{r_i} \simeq 625 \text{ GeV} \qquad (2.5)$$

In order to obtain such large scales Marciano[4] assumed, that chiral symmetry breaking depends entirely on the strength of the effective quark-antiquark binding potential, based on the one gluon approximation. Effective potential calculations[14] of chiral symmetry breaking yield that chiral symmetry breaking is induced for

$$C_2(r) \, \alpha_s(\mu_r) \gtrsim \text{const.} \sim \frac{\pi}{3} \qquad (2.6)$$

Also non-perturbative analytic methods support the validity of eq. (2.6)[15].

Next one can integrate the renormalization group equation and finds, neglecting the contribution of the high colour fermions to the β-function, the so-called exponential "Casimir scaling":

$$\frac{\mu_r}{\mu_3} \simeq \exp\left[\frac{2\pi}{7\alpha_s(\mu_3)} \left(\frac{C_2(r)}{C_2(3)} - 1 \right) \right] \qquad (2.7)$$

Then, it can be easily seen, that with $r = 6$ scales $\mu_r \approx 0(100 \text{ GeV})$ can be reached. The behaviour at $\alpha_s(\mu)$ is summarized in figure 1. We have introduced a cut-off Λ where the theory changes and $SU(3)_C$ becomes again asymptotically free (see also ref. 10 and I. Bars ref. 12).

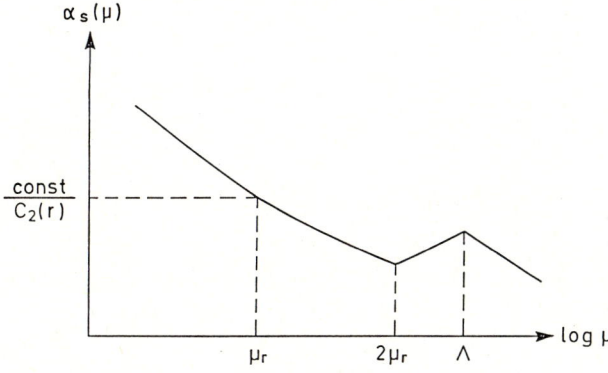

Fig. 1

3. DYNAMICAL FERMION MASS GENERATION

The basic mechanism for fermion mass generation, as we have already briefly outlined, consists of the breaking of a strong gauge group G_s in one or more steps to a $SU(3)_C$ subgroup under which irreducible representations of G_s decompose to colour-singlets, colour-triplets and high-colour representations of $SU(3)_c$. Then, the massive gauge bosons V_μ of $G_s/SU(3)_c$ give rise to radiative diagrams like the ones shown in Fig. 2.

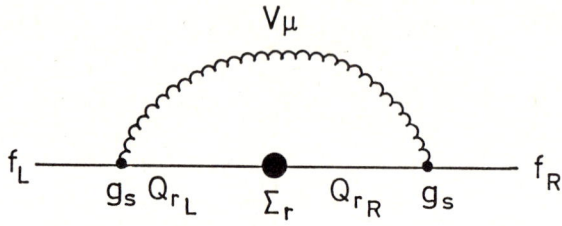

Fig. 2

We can calculate this diagram and obtain:

$$m_f \simeq \left(\frac{3\alpha_s}{4\pi}\right) \frac{3\, C_2(r)\, \mu_r^3}{M_V^2} \ell n \left(\frac{M_V^2}{(3C_2(r))^{2/3} \mu_r^2}\right) \qquad (3.1)$$

The models that can be constructed within this framework can be divided into two classes with the main characteristic feature whether quarks and leptons belong to separate or common representations of G_s.

One can realize quite easily models of the class 1 by enlarging $SU(3)_c$ to $G_s = [SU(3)]^n$. Then G_s can be broken down to $SU(3)_c$ dynamically by the introduction of a new hypercolour force as it was described in ref. 9. Another possibility for the breaking can arise from Higgs fields which have geometrical origin in a dimensional reduction scheme[8]. These breaking mechanisms reduce the gauge symmetry of G_s to an unbroken diagonal $SU(3)$ gauge group which we identify with the colour group. The rest n-1 octets of vector bosons V_{8i} (i=1...n-1) receive masses which are ordered $M_{V8i} > M_{V8i+1}$. They connect ordinary quarks and leptons with the corresponding echos, but they never couple identical representations to each other. To avoid tree order FCNC each quark or lepton family must be assigned to different G_s-representations. Taking into account this requirement it turns out that every colour triplet quark family f (every colour singlet lepton family f) is accompanied by one or several high colour representations r_i^f ($\tilde{r}_i^f$) (for more involved discussion see ref. 10). To obtain the correct $W^\pm$ and Z° masses all the μ_r should lie in the following range:

$$50 \text{ GeV} \leq \mu_r \leq 200 \text{ GeV} \qquad (3.2)$$

Then the mass splitting between different quarks and leptons arises because different high colour condensates are responsible for the radiative mass of different fermions. Also different heavy vector bosons V_{8i} with independent masses can contribute to the fermion mass hierarchy.

The mixing between different quark families arises because some echo-quarks of one family can condense with echo-quarks of another family. Thus, these family number violating condensates produce off diagonal elements in the quark mass matrix.

In ref. 7,10 explicit models realizing these ideas had been constructed. They are based on the gauge groups G_s=SU(3) x SU(3) and G_s=SU(3) x SU(3) x SU(3) and describe two families of quarks and leptons with realistic mass values and family mixing.

In the Pati-Salam type models quarks and leptons are unified in one G_s-representation. Their structure has been discussed in refs. 6,7. The main difference, as compared to the $[SU(3)]^n$ type models, is that here heavy colour triplet gauge bosons V_3 exist which connect not only quarks with echo-quarks but also quarks with leptons. Therefore the leptons obtain masses from the same high colour representations as the quarks, but their radiative diagrams always involve the exchange of one more heavy gauge boson which then leads to the relation

$$m_\ell \sim \frac{\alpha_s}{4\pi} m_q \qquad (3.3)$$

4. HIGH COLOUR BOUND STATES

In this chapter we shall discuss some properties of the $SU(3)_c$-singlet bound states involving high colour quarks (see also ref. 16).

First we consider boundstates which contain only one high colour fermion and several light quarks. We estimate the masses of these high colour baryons in analogy to the mass of the ordinary baryons i.e. by adding the constituent masses of the echo-quarks and the ordinary quarks:

$$M_{B_r} \approx 1.3 \; \mu_r = 100\text{-}200 \text{ GeV} \qquad (4.1)$$

Now the question is if the high colour baryons are stable or if they decay finally into ordinary quark matter. The answer depends crucially on the structure of the considered model. In the first option ($G_s = [SU(3)]^n$) the baryons are absolutely stable. However the situation is totally different in the Pati-Salam type models where the heavy gauge boson V_3 couple also quarks to leptons. In this scheme the echo-quarks can decay into two quarks and one antilepton.

High colour baryons should be detected in future high energy accelerators especially in high energetic hadron-hadron collisions like in the Tevatron. The exact cross sections are presented in ref. 10. But it is worth to mention that high colour baryons could have been already copiously produced in the present CERN collider if their mass is less than about 80 GeV.

The importance of the high colour Goldstone bosons π_r which are composed out of an echo-quark anti-echo-quark pair lies in the fact that their mass could be already in the range of present accelerators. As we shall see, some of them can be even totally massless, a fact which has unpleasant consequences for rare decays.

In order to determine the spectrum of the high colour Goldstone bosons one has to consider the global symmetries of our model. Because every fermion family was accompanied by several high colour representations the global symmetry G of QCD in the high colour sector is rather huge (for explicit discussion see ref. 10). Therefore the spontaneous breakdown of this group by the echo-quark condensates produces a large number of high colour Goldstone bosons. However not all of them remain massless. Most of them obtain mass by the broken $G_s/SU(3)_c$ interactions in the following range:

$$30 \text{ GeV} \leq m_{\pi_r} \leq 200 \text{ GeV} \tag{4.2}$$

In addition electromagnetic effects contribute to the mass of the charged Goldstone bosons:

$$m_{\pi_r^\pm} \sim 10\text{-}20 \text{ GeV} \tag{4.3}$$

So eventually one is left with some neutral true Goldstone bosons.

The massive pseudo Goldstone bosons decay mainly into two gluons or into a fermion antifermion pair with comparable rates (see ref. 10). We also expect clear signals from the high colour pseudo Goldstone boson in high energetic hadron-hadron collisions. (For the production of the Goldstone bosons in e^+e^--collisions see ref. 17). In the present CERN collider they could be seen if their mass is smaller than about 30 GeV.

5. LOW ENERGY PHENOMENOLOGY

Contrary to ETC models where the ETC gauge bosons induce horizontal interactions, in the present scheme the heavy vector bosons couple only flavour diagonally. The reason is that every representation of G_s contains only one quark or lepton family. Therefore the heavy vector boson is flavour blind just like the ordinary gluon. So in all of the present schemes, there are no induced transitions contributing to K_L-K_S mass difference etc. by single gauge boson exchange which was the major difficulty of the ETC models.

However in the Pati-Salam models, semileptonic flavour changing neutral interactions are possible due to exchange of the vector boson V_3. The most stringent limits on the mass of V_3 are provided by the decay of $K_L \to e\mu$:

$$M_{V_3} \gtrsim 50 \text{ TeV} \tag{5.1}$$

In the $[SU(3)]^n$ type models FCNC can be only induced by high order diagrams involving the flavour violating condensates of the high colour fermions. For example to respect the limit on the K_L°-K_S mass difference one finds[10]:

$$M_{V_8} \gtrsim 10 \text{ TeV} \tag{5.2}$$

This bound and also all other bounds of different processes[10] are satisfied in the models of ref. 10.

As we have already discussed there exist some neutral massless Goldstone bosons in the quark sector of our models. They behave like axions. Therefore, one can derive limits on the corresponding decay constant F_{π_r} using the bounds obtained in the searches for axions[18]. The massless Goldstone bosons contribute also to FCNC because of the existence of flavour violating condensates.

The most dangerous process is the decay of K° into an ordinary pion π° and a high colour pion π_r°.

The bounds on the decay rate leads to

$$\mu_r \gtrsim 10^5 \text{ TeV} \tag{5.3}$$

in complete disagreement with any of the values of eq. (3.2).

In conclusion there is an apparent serious difficulty related to the existence of truely massless high colour Goldstone bosons in the present scheme.

6. SUMMARY

We have discussed the phenomenological consequences of models involving high colour fermions. High colour fermions were introduced in an attempt to understand dynamically the electroweak symmetry breaking, the fermion mass generation mechanism and the flavour mixing. Such an attempt is supported by recent lattice calculations.

The advantage of the discussed models, compared to technicolour models, is the lack of tree order FCNC by heavy gauge boson exchange. However the heavy vector boson do induce flavour changing neutral currents in higher orders which puts limits on their masses. These limits are easily satisfied in our models when they are confronted with the values which are needed in order to reproduce the fermionic spectrum.

Models that can be constructed within the presented scheme suffer from the presence of true massless Goldstone bosons which induce rare processes in unacceptably high rates. Therefore better models are needed in which the dangerous

Goldstone bosons are either absorbed by vector bosons or become massive in some way.

Although our specific models should not be taken as totally realistic, they have a number of attractive points and definite predictions. Therefore one should search for the signals of high colour fermions and keep in mind that the existence of these particles with the described or similar properties will be a big step towards the understanding of the origin of the $W^{\pm}$, Z°, quark and lepton masses.

ACKNOWLEDGEMENTS

It is a pleasure to thank E. Papantonopoulos and K.H. Streng for a very fruitful collaboration on this subject. We would like also to thank A.J. Buras, U. Baur, R. Decker, H. Fritzsch and S. Narison for useful discussions. One of us (D.L.) would like to thank Prof. G. Lazarides for the invitation to the very interesting Symposium on "Particles and the Universe" in Thessaloniki.

REFERENCES

1. S. Weinberg, Phys.Rev.Lett. 19 (1967) 1264;
   A. Salam, Proc. 8th Nobel Symposium, ed. N. Svartholm, Stockholm (1968);
   S.L. Glashow, J. Iliopoulos and L. Maiani, Phys.Rev. D2 (1970) 1285.
2. For a review, see:
   E. Farhi and L. Susskind, Phys.Rep. 74C (1981) 277;
   J. Ellis, Lectures given at Les Houches Conference (1981);
   R.K. Kaul, Rev.Mod.Phys. 55 (1983) 449.
3. S. Dimopoulos and J. Ellis, Nucl.Phys. B182 (1981) 505.
4. W.J. Marciano, Phys.Rev. D21 (1980) 2425.
5. G. Zoupanos, Phys.Lett. 129B (1983) 315.
6. P.Q. Hung, "Dynamical generation of fermion masses", Fermilab preprint 80/78-THY (1980), unpublished.
7. D. Lüst, E. Papantonopoulos and G. Zoupanos, Z.Phys. C25 (1984) 81.
8. P. Forgacs and G. Zoupanos, Phys.Lett. 148B (1984) 99.
9. G. Zoupanos, J.Phys.G: Nucl.Phys. 11 (1985) L31;
   D. Lüst, E. Papantonopoulos and G. Zoupanos, Phys.Lett. 158B (1985) 55.
10. D. Lüst, E. Papantonopoulos, K.H. Streng and G. Zoupanos, "Phenomenology of High Colour Fermions "Max-Planck preprint MPI-PAE/PTh 35/85 (1985).
11. G. Zoupanos, Proceedings of International Conference for High Energy Physics, Leipzig (1984).
12. C.H. Albright, B. Schrempp and F. Schrempp, Phys.Rev. D26 (1982) 1737;
    K. Konishi and R. Tripiccione, Phys.Lett. 121B (1983) 403; 132B (1983) 347;
    A.R. White, "Quark-condensate, Higgs scalars and chiral limit of QCD", Argonne preprint ANL-HEP-PR-84-31 (1984);
    I. Bars, "Can the preon scale to be small?", USC 84/033 preprint (1984).
13. J.B. Kogut, J. Shigemitsu and D.K. Sinclair, Phys.Lett. 138B (1984) 283; Phys.Lett. 145B (1984) 239.
    J.B. Kogut et al., Phys.Lett. 50 (1983) 393; Nucl.Phys. B225 (1983) 326.

14. M. Peskin, "Chiral Symmetry and Chiral Symmetry Breaking", Les Houches Lectures, SLAC-PUB-3021 (1982);
    R. Casalbuoni, S. de Curtis, D. Dominici and R. Gatto, Phys.Lett. 140B (1984) 357.
    R.W. Haymaker and J. Perez-Mercader, "Natural, large, dynamically generated gauge hierarchies", Los Alamos preprint (1984) and reference therein.
15. M. Floratos, E. Papantonopoulos and G. Zoupanos, Phys.Lett. 151B (1985) 433; J. Hosek, "Dynamical breakdown of electroweak gauge symmetry", Dubna preprint E2-83-657 (1983).
16. F. Wilczek and A. Zee, Phys.Rev. D16 (1977) 860.
17. S. Narison and J.C. Wallet, Phys.Lett. 158B (1985) 355.
18. G.B. Gelmini, S. Nussinov and T. Yanagida, Nucl.Phys. B219 (1983) 31.

THE PROBLEM OF FAMILIES

Goran SENJANOVIĆ

Physics Department, Brookhaven National Laboratory, Upton, New York, 11973, U.S.A.

I review attempts to unify fermionic families and discuss the question of quark and lepton masses.

In memory of my brother Pavle Senjanović

1. INTRODUCTION

Why do quarks and leptons come in repetitive structures (families)? How many families are there? Are families grouped in larger structures?

Nobody knows. These are the central issues of the weak interaction physics. I don't think it is an accident of semantics that they are jointly considered to constitute "the problem of families". The answers to these questions may require a radical change in our thinking. It could be that quarks and leptons are not elementary. It could be that the underlying objects are strings and all the energy phenomena will be determined by the physics at the Planck scale. It could be ... .

On the other hand, for all that we know the answers may still be provided in the context of the physics of weak interactions or its extensions to grand unified theories. In some sense, the success of perturbative unification demands that we try this direction. This is what I focus on in my talk, I assume that quarks and leptons are elementary particles and study the consequences of the perturbative unification of families. The main stress will be on model independent features that lead to testable low energy predictions. In other words families <u>are</u> grouped in larger structures: they form <u>communes</u>. The question then becomes how to verify this.

In the usual language we study general aspects of the horizontal gauge family symmetry, both global and local. In either case the outcome is quite interesting, assuming that this symmetry is broken spontaneously (it must be broken, since $m_e \neq m_\mu$). In the global symmetry case this implies the existence of Goldstone bosons, <u>familons</u> with some interesting phenomenological consequences. We will discuss this at length. Furthermore, if this symmetry

---

*This work supported in part by the Department of Energy under Contract No. DE-AC02-76CH00016.

is chiral and the number of families different from four, the axion sneaks in beside the familons, leading to even more exciting physics.

On the other hand, the case for local horizontal symmetry is even stronger, after all, other interactions in nature stem from <u>local</u> gauge symmetries. And if, more ambitiously, horizontal and vertical interactions are unified in some grand scheme, it will, in my opinion, be based on large orthogonal groups and <u>mirror fermions</u>, the complete, opposite helicity replica of our world.

The idea of family symmetry is an attempt to do away with a plethora of arbitrary Yukawa couplings of the standard theory. To be completely successful it should eventually predict the quark and lepton masses; it should tell us why the muon is heavier than the electron. Better to say, we ought to compute such ratios as m_e/m_μ. The above approaches, as all others up to now, have failed to achieve this as yet. I will say a few words regarding this problem.

The material discussed here is divided naturally into the following sections: 2. Global family symmetry; 3. Local family symmetry; 4. Quark and lepton masses; 5 Concluding remarks.

Finally, I should mention that this talk is a slight extension of my recent Moriond review of family problem. You will find nothing relevant here not discussed there, just maybe somewhat differently and in more detail.

2. GLOBAL FAMILY SYMMETRY

It is probably more appealing to believe in local symmetry, but our ignorance demands that we study this possibility too. At least a partial motivavation could be to incorporate otherwise ad hoc Peccei-Quinn symmetry into a more natural scheme. If a global horizontal symmetry is broken spontaneously, there <u>will be a plethora of Goldstone bosons</u>, so-called <u>familons</u>[1]. This has exciting phenomenological consequences which are worth going through.

First, Goldstone bosons, as you know, have only spin-dependent couplings. Let me remind you of that. Imagine a global symmetry group G broken down to H. Perform a local rotation

$$U = \exp[i\, T_a G_a(x)/M] \tag{2.1}$$

where M is the scale of symmetry breaking (vacuum expectation value of some scalar field) and $G_a(x)$ are the corresponding Goldstone bosons (T_a are broken generators). This will clearly leave the potential invariant, but <u>not the kinetic terms</u>. Therefore, in this gauge the interactions of matter with Goldstone bosons are necessarily proportional to $\partial_\mu U$. The leading fermion-boson interaction will then be (a, b are some constants)

$$L_{eff} = \frac{1}{M} \partial_\mu G_a\, \bar\psi\, T_a \gamma^\mu (a + b\, \gamma_5)\, \psi \tag{2.2}$$

which on mass shell becomes

$$L_{eff} = \frac{1}{M} G_a \bar{\psi} T^a{}_b \gamma_5 \psi \qquad (2.3)$$

Goldstone boson couplings are always axial (flavor-diagonal part); this means only spin-dependent nonrelativistic terms. This is why gravity is correctly described by the general theory of relativity.

Now, familons are peculiar beasts. Since they are associated with the breaking of family symmetry, their main task - the essence of their existence - is to change flavor through such terms as

$$L_{familon} = \frac{\phi_F}{M_F} (m_\mu \bar{e} \mu + m_s \bar{d} s + \ldots) \qquad (2.4)$$

where M_F is the scale of family symmetry breaking and ϕ_F stands for familons. The branching ratios for the resulting rare decays come to be

$$B = \frac{\Gamma(K^+ \to \pi^+ \phi_F)}{\Gamma(K^+ \to \pi^+ \pi^0)} \simeq \frac{10^{14}}{M_F^2 (GeV)} \qquad (2.5)$$

and similarly $\mu \to e \phi_F$. From experiment[2] $B \leq 10^{-7}$, implying

$$M_F > 10^{10} \text{ GeV} \qquad (2.6)$$

This limit will be pushed by the BNL experiment[3] currently underway to 10^{-10} or even 10^{-11}. This will probe M_F to 10^{12} GeV! Remarkably enough it is not impossible that M_F is limited from above by precisely 10^{12} GeV. The only requirement seems to be a <u>chiral family group</u>[4], as long as the <u>number of families is larger than four</u>.

Let me describe now how it works, without going into detail. In the absence of interaction terms, we have a large chiral flavor symmetry. It may not be crazy then to postulate the underlying theory to possess a <u>chiral</u> family symmetry G_F, subsequently broken at high energies. Example

$$G_F = SU(N)_L \times SU(N)_R \qquad (2.7)$$

N being the number of families and L and R denotes the groups acting on left-handed and right-handed particles, respectively (actually, $SU(N)_R$ could be $SU(N)_\uparrow \times SU(N)_\downarrow$, acting separately on up and down quarks). This forces Yukawa couplings to take the form

$$L_Y = h(\bar{Q}_L^i \phi_d d_R^i + \bar{Q}_L^i \phi_u^* u_R^i) + h.c. \qquad (2.8)$$

where $i = 1, \ldots, N$ counts families. Notice the necessity of different doublets in (2.8) with quantum numbers under $SU(N)_L \times SU(N)_R$

$$\phi_d(N, N^*); \quad \phi_u(N^*, N) \qquad (2.9)$$

as long as $N > 2$. <u>For more than two families we get Peccei-Quinn symmetry!</u> Is this why there are at least three families?

This is not the whole story, of course. You have to break G_F down to nothing at $M_F \gg M_W$. For this you need a set of $SU(2) \times U(1)$ singlets, commonly denoted H. Now you can see the danger immediately. Couple H to ϕ_d and ϕ_u, through, say

$$L_{mix} = M\phi_d^* \phi_u H + h.c. \tag{2.10}$$

so that the $SU(N) \times SU(N)$ quantum numbers of H are $(N^* \times N^*, N \times N)$.

a) <u>$N = 3$</u>. The dangerous term is H^3. Forbidding this term fixes H uniquely to be $(3,3^*)$.

b) <u>$N = 4$</u>. Notice that H^4 is always allowed. No $U(1)_{PQ}$ symmetry in this case.

c) <u>$N > 4$</u>. You have to go to $d = 5$ terms (or larger) to break $U(1)_{PQ}$. For more than four families Peccei-Quinn symmetry is automatic.

<u>Let there be more than four families!</u> The consequence is dramatic. Since[5] $M_{PQ} < 10^{12}$ GeV, this implies in our case $M_F < 10^{12}$ GeV, too. In other words, BNL ought to see a rare decay $K^+ \to \pi^+ + $ nothing!

3. LOCAL FAMILY SYMMETRY

Weak and strong interactions are local gauge theories. Perhaps the horizontal symmetries are local too. This is particularly appealing if you believe in grand unification. Now, the scale of local family symmetry does not have to be that large; all you need is to suppress strangeness changing interactions, which requires $M_F > 10^5$ GeV. It could be that this horizontal symmetry is not unified with other interactions, and that M_F is quite low. You would then see various rare decays such as say $K_L \to \mu e$, which will be probed by experiments at BNL. With big enough accelerators, you could even hope to produce the horizontal gauge bosons. I don't believe in this possibility, but in all fairness, this is an experimental - not theoretical issue. My prejudice is that local horizontal symmetry is actually unified in some grand scheme and that is what I wish to pursue here. We shall try to search for some prinicples that could lead us to such a scheme.

Let us recall what we know of GUTS. Could we extend SU(5) to some large SU(N) group containing family symmetry? I find this ugly. The problem is that SU(5) is inherently sick as far as the family problem is concerned: <u>it does not even unify the fermions of the single family</u>; up and down quarks belong actually to different representations of SU(5)![6] The theory that does this, <u>the minimal theory</u>, is based on SO(10) group. Here all sixteen Weyl fields of one generation (include ν_R please) belong to a single spinor of SO(10). SO(10) is a perfect one-family GUT. The strategy is now obvious: find a large enough group (SO(10+?)) that can unify all the families into a single spinorial representation[7]. If you are not convinced, there is even stronger reason to do this and it has to do with a unique property of spinors: <u>when decomposed under subgroups, they break into spinors only</u>.

The proposal is the following: (i) <u>the low energy world contains only matter with the standard quantum numbers (no exotics)</u>; and (ii) <u>the matter, separated into families, is unified into a large grand unified theory</u>. This singles out orthogonal groups. The consequences, you will see, are quite dramatic.

But first, we have to review a few things about spinors, an easy task since they are just Euclidian analogs of Dirac spinors[8]. I will be very brief: you start with a Clifford algebra

$$\{\gamma_i, \gamma_j\} = 2\delta_{ij} \quad ; \quad i,j = 1, \ldots, 2N , \tag{3.1}$$

construct O(2N) generators in the usual manner

$$T_{ij} = \frac{1}{2} \sigma_{ij} = \frac{1}{4i} [\gamma_i, \gamma_j] \tag{3.2}$$

as in the case of Lorentz group define γ_{five}

$$\gamma_{five} = i^N \gamma_1 \cdots \gamma_{2N} \quad ; \quad \gamma_{five}^2 = 1$$

$$[\gamma_{five}, T_{ij}] = 0 \tag{3.3}$$

In other words, irreducible spinors possess internal helicity $L(R) \equiv (1\pm\gamma_5/2)$; Their dimension is 2^{N-1}. We set out conventions by choosing the N neutral generators of SO(2N) to be

$$\hat{\varepsilon}_i = 2 T_{2i-1, 2i} \qquad i = 1, \ldots, N \tag{3.4}$$

with eigenvalues $\varepsilon_i = \pm 1$. The physical state is given by

$$|\varepsilon_1 \varepsilon_2 \cdots \varepsilon_N\rangle \tag{3.5}$$

Furthermore, $\gamma_{five} = \varepsilon_1 \cdots \varepsilon_N$; and we take $\gamma_{five} = 1$. We now derive our major results.

3.1 Chiral fermions and SO groups

Good theories must forbid bare mass term. Define

$$U = \exp[i \pi/2 (\hat{\varepsilon}_1 + \ldots + \hat{\varepsilon}_N)] = i^N \hat{\varepsilon}_1 \cdots \hat{\varepsilon}_N = i^N \tag{3.6}$$

therefore

$$U(\psi\psi) = i^{2N}(\psi\psi) \tag{3.7}$$

and so we demand $\underline{2N = 4k + 2}$, in order to allow for chiral fermions. SO(4k) groups must be discarded; and similarly SO(odd) groups.

3.2 Decomposition of spinors

The minimal physical group is obviously SO(10); now popular GUT. Its basic spinor is $\underline{16}$ dimensional and contains a full family of quarks and leptons; defined with γ_{five} (SO(10)) = 1. Since γ_{five} (SO(10+4k)) = 1, we get

(a) $\varepsilon_1 \ldots \varepsilon_5 = 1$ or (b) $\varepsilon_1 \ldots \varepsilon_5 = -1$

$\varepsilon_6 \ldots \varepsilon_{5=2k} = 1$ $\varepsilon_6 \ldots \varepsilon_{5+2k} = -1$ (3.8)

<u>Spinors are decomposed only into spinors</u> (or antispinors). Furthermore, we predict for each fermion (a) in the world its opposite helicity <u>mirror</u> (b) (helicity defined through weak interactions). A remarkable outcome, completely independent of the precise theory or the number of families.

3.3 SO(18)

You can easily convince yourself that the minimal possible GUT is SO(18).[7] These theories are of the form SO(10+4k), k = 1,2 ... N ; k =1 or SO(14) is not big enough, it contains only two families and two mirror families. So be it SO(18). It seems to have an opposite problem at the first glance: it contains far too many particles, 8 families and 8 mirror families. If they are all light, asymptotic freedom is gone and we cannot talk of perturbative unification. The question is what, if anything, of the original $SO(8)_H$ (SO(18) $\supset$ SO(10) × $SO(8)_H$) family symmetry remains unbroken down to low energy, or better, what happens to the mirror symmetry in the process of symmetry breaking[9]? It should not remain unbroken forever, since we don't want stable mirror particles.

The physics of mirror fermions is to a large degree independent of the fine details of the theory[10].

a) The number of families must be less than five for the sake of asymptotic freedom. It is worthwhile mentioning that for four families (and four mirror families) proton lifetime in the minimal SU(5) theory is as large as 10^{32} yr. or more[11]! We could still descend through SU(5) theory.

b) The effective fermion-mirror fermion mixing comes from dimension 5 terms[9]

$$L_{eff} = \frac{1}{M_X} \bar{f} \phi^2 F \qquad (3.9)$$

where f denotes ordinary fermions, F mirror fermions and ϕ^2 indicates the necessity of two doublets needed to form a weak singlet. Therefore the mixing is

$$\theta_{fF} \simeq \frac{M_W}{M_X} \lesssim 10^{-13} \qquad (3.10)$$

induced through weak breaking. For laboratory purposes mirror particles are essentially decoupled!

c) Where are mirror particles today? We don't see any on earth or elsewhere. Well, they presumably all decayed a long time ago. From (3.10) the lifetime of a lightest mirror fermion (through ordinary weak decays) is

$$\tau_F = \frac{M_X^2}{m_F^3} \simeq 1 \text{ sec.}$$

for $M_X \simeq 10^{15}$ GeV

$m_F \simeq 100$ GeV (3.11)

All the mirror baryons and leptons could have decayed prior to nucleosynthesis.

Thus the phonomenology of mirror fermions is completely consistent. The bare mass terms are forbidden by mirror symmetry, a part of the gauge theory itself.

d) We still don't know the number of families; but its either <u>three</u> as in our case and in the case of Chang et al., or <u>four</u> as in the program of Bagger and Dimopoulos. Contrast this with the global symmetry analysis in which case one had a rationale for Peccei-Quinn symmetry only if there were more than four families.

e) I shall not discuss here the issue of neutrinos, since I feel strongly that we ought to have some patience and wait for the limit on the number of neutrinos from the Z^0 decays. Cosmological limit depends unfortunately on some assumption, in particular on the <u>belief</u> that the lepton number of the universe is small.[12] I don't know why this should be so. Suffice it to mention the point of Chang et al.[13] who claim that neutrinos in SO(18) are OK, as long as the scale of B-L breaking is not larger than 10^4 TeV. They also study the constraints of grand unification and find the theory consistent if the horizontal $SO(8)_H$ scale is above the SO(10) scale.

3.4 Baryon asymmetry

Since a spinor of SO(18) contains both f's and F's, it is perhaps not too surprising that there is a discrete part of SO(18) D_M which transforms D_M f = F (for each particle). This is similar to charge conjugation being embedded in SO(10). Such symmetry must be broken in order that a nonvanishing baryon asymmetry is generated (mirrors are like antiparticles with the same helicity). Providing enough of a baryon number requires the scale of this symmetry to be close to the SO(18) scale.[13]

4. QUARK AND LEPTON MASSES

So families may form communes with some nice things resulting from the life in a commune. But, so what, you may say. What about the even more important question, the calculability of quark and lepton masses? In the standard theory, GUT or not, particle masses are just arbitrary parameters in the basic Lagrangian. Family symmetry on the other hand restricts Yukawa couplings (or even picks a unique coupling) so that, at the first glance, it ought to be fairly easy to predict relationships between particle masses. Unfortunately, the store is more complicated.

4.1 Global family symmetry

Take, for example, the case of $SU(N)_L \times SU(N)_R$ global horizontal symmetry discussed in section 2. The Yukawa interaction is

$$L_Y = h(\bar{Q}_L^i \phi_d d_R^i + \bar{Q}_L^i \phi_u^* u_R^i) + h.c. \qquad (4.1)$$

Upon symmetry breaking, you could conclude

$$(M_d)_{ij} = h \langle\phi_d\rangle \delta_{ij}$$

$$(M_u)_{ij} = h \langle\phi_u\rangle \delta_{ij} \qquad (4.2)$$

which is nonsense. Of course, there is a weak singlet (s) H which breaks family symmetry, but since $\langle H \rangle = M_F \gg M_W$, its effects should decouple from low energies. Right? <u>Wrong!</u> The decoupling argument does not apply here, since (4.1) is not the most general $SU(2) \times U(1)$ Lagrangian. Unfortunately, (or fortunately) dimension 4 terms are equally important, e.g. d = 5 operator $1/M \; \bar{q}q \; \phi H$, coming from the figure below:

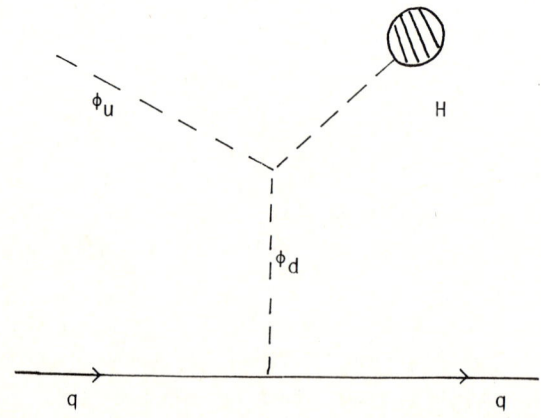

In other words, computing a fermion mass matrix requires summing up all the higher dimension operator; a monumental task. Hopefully, one day it will be performed.

4.2 Local family symmetry

The problems are obviously similar. No determination of quark and lepton masses have yet been achieved. One important point: you can achieve mirror fermions heavier than the ordinary ones in the following manner. Under $SO(10) \times SO(8)_H$, the $\underline{256}$-dimensional spinor of $SO(18)$ is decomposed as

$$\underline{256} = \underset{f}{(16,8)} + \underset{F}{(\overline{16},8')} \qquad (4.3)$$

In $SO(8)$

$$8 \times 8 = 1 + 28 + 35$$

$$8' \times 8' = 1 + 28 + 35' \qquad (4.4)$$

with 35 and 35' not equivalent to each other. The strategy is then simple: choose only $(10,35')$ of $SO(10) \times SO(8)_H$ to break $SU(2) \times U(1)$; this would make mirrors heavy and keep ordinary fermions massless. However, there is more to it. Chang et al.[13] show that unless D_M is broken

$$\langle 10,35 \rangle = \langle 10,35' \rangle \qquad (4.5)$$

a clear disaster. Assuming D_M broken at the GUT level M_X, they get

$$\langle 10,35 \rangle \simeq \left(\frac{M_8}{M_X}\right)^2 \langle 10,35' \rangle \qquad (4.6)$$

where M_8 is the $SO(8)_H$ scale. This is remarkable, since it trades the smallness at arbitrary Yukawa couplings in the standard model for the ratio of mass scales in a family GUT.

This, of course, is not the calculation of masses. As we know, the deep problem is electron to muon mass ratio or the mass relations between different families. This has been an outstanding problem and has been pursued for years. Unfortunately, it is still as much a myster as it ever was.

5. CONCLUDING REMARKS

There is not much more to be said. The question of quark and lepton masses is too deep to be answered today. The program described here is somewhat pragmatic in that it puruses the established ideas of perturbative unification. Its major hope in a sense is that nature eventually runs out of imagination. While the program may be lacking in depth, at least it is predictive and as described in the course of the paper, it will be test in the near future.

The familons (massless Goldstone scalars) of the global symmetry manifest themselves in rare decays that change K^+ into π^+ or μ into e. When the symmetry is chiral, they automatically include the axion as well (at least for $N_F > 4$) and furthermore, the above rare decays ought then to show up in the experiment under way.

The grand unified version of the local horizontal symmetry points towards large orthogonal groups SO(10+4n) (SO(18)?). This results in the existence of mirror fermions which should weigh less than 250 GeV, the lightest of them being strikingly long-lived (~ 1 sec.). This idea, if it works, should explain both the proliferation of families and the proliferation of arbitrary parameters of the standard theory. All the families (together with mirrors) constitute a large, irredicible representation of a unifying group SO(18)?), so that the particle masses ought to be predicted from a single starting parameter. Hopefully, one day this will happen.

In all honesty, I, for one, wouldn't mind a more fundamental explanation for the existence of families. Certainly a possibility that the underlying unique string theory in higher number of dimensions, eventually determines all the low energy parameters is far more appealing than the program I have outlined, especially in view of the complete omission of gravity in the usual GUT program.

ACKNOWLEDGMENTS

I am grateful to the organizers of this impressive conference, George Lazarides, Quiser Shafi, and all the local people for the warm hospitality they extended to us. I hope enjoyed my stay in Thessaloniki greatly.

REFERENCES

1) G. Gelmini, S. Nussinov and T. Yanagida, Nucl. Phys. B219 (1983) 31;
   F. Wilczek, Phys. Rev. Lett. 49 (1982) 1549.

2) Y. Asano et al., Phys. Lett. 107B (1981) 159.

3) I.-H. Chiang et al., BNL-AGS Exp. is searching for $K^+ \to \pi^+$ + nothing, with sensitivity up to 2×10^{-10}.

4) D. Chang, P. Pal and G. Senjanović, Phys. Lett. 153B (1985) 407.

5) L. Abbott and P. Sikivie, Phys. Lett. B120 (1983) 133;
   M. Dine and W. Fischler, Phys. Lett. B120 (1983) 137;
   J. Preskill, M. Wise and F. Wilczek, Phys. Lett. B120 (1983) 127.

6) For an ingenious attempt to pursue SU(N) groups, see H. Georgi, Nucl. Phys. B156 (1979) 126.

7) M. Gell-Mann, P. Ramond and R. Slansky, in: Supergravity, eds. P. von Nieuwenhuizen and D.Z. Freedman (North-Holland, Amsterdam, 1979).

8) For a nice expose on spinors, see F. Wilczek and A. Zee, Phys. Rev. D25 (1982) 553.

9) The physics of SO(18) and mirrors has recently been revived by:
   G. Senjanović, F. Wilczek and A. Zee, Phys. Lett. 141B (1984) 389;
   J. Bagger and S. Dimopoulos, Nucl. Phys. B244 (1984) 247.

10) Mirrors have been introduced before in the context of SU(N) groups by J.C. Pati and A. Salam, Phys. Lett. 58B (1975) 333. For a large literature on mirrors, see G. Senjanović, Proc. of Inner Space/Outer Space Conference, Fermilab (1984), eds. E. Kolb et al.

11) J. Bagger, S. Dimopoulos and E. Masso, Phys. Lett. 145B (1984) 211.

12) See, e.g. J. Harvey and E.W. Kolb, Phys. Rev. D24 (1981) 2090, and references therein.

13) D. Chang and R.N. Mohapatra, Phys. Lett.
D. Chang, A. Kumar and R.N. Mohapatra, Univ. of Maryland preprint #86-18 (1985);
See also, J. Bagger, S. Dimopoulos, E Masso and M.H. Reno, SLAC-Pub-3586 (1985).

THE NEUTRINO MASS, LEPTON FLAVOR AND LEPTON NUMBER NON CONSERVATION

J.D. VERGADOS

Theoretical Physics Division, The University of Ioannina, GR 453 32
Ioannina - Greece

The properties of neutrino are discussed in the context of modern gauge theories. The connection between the nature of the neutrino and family and lepton number conservation is examined. The possible neutrino mass combinations which can be measured in various processes are examined and the consistency between the various experiments is discussed.

1. INTRODUCTION

There is a familiar to all physicists relationship between symmetries on one hand and conservation laws on the other. If a symmetry is known to govern the laws of physics one expects a conserved quantity associated with it. And conversely if a conservation law is known to exist one is not completely satisfied until one finds the symmetry behind it.

Electric charge conservation guarantees the absolute stability (in free space) of the electron which is the lightnest singly charged particle. As of last year the experimental limit [1] on its half-life was $T_{1/2} > 2 \times 10^{22}$y. A new limit has been obtained in the neutrinoless ββ-decay experiment of the South Carolina group [2] which also looked for the decay $e^- \to \gamma \nu_e$. This is

$$T_{1/2} > 3 \times 10^{25} y \tag{1}$$

Electric charge by itself, however, cannot explain the stability of the proton, since many decays are possible e.g.

$$p \to e^+ e^- e^+, \quad p \to e^+ \pi^0, \quad p \to \mu^+ K^0, \quad p \to \tilde{\nu}_\mu K^+ \text{ etc.} \tag{2}$$

Furthermore the neutrons inside nuclei as well as the nuclei themselves appear stable against processes like

$$(A,Z) \to (A-2,Z-2) + e^+ + e^+, \quad (A,Z) \to (A-2,Z-2) + e^+ + e^+ + \text{mesons} \tag{3}$$

The experimental limits are [1]

$$T_{1/2} > 10^{30} y; \quad T_{1/2}(p \to e^+ \pi^0) > 10^{32} y \tag{4}$$

The nonoccurence of processes (2) and (3) is understood if one introduces an additional charge B, the Baryon quantum number, and postulates its conservation.

We also know that the lightest fermions, the leptons (e,μ,τ and their neutrinos), always appear in pairs. Furthermore processes like

$$(A,Z) \to (A,Z\pm 2) + e^{\mp} + e^{\mp} \tag{5}$$

$$e_b^- + (A,Z) \to (A,Z-2) + e^+ \tag{6}$$

$$\mu_b^- + (A,Z) \to (A,Z-2) + e^+ \tag{7}$$

have never been observed. The longest experimental limit ever set for reactions (5) and (6) has recently been achieved by the South Carolina group [2]

$$T_{1/2} > 1.7 \times 10^{23} \, y \tag{8}$$

which is an order of magnitude improvement over that of Bellotti et al [2]. This is much longer than expected even if such processes are viewed as second order weak interactions. The branching ratio for process (7) is also very small

$$R_{exp} < 9 \times 10^{-10} \tag{9}$$

(see e.g. A. Badertscher et al., ref. 3).

The above phenomena are understood if one assumes the existence of a new charge L, the lepton quantum number, and postulates its absolute conservation.

Even the introduction of the lepton quantum number L is not enough. Processes like

$$\mu \to e e^+ e^-, \quad \mu \to e\gamma, \quad \mu \to e\gamma\gamma \tag{10}$$

$$\mu_b^- + (A,Z) \to e^- + (A,Z) \tag{11}$$

have not been observed. The branching ratios for (10) - (11) are
$R(\mu \to 3e) < 2.4 \times 10^{-12}$ (ref. 4), $R(\mu \to e\gamma) < 1.7 \times 10^{-10}$ (ref. 5),
$R(\mu \to e\gamma\gamma) < 8.9 \times 10^{-9}$ (ref. 6), $R(\mu\text{-}e) < 1.6 \times 10^{-11}$ (ref. 7).
Also the decays

$$\tau \to e\gamma, \quad \tau \to \mu\gamma, \quad \tau \to \mu e^+ e^- \tag{12}$$

have not been observed. The obtained limits are [1]
$$R < 5 \times 10^{-4}$$

These facts are explained if one introduces three additive family (lepton) numbers, L_α, $\alpha = \mu, e, \tau$ as follows:

$$\begin{aligned} e^- &: L_e = 1, \ L_\mu = 0, \ L_\tau = 0 \\ \mu^- &: L_e = 0, \ L_\mu = 1, \ L_\tau = 0 \\ \tau^- &: L_e = 0, \ L_\mu = 0, \ L_\tau = 1 \end{aligned} \tag{13}$$

and postulates that they are conserved. Then the lepton number $L = L_e + L_\mu + L_\tau$ is also conserved. The introduction of the family quantum number has the following implications: i) There exist three different families of neutrinos. Which differ from each other in lepton flavor charge.
ii) In conjuction with the absence of right-handed neutrinos it guarantees

that the neutrinos are massless. Indeed the only possible Lorentz scalars which one can write down are

(a) $\bar{\nu}_L \nu_R^c$: (Majorana mass term) (14)

(b) $\bar{\nu}_L N_R$: (Dirac mass term) (15)

where $\nu_R^c = \frac{1}{2}(1+\gamma_5)\nu^c$; $\nu^c = C(\bar{\nu})^T$; C = charge conjugation.

The Majorana mass term violates charge by two units. The Dirac mass term is absent if the right-handed neutrino N_R does not exist.

If the neutrinos are massless one cannot tell whether they are Dirac of Majorana Particles. (There is a unitary transformation which transforms one to the other).

2. THE NEUTRINO MASS

The standard model proposed by Glashow, Salam and Weinberg [8-10] was the successful culmination of attempts to construct a renormalizable theory of weak and electromagnetic interactions. It explains all presently known experimental data. The standard model, however, is currently viewed more as a phenomenologically successful description rather than the ultimate theory of nature [11].

If one goes beyond the standard model the neutrino naturally acquires a mass by introducing one or both terms given by eqns (14) and (15). Thus the most general neutrino mass matrix [12] m is

	ν_R^{oc}	N_R^o
$\bar{\nu}_L^o$	(m_ν)	(m_D)
$\bar{N}_L^{oc}$	(m_D^T)	(m_N)

(16)

where ν_L^o, N_R etc. is a short hand notation for

$$\nu_L^o = (\nu_e^o, \nu_\mu^o, \nu_\tau^o \ldots)_L, \quad N_R = (N_e^o, N_\mu^o, N_\tau^o, \ldots)_R \text{ etc} \quad (17)$$

The following special cases will be examined seperately.

a) Extension of the fermion sector only. If the standard model is extended to contain the right-handed neutrino the following Yukawa coupling is allowed

$$L_Y = \sum_{\alpha,\beta} \Gamma_D^{\alpha,\beta} (\bar{\nu}_\alpha \; \bar{e}_\alpha)_L \begin{pmatrix} \phi^{o*} \\ -\phi^- \end{pmatrix} N_{\beta R} \quad (18)$$

which in a way analogous to the up-quark case yields the Dirac mass

$$m^{\alpha\beta} = \Gamma_D^{\alpha,\beta} \frac{\upsilon}{\sqrt{2}}, \quad \upsilon = \langle \phi^o \rangle \quad (19)$$

In such a model, assuming natural Higgs couplings, it is very difficult to see

why the neutrino mass spectrum should be different from that of the up quarks.
b) Extension of the Higgs sector only. Without the right-handed neutrino one can only generate the Majorana mass matrix (m_ν). Since the fermion bilinear $\bar{\nu}_L \nu_R^c$ has the quantum numbers $I = -I_3 = 1$ and $Y = 2$, to get a gauge invariant Yukawa coupling one needs to introduce an isotriplet Higgs field with $Y = -2$ i.e. with charges $Q = I_3 - Y/2 = 0, -1, -2$, i.e.

$$L_Y = \Sigma \frac{g_1^{\alpha,\beta}}{\sqrt{2}} (\bar{\nu}_\alpha, \bar{e}_\alpha)_L \begin{pmatrix} \Delta^-/\sqrt{2} & \Delta^0 \\ \Delta^{--} & -\Delta^-/\sqrt{2} \end{pmatrix} \begin{pmatrix} e_\beta^c \\ -\nu_\beta^c \end{pmatrix}_R \quad (20)$$

Lepton conservation demands that the isotriplet has $L = 2$. After the spontaneous symmetry breaking, i.e. when the isotriplet acquires a vacuum expectation value

$$(\Delta^0, \Delta^-, \Delta^{--}) \to (\upsilon^T/\sqrt{2}, 0, 0) \quad (21)$$

one gets the Majorana mass matrix

$$m_\nu^{\alpha,\beta} = (g_1^{\alpha,\beta}/\sqrt{2})(\upsilon^T/\sqrt{2}) \quad (22)$$

c) Extension of both the Higgs and fermion sector. The most economic way to do this (without introducing the isotriplet or preventing it form acquiring a vacuum expectation value), is via an isosinglet Higgs scalar (majoron) together with the right-handed neutrino. The new Yukawa coupling is

$$L_Y = \Sigma\, g_2^{\alpha\beta}\, \overline{N_{\alpha L}^{0c}}\, N_{\beta R}\, \phi_S \quad (23)$$

which Yields

$$m_N^{\alpha,\beta} = g_2^{\alpha,\beta}\, \upsilon_S \quad \text{with} \quad <\phi^S> \to \upsilon^S \quad (24)$$

then we get the matrix of equ. (16) without (m_ν). One can, however, obtain an effective light neutrino Majorana matrix which in perturbation theory becomes [12]

$$(m_\nu) \approx -m_D(m_N)^{-1} m_D^T \quad (25)$$

In this case the neutrino mass can be much smaller than the up-quark mass by a suitable choice of the scale for M_N.
d) Radiatively induced masses [12]. It is possible to construct more economic extensions of the minimal model in which the masses are induced radiatively via higher order diagrams. For more details see ref. 13.

The matrix m can be diagonalized by separate left and right unitary transformations. Assuming that the masses of the left and right-handed neutrinos are vastly different we label the eigenstates as follows:

$$\nu'_{jL},\ j = 1,2,\ldots,n_f \quad \text{and} \quad N'_{jL},\ j = 1,2,\ldots,N_f \quad (26)$$

where the subscript L here indicates left eigenvectors. Then

$$\begin{pmatrix} \nu^0 \\ N^{0c} \end{pmatrix}_L = \underbrace{\begin{pmatrix} S_L^{(11)} & S_L^{(12)} \\ S_L^{(21)} & S_L^{(22)} \end{pmatrix}}_{S_L'} \begin{pmatrix} \nu'_L \\ N'_L \end{pmatrix} \qquad (27)$$

The right eigenvectors are obtained analogously ($S_R' = S_L'{}^*$). The submatrices $S^{(11)}$ and $S^{(22)}$ are not unitary but they may be approximately so. Then

$$m = \sum_{j=1}^{n_f} m_j \bar{\nu}'_{jL} \nu'_{jR} + \sum_{j=1}^{N_f} M_j \bar{N}'_{jL} N'_{jR} + H \cdot C \qquad (28)$$

Equ. (28) is cast in more familiar form by defining

$$\nu_j = e^{i\Lambda_j(\nu)} (\nu'_{jL} + e^{i\lambda_j(\nu)} \nu'_{jR}), \quad N_j = e^{i\Lambda_j(N)} (N'_{jL} + e^{i\lambda_j(N)} N'_{jR}) \qquad (29)$$

where $\Lambda_j(\nu)$ and $\Lambda_j(N)$ are arbitrary and $\lambda_j(\nu)$ and $\lambda_j(N)$ are such that

$$m_j = |m_j| e^{i\lambda_j(\nu)}, \quad M_j = |M_j| e^{i\lambda_j(N)} \qquad (30)$$

Then

$$m = \sum_{j=1}^{n_f} |m_j| \bar{\nu}_j \nu_j + \sum_{j=1}^{n_f} |M_j| \bar{N}_j N_j \qquad (31)$$

Equation (27) can be rewritten in terms of the eigenfields as follows:

$$\begin{pmatrix} \nu^0 \\ N^{0c} \end{pmatrix}_L = \underbrace{\begin{pmatrix} S_L^{(11)} e^{-i\Lambda(\nu)} & S_L^{(12)} e^{-i\Lambda(N)} \\ S_L^{(21)} e^{-i\Lambda(\nu)} & S_L^{(22)} e^{-i\Lambda(N)} \end{pmatrix}}_{S_L} \begin{pmatrix} \nu \\ N \end{pmatrix}_L \qquad (32)$$

The right-handed fields are related by

$$\begin{pmatrix} \nu^{0c} \\ N^0 \end{pmatrix}_R = \begin{pmatrix} S_L^{*(11)} e^{-i\alpha} & S_L^{*(12)} e^{-i\phi} \\ S_L^{*(21)} e^{-i\alpha} & S_L^{*(22)} e^{-i\phi} \end{pmatrix} \begin{pmatrix} \nu \\ N \end{pmatrix}_R \qquad (33)$$

where

$$\alpha_j = \lambda_j(\nu) + \Lambda_j(\nu), \; j = 1,2,\ldots,n_f; \; \phi_j = \lambda_j(N) + \Lambda_j(N), \; j=1,2,\ldots,N_f \qquad (34)$$

The charged leptonic left-handed current becomes

$$J_\mu^L = 2(\bar{e}_L^0 \gamma_\mu \nu_L^0) + H \cdot C = 2(\bar{e}_L \gamma_\mu U^{(11)} \nu_L + \bar{e}_L \gamma_\mu U^{(12)} N_L) + H \cdot C \qquad (35)$$

where

$$U^{(11)} = e^{i\Lambda(e)} (S_L^{(e)})^+ S_L^{(11)} e^{-i\Lambda(\nu)}, \quad U^{(12)} = e^{i\Lambda(e)} (S_L^{(e)})^+ S_L^{(12)} e^{-i\Lambda(N)} \qquad (36)$$

$S_L^{(e)}$ is the familiar lepton matrix and $\Lambda(e)$ arbitrary phase (see ref. 12). We note, however, that the matrix $U^{(11)}$ is not unitary. If however $S_L^{(11)}$ and $S_L^{(22)}$ are approximately unitary then the matrix $U^{(11)}$ can be brought into the Kobayashi-Maskawa parametrization.

It is possible to show [11] that the fields ν_j and N_j obey the transformations

$$CP \; \nu_j (CP)^{-1} = e^{-i\alpha_j} \nu_j \;, \quad CP \; N_j \; CP^{-1} = e^{-i\phi_j} N_j \tag{37}$$

where α_j and ϕ_j are defined by eqn. (34). Thus the phases α_j and ϕ_j are related to the CP eigenvalues of the neutrino eigenstates. In particular, if the mass matrix is real, i.e. Hermitian, these phases can be chosen to be +1 or -1. They cannot all be chosen to be +1 since some of the eigenvalues in eqn. (35) may be negative. This has profound implications in ββ-decay [12].

We stress that the above phases α_j and/or ϕ_j are in principle measurable since the charged current involving the conjugate fields becomes [12]

$$J_\mu^L = 2(\bar{\nu}_R \gamma_\mu e^{i\alpha} (U^{11})^T e_R^c + \bar{N}_R \gamma_\mu e^{i\phi} (U^{12})^T e_R^c) \tag{38}$$

These phases appear in the virtual neutrino-antineutrino vertices (e.g. ββ-decay, (μ^-, e^+) conversion, $\nu \leftrightarrow \bar{\nu}$ oscillation etc.). Also the neutral currents involving the neutrinos are no longer exactly diagonal [12].

From the above discussion we conclude that in purely left-handed theories the matrices $S_L^{(21)}$ and $S_L^{(22)}$ are not observable.

In the presence of right-handed currents we find that

$$J_\mu^R = 2(\bar{e}_R^{-0} \gamma_\mu N_R^0) + H \cdot C = 2(\bar{e}_R \gamma_\mu U^{(21)} \nu_R + \bar{e}_R \gamma_\mu U^{(22)} N_R) + H \cdot C \tag{39}$$

where

$$\begin{aligned} U^{(21)} &= e^{i(\lambda(e) + \Lambda(e))} (S_L^{(e)})^T S_L^{*(21)} e^{-i\alpha} \\ U^{(22)} &= e^{i(\lambda(e) + \Lambda(e))} (S_L^{(e)})^T S_L^{*(22)} e^{-i\phi} \end{aligned} \tag{40}$$

Thus now the matrices $U^{(21)}$ and $U^{(22)}$ are, in principle, measurable. The current involving the conjugate fields becomes [12]

$$J_\mu^R = 2(\bar{\nu}_L \gamma_\mu e^{i\alpha} (U^{21})^T e_L^c + \bar{N}_L \gamma_\mu (U^{22})^T e^{i\phi} e_L^c) + H \cdot C \tag{41}$$

The special case $(m_\nu) = (m_N) = 0$ is of special interest. Now the non zero eigenvalues come in pairs which are opposite. Suppose that the states ν_{jL}' and N_{jL}' belong to such a pair. Then one can write [12]

$$m = \sum_j |m_j| (\bar{\nu}_j^{(+)} \nu_j^{(+)} + \bar{\nu}_j^{(-)} \nu_j^{(-)}) \tag{42}$$

with

$$\nu_j^{(\pm)} = (e^{i\Lambda_j}/\sqrt{2}) [\nu_{jL}' + N_{jL}' \pm e^{i\lambda_j}(\nu_{jR}' + N_{jR}')] \qquad (43)$$

which are majorana particles. A more convenient choice can be made however. E.g.

$$\nu_{Dj} = \frac{1}{\sqrt{2}}(\nu_j^{(+)} + \nu_j^{(-)}) = e^{i\Lambda_j}(\nu_{jL}' + e^{i\lambda_j} N_{jR}') \qquad (44)$$

$$\nu_{Dj}^c = (e^{i\alpha_j}/\sqrt{2})(\nu_j^{(+)} - \nu_j^{(-)}) = e^{-i\Lambda_j}(\nu_{jR}' + e^{-i\lambda_j} N_{jL}') \qquad (45)$$

which are Dirac neutrinos. The mass matrix in this case becomes

$$m = \sum_j |m_j|(\bar{\nu}_{Dj}\nu_{Dj} + \bar{\nu}_{Dj}^c \nu_{Dj}^c) \qquad (46)$$

i.e. two majorana neutrinos combine to form a Dirac neutrino. Note that

$$\bar{e}_L \gamma_\lambda \nu_L = \bar{e}_L \gamma^\lambda \nu_D \qquad (47)$$

3. SOME NEUTRINO MEDIATED PROCESSES

We are now going to present a brief summary of various processes mediated by neutrinos. For more details see ref. 12.

3.1. Decay experiments.

In a decay experiment the weak eigenstate β is produced which is a linear combination of mass eigenstates

$$\nu_\beta = \sum_j U_{\beta j}^{(c)} |\nu_j\rangle, \quad U^c \approx U^{(11)} \qquad (48)$$

Thus the decay width of a particle with mass m is

$$\Gamma(\nu_\beta) = \sum_j |U_{\beta j}|^2 \Gamma(m_j)\Theta(m-m_j)$$

For a review of the experimental situation see ref. 14.

a) Muon decay ($\beta = \mu$). In this case one expects various monochromatic lines of momentum P_μ given by

$$m = \sqrt{P_\mu^2 + m_\mu^2} + \sqrt{P_\mu^2 + m_j^2} \qquad (49)$$

each with probability $|U_{\mu j}|^2$. The best experimental limit set this way is

$$m_\mu < 0.25 \text{ MeV} \quad (90\% \text{ C.L.; Abela et al ref. 15}) \qquad (50)$$

b) The τ-neutrino mass is measured analogously in the reaction $\tau^- \to \pi^- \nu_\tau$ from which the best limit obtained is

$$m_{\nu_\tau} < 250 \text{ MeV } (90\% \text{ C.L.; Blocker et al ref. 16})$$
$$m_{\nu_\tau} < 164 \text{ MeV } (95\% \text{ C.L.; A. Mateuzzi et al ref. 17}) \qquad (51)$$

c) The $\bar{\nu}_e$ mass is obtained from a detailed study of the β-spectrum of nuclei.

The process $H^3 \rightarrow {}^3He + e^- + \vec{\nu}_e$ is preferred due to the small energy release $E_{max} = 0.529559$ MeV. The β-spectrum is usually presented in the form of a curie plot

$$K(E) = \frac{dN/dE}{C F (E/P_1 Z) PE} = [\sum_j |U_{ej}|^2 (E_{max}-E)^2 - m_j^2 (E_m-E)]^{1/2} \quad (52)$$

From end point experiments the following limits have emerged

$$m_{\nu_e} \ll 55 \text{ eV } (90\% \text{ C.L., ref. 18}), \quad m_{\nu_e} < 33 \text{ eV } (90\% \text{ C.L. ref. 19}) \quad (53)$$

and more recently

$$14 \text{ eV} < m_{\nu_e} < 46 \text{ eV } (99\% \text{ C.L., ref. 20}) \quad (54)$$

This last important result has not yet been univerally accepted due to uncertainties in the analysis of the data which result mainly from the molecular structure of triton compounds employed. Many experiments are planned [14], some of them involving atomic triton, which aim at the confirmation of the above result.

In another recent experiment not at the end but at the beginning of the β-spectrum [21] the following result was reported

$$m_j = 17.1 \text{ KeV}, \quad |U_{ej}|^2 = 0.03 \quad (55)$$

This result has not been confirmed in the ${}^{35}S \rightarrow {}^{35}C\ell + e^- + \vec{\nu}_e$ decay (see F. Calaprice, ref. 22).

3.2. Neutrino Oscillations.

Since the neutrino of equ. (48) produced at time $t = 0$ is not a stationary state it will evolve according to the laws of quantum mechanics. Thus at $t > 0$ it will look like

$$|\nu_\beta(t)> = \sum_j U^{(c)}_{\beta j} e^{-iE_j t} |\nu_j> \quad (56)$$

where $E_j = \sqrt{p^2 + m_j^2}$ with p the neutrino momentum. Thus the oscillation probably becomes

$$P(\nu_\gamma \rightarrow \nu_\beta) = |<\nu_\beta(t)|\nu_\gamma(0)>|^2 = \sum_j |U^{*(c)}_{\beta j} U^{(c)}_{\gamma j}|^2$$
$$+ 2 \sum_{j<k} \text{Re}\{U^{(c)}_{\beta k} U^{*(c)}_{\gamma k} U^{*(c)}_{\beta j} U^{(c)}_{\gamma j} e^{-i(E_k - E_j)t}\} \quad (57)$$

If we define

$$L = c t, \quad \Delta_{kj} = \frac{1}{2}(\sqrt{P_\nu^2 + m_k^2} - \sqrt{P_\nu^2 + m_j^2})L \quad (58)$$

and use the unitarity of the matrix $U^{(c)}$ we can write Equ. (57) as follows:

$$|<\nu_\beta(L)|\nu_\gamma(0)>|^2 = \delta_{\beta\gamma} - 4 \sum_{j<k} \text{Re}(U^{*(c)}_{\beta k} U^{(c)}_{\gamma k} U^{(c)}_{\beta j} U^{*(c)}_{\gamma j}) \sin^2 \Delta_{kj}$$

$$+ 2 \sum_{j<k} \mathrm{Im}(U^{*(c)}_{\beta k} U^{(c)}_{\gamma k} U^{(c)}_{\beta j} U^{*(c)}_{\gamma j}) \sin 2\Delta_{kj} \qquad (59)$$

the second term appears if the matrix $U^{(c)}$ is complex and it is absent in a CP conserving theory. From equ. (59) immediately follows:

$$|\langle \nu_\beta(L)|\nu_\beta(0)\rangle|^2 = 1 - \sum_{\gamma \neq \beta} |\langle \nu_\gamma(L)|\nu_\beta(0)\rangle|^2 \qquad (60)$$

where we made use of the unitarity of the matrix $U^{(c)}$. The above formulas do not distinguish between Dirac and Majorana neutrinos.

For neutrino masses small compared with the neutrino momentum we get

$$\Delta_{kj} = \delta m^2 L/4 E_\nu, \quad \delta m^2 = m_k^2 - m_j^2 \qquad (61)$$

measuring Δm^2 in eV^2, L in meters and E_ν in MeV we get

$$\Delta_{kj} = 1.269 \, (\delta m^2/(eV)^2) \, (L/m)/(E_\nu/MeV) \qquad (62)$$

This equation is also sometimes written as

$$\Delta_{kj} = 1.269 \, (\delta m^2/(eV)^2) \, (L/km)/(E_\nu/GeV) \qquad (63)$$

Sometimes we exhibit the periodic nature of (59) as a function of the distance L between the source and the detector by writing

$$\Delta_{kj} = \pi \frac{L}{L_0}, \quad L_0 = \frac{4\pi E}{\delta m^2} = 2.476 \text{ km} \frac{E_\nu/GeV}{\delta m^2/(eV)^2} = 2.476 \text{ m} \frac{E_\nu/MeV}{\delta m^2/(eV)^2} \qquad (64)$$

The quantity L_0 is called the oscillation length. For sufficiently small δm^2, L_0 can be of the order of the Sun-Earth distance. The sensitivity of the various experiments in a two state model (see eqns. 66-68) is given in fig.1.

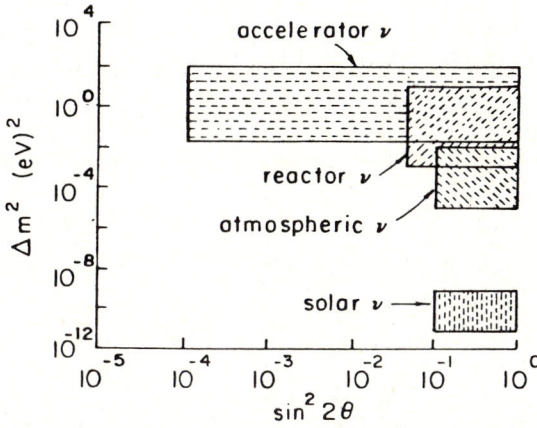

FIGURE 1. Regions in the δm^2 and $\sin^2 2\theta$ plane which are accessible to various neutrino oscillation experiments.

For illustration purposes we will consider a two generation model (e.g. $\nu_\beta = \nu_e$ and $\nu_\alpha = \nu_\mu$). In this case

$$\begin{pmatrix} \nu_e \\ \nu_\mu \end{pmatrix}_L = \begin{pmatrix} \cos\theta & \sin\theta \\ -\sin\theta & \cos\theta \end{pmatrix} = \begin{pmatrix} \nu_1 \\ \nu_2 \end{pmatrix}_L \tag{65}$$

$$|<\nu_e(L)|\nu_e(0)>|^2 = 1 - (\sin 2\theta \sin\Delta)^2 \tag{66}$$

$$|<\nu_\mu(L)|\nu_e(0)>|^2 = |<\nu_e(L)|\nu_\mu(0)>|^2 = (\sin 2\theta \sin\Delta)^2 \tag{67}$$

with

$$\Delta = 1.269 \frac{\delta m^2}{eV^2} \frac{L/m}{E_\nu/MeV} \quad \text{and} \quad L_0 = 2.476 \frac{E_\nu/GeV}{\delta m^2/(eV)^2} \tag{68}$$

Some fashionable 3-generation models are examined in ref. 12.

There exists a variety of neutrino oscillation experiments

a) Reactor Experiments. These utilize low energy $\tilde{\nu}_e$'s from reactors. Thus oscillation to other flavors cannot be detected (the threshold for their detection is the muon mass). Thus such experiments attempt to measure the depletion of the $\tilde{\nu}_e$ flux due to oscillation (disappearance experiments). Since the neutrino flux is not accurately known the experimental set up is arranged to eliminate such uncertainties. Two possibilities have been employed to this end:

i) Experiments on deuteron target. Such experiments have been performed in the Savanah reactor by the Irving group [23]. They search for charged current (cc) and neutral current (nc) events in the reactions

$$\begin{aligned} \tilde{\nu}_e + d &\to e + n + n \quad \text{(ccd-reaction)}, \; E_{th} = 4.0 \text{ MeV} \\ \tilde{\nu}_e + d &\to \tilde{\nu}_e + p + n \quad \text{(ncd-reaction)}, \; E_{th} = 2.2 \text{ MeV} \end{aligned} \tag{69}$$

Since the neutral currents are diagonal in flavor space in the ratio

$$r = \sigma(\tilde{\nu}_e d \to e^+ n n)/\sigma(\tilde{\nu}_e d \to \tilde{\nu}_e pn) = R(ccd)/R(ncd) \tag{70}$$

the uncertainties in the neutrino flux drop out. Then

$$\rho = \frac{r(\text{with oscillation})}{r(\text{no oscillation})} = |<\tilde{\nu}_e(L)|\tilde{\nu}_e(0)>|^2 \tag{71}$$

Unfortunately, however, the analysis of such experiments is difficult since r depends somewhat on the shape of the neutrino spectrum due to the different thresholds for the two reactions. No evidence for neutrino oscillations has been found.

ii) The ccp experiments. These include the Grenoble [24], Gosgen [25] and Bugey [26] experiments which study the reaction

$$\tilde{\nu}_e + p \to e^+ + n \quad \text{(ccp reaction)}, \quad E_{th} = 1.806 \text{ MeV} \tag{72}$$

The yield with oscillation at distance L is related to that without oscillation as follows

$$Y_{osc}(E_e, L) = Y_0(E_e, L) <\tilde{\nu}_e(L)|\tilde{\nu}_e(0)>|^2 \tag{73}$$

To avoid uncertainties in the neutrino flux, inherent in $Y_0(E_e,L)$, it is desirable to perform the experiment at two distances. Then

$$\frac{Y_{osc}(E_e,L_1)}{Y_{osc}(E_e,L_2)} = \frac{|<\tilde{\nu}_e(L_1)|\tilde{\nu}_e(0)>|^2}{|<\tilde{\nu}_e(L_2)|\tilde{\nu}_e(0)>|} \tag{74}$$

This ratio can be energy dependent if Δ is not small. The Grenoble and Gosgen experiments, which are single position experiments, see no evidence for neutrino oscillations but the two position Bugey experiment [26] sees evidence for neutrino oscillations. It is consistent with the Gosgen experiment for $\delta m^2 = 0.2$ (eV)^2 and $\sin^2 2\theta = 0.2$ and 0ν ββ-decay (see below) but not consistent with the Grenoble experiment. It needs confirmation before it is accepted as evidence for neutrino oscillation. A recent review of the experimental situation has been given by Boehm [27].

b) Neutrinos from low energy accelerators. Such an experiment has been carried out at LAMPF [28]. The experiment searches for $\tilde{\nu}_e$'s in the π^+ decay

$$\pi^+_{rest} \to \mu^+ \nu_\mu \tag{75}$$
$$\hookrightarrow \mu^+_{rest} \to e^+ \nu_e \tilde{\nu}_\mu$$

For a clean beam the sought $\tilde{\nu}_e$'s can only arise from the oscillation of $\tilde{\nu}_\mu$'s. No evidence for oscillation has been seen but useful limits have been set [28].

c) Neutrinos from high energy accelerators. These are suitable for oscillations involving the τ-family.

They involve traditional reactions like

$$\tilde{\nu} + N \to \ell^+ + X, \quad \nu + N \to \ell^- + X \tag{76}$$

(where N is the nucleon and X stands for anything) or prompt neutrinos produced in the beam dump from short lived (charmed) particles:

$$D^+ \to \ell^+ \nu \bar{K}^0, \quad D^0 \to \ell^+ \nu K^-, \quad F^+ \to \ell^+ \nu K^+ K^- \quad \text{etc.} \quad \ell = e \text{ or } \mu$$

A deviation from unity of the following measured quantity:

$$R_{pr} = N_{pr}(e^+) + N_{pr}(e^-)/N_{pr}(\mu^+) + N_{pr}(\mu^-) \tag{77}$$

will indicate 3-generation oscillations. For a review of such experiments see 28 and 12. No evidence for such oscillations has yet been seen.

d) Atmospheric neutrinos [29] and neutrinos from the SUN [30]. Experiments of the first category search for neutrinos produced on the opposite side of the Earth, i.e. $L/E_\nu \simeq 2\times 10^4$ (Km/GeV) or $\delta m^2 \simeq 10^{-4}$ eV^2. For the solar neutrinos, since L/E_ν is very large (10^{12} Km/GeV), it is, in principle, possible to measure $\delta m^2 \simeq 10^{-12}$ (eV)^2.

3.3. Neutrinoless double β-decay.

This process, given by equ. (5) and (6), is the oldest lepton violating reaction suggested by Furry 50 years ago as an experimental test of whether the neutrinos are Majorana particles [12]. Of its aspects we are going to summarize here only those which are related to the neutrino properties and refer the reader to recent reviews [12,31] for more details.

The analysis of 0ν ββ-decay requires the solution of a number of complex nuclear and particle problems. For neutrinos which are very light or very heavy compared to the proton mass the nuclear physics can be separated from the gauge parameters so that the expression for the life-time can be cast in the form [12]

$$T_{1/2}(0\nu) = K_{0\nu}(\varepsilon_0, Z, A)/|\eta|^2 |ME|^2_{0\nu} \qquad (78)$$

where η is a lepton violating parameter, which contains all the essential information about the neutrino, and $|ME|^2_{0\nu}$ is the relevant nuclear matrix element varying in the range 1-30 for various nuclei. The quantity $K_{0\nu}(\varepsilon_0, Z, A)$ essentially a kinematical function which depends on the energy released in the decay and the bulk nuclear parameters (the charge Z and the mass number A). It varies between 10^{13} y and 10^{15} y for various nuclei [12]. The interesting parameter, which one hopes to extract from the data, once the nuclear matrix element is known, is η.

The available experiments belong to two classes
a) Indirect experiments [32]. In these one tries to detect the daughter nucleus (A,Z+2) in a sample with high concentration of the parent nucleus provided that all other possible mechanisms for producing the daughter nucleus have been eliminated. A recent review of such experiments has been given by Kirsten [33]. Such experiments cannot distinguish between the neutrinoless (0ν) mode and the usual lepton allowed 2ν mode in which processes (5) and (6) are accompanied by neutrinos. The lepton violating parameter is extracted from the data via a relation of the type:

$$|\eta|^2 = \frac{K_{0\nu}}{|ME|^2_{0\nu}} \left| \frac{1}{T_{1/2}(\exp)} - \frac{1}{T_{1/2}(2\nu)} \right| \qquad (79)$$

once $T_{1/2}(2\nu)$ is calculated. Unfortunately it requires rather precise nuclear calculations. 2ν ββ-decay has been seen this way. The measurements are con-

sistent with no 0ν decay [32].
b) Laboratory Experiments [2,12,31]. These measure the characteristic properties of the 0ν ββ-decay i.e.
i) Simultaneous emission of two electrons (on positrons)
ii) Emission of two electrons (positrons) from one point inside the source.
iii) Constancy of the energy of the emitted leptons. No evidence of 0ν ββ-decay has yet been seen. The best limit obtained [2] is given by equ. (8).

The lepton violating parameter depends on the gauge model [12]. We will consider only those mechanisms which involve intermediate neutrinos. We distinguish the following cases:

A. Left-handed theories. The basic diagrams are shown in fig. 2.

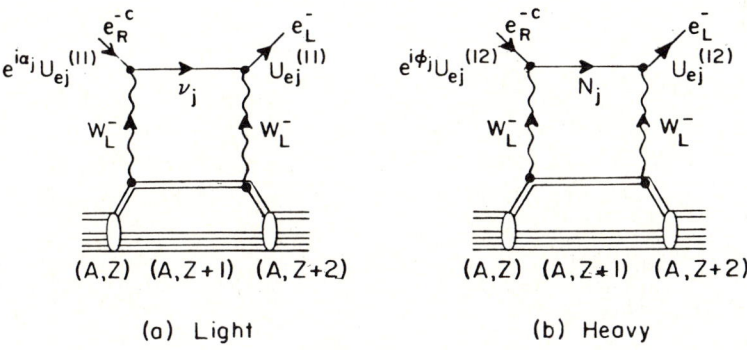

(a) Light   (b) Heavy

FIGURE 2. Diagrammatic representation of neutrinoless double β-decay in left-handed theories (a) Light intermediate neutrinos (b) Heavy intermediate neutrinos. In addition to the couplings, $U^{(11)}$ and $U^{(12)}$ resulting from the mixing mass matrix, the CP eigenvalues $e^{i\alpha_j}$ and $e^{i\phi_j}$ also appear in the antiparticle vertices. A summation over all possible mass eigenstates is implied. For light neutrinos the amplitude is proportional to the neutrino mass.

In this case the lepton violating parameter is

$$\eta = \begin{cases} \eta^{(L)}_\nu = \dfrac{<m_\nu>}{m_e}, \quad <m_\nu> = \sum_j (U^{(11)}_{ej})^2 e^{i\alpha_j} m_j & \text{(light neutrino)} \\ \\ \eta^{(L)}_N = \sum_j (U^{(12)}_{ej})^2 e^{i\alpha_j} \dfrac{m_p}{M_j} & \text{(heavy neutrino)} \end{cases} \quad (80)$$

see equs. (36) and (38) for the required definitions. Note that η is pro-

portional to m_j or $1/M_j$. In most gauge models $U^{(12)}_{ej}$ is very small.
For two neutrino eigenstates we have

$$\eta^{(L)}_\nu = \frac{m_1}{m_e} \cos^2\theta + \frac{m_2}{m_e} \sin^2\theta \, e^{i\alpha}, \quad \alpha = \alpha_1 - \alpha_2 \tag{81}$$

We thus see that $o\nu$ $\beta\beta$-decay is suppressed even if the neutrino are massive Majorana particles provided that

$$\frac{m_2}{m_1} = -e^{-i\alpha}\cos^2\theta \tag{82}$$

In particular we get $\eta_\nu = 0$ if the neutrino is a Dirac particle ($m_1 = m_2$, $\alpha = \pi$, $\theta = \frac{\pi}{4}$). Thus the absence of $o\nu$ $\beta\beta$-decay is consistent with equs. (54) or (55).

B. Right-Left symmetric theories. There exist now two possibilities.

a) Both leptonic currents are right-handed. The new diagrams are analogous to those of fig. 2 with L ↔ R and (11) ↔ (22), (12) ↔ (21). Then

$$\eta = \begin{cases} \eta^{(R)}_\nu = \left[\sum_j (U^{(21)}_{ej}) \, e^{i\alpha_j} \, m_j/m_e\right] (\kappa^2 + \varepsilon^2) & \text{(Light neutrino)} \\ \eta^{(R)}_N = (\kappa^2 + \varepsilon^2) \sum_j (U^{(22)}_{ej})^2 \, e^{i\varphi_j} \frac{m_p}{M_j} & \text{(Heavy neutrino)} \end{cases} \tag{83}$$

where $\kappa = (M_L/M_R)^2$ with $M_L = M_W$ and M_R the mass of the vector boson which mediates the right-handed interaction. $\varepsilon \simeq \tan\zeta$, ζ being the mixing angle between W_L and W_R. $\eta^{(R)}_\nu$ is expected to be negligible compared to $\eta^{(L)}_\nu$ while $\eta^{(R)}_N$ may dominate over $\eta^{(L)}_N$.

b) Interference between the left-handed and right-handed leptonic currents (see fig. 3).

Only the light neutrino component is relevant. Then

$$\eta = \eta_{RL} = \sum_j U^{(21)}_{ej} U^{(11)}_{ej} e^{i\alpha_j} \times \begin{array}{l} \kappa \text{ (fig. (2a))} \\ \varepsilon \text{ (fig. (2b))} \end{array} \tag{84}$$

Note that this amplitude is, to leading order, independent of the neutrino mass. It also vanishes if the neutrino is not a Majorana particle.

Form the geochemical measurements one finds

$$|<m_\nu>| < 5.5 \text{ eV} \quad |\eta_{RL}| < 3.0 \times 10^{-5} \tag{85}$$

The corresponding limits obtained from equ. (8) are

$$|<m_\nu>|^2 < 3.2 \text{ eV} \, (1 \, \sigma \text{ level}), \, |\eta_{RL}| < 1.0 \times 10^{-5} \tag{86}$$

The other lepton violating parameters have been extracted only from the data on the theoretically simple $^{48}Ca \to {}^{48}Ti$ nuclear decay [12] which, however, are 10 years old. These are [12]

$$|\eta_N| = |\eta_R| < 5.6 \times 10^{-7}, \quad |(g_1)^{ee}| < 7 \times 10^{-4} \tag{87}$$

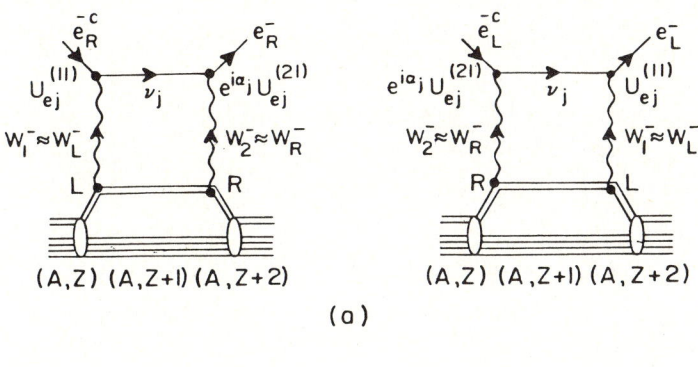

FIGURE 3. 0ν ββ-decay in the presence of right-handed currents. Light intermediate neutrinos are considered. The interference between the left-handed and right handed leptonic currents j_L - j_R leads to an amplitude independent of the neutrino mass. In case (a) the hadronic currents are also J_L - J_R. In case (b) both hadronic currents are left-handed. This is possible due to the mixing of the vector Bosons W_L and W_R.

3.4. Muon number Violating Processes.

One such is the lepton violating (μ^-, e^+) process (see equ. 7). Except for kinematics this process is analogous to 0ν ββ-decay. The lepton violating parameters η' are obtained from those of the previous subsection with the obvious substitution $U_{ej} U_{ej} \to U_{ej} U_{\mu j}$ and complex conjugation. In the special case of two generations we get

$$\eta_\nu^{'(L)} = \frac{1}{2} \sin 2\theta \left| \frac{m_1}{m_2} - \frac{m_2}{m_e} e^{i\alpha} \right| \tag{88}$$

we stress that $\eta_\nu^{(L)}$ and $\eta_\nu^{'(L)}$ cannot simultaneously vanish by accident.

The situation for the lepton conserving but flavor violating processes (equs. (10) - (12)) is somewhat complicated [12]. The relevant lepton violating parameters are approximately given as follows:

A. Left-handed theories

$$\tilde{\eta} = \begin{cases} \tilde{\eta}_\nu^{(L)} = \sum_j U_{ej}^{(11)} U_{ej}^{*(11)} m_j^2/m_e^2 & \text{(Light neutrino)} \\ \tilde{\eta} \approx \sum_j U_{ej}^{(12)} U_{\mu j}^{*(12)} M_W^2/M_j^2 & \text{(Heavy neutrino)} \end{cases} \tag{89}$$

B. Right-Left Symmetric Models

$$\tilde{\eta} = \tilde{\eta}_N^R = (\epsilon^2 + \kappa^2) \sum_j U_{ej}^{(22)} U_{\mu j}^{*(22)} M_W^2/M_j^2 \tag{90}$$

$$\tilde{\eta} = \tilde{\eta}_\nu^{(\pm)} \approx \epsilon \sum_j (U_{ej}^{(11)} U_{\mu j}^{*(21)} \pm U_{ej}^{(21)} U_{\mu j}^{(11)}) m_j/m_e \tag{91}$$

$$\tilde{\eta} = \tilde{\eta}^\pm \approx \epsilon \sum_j (U_{ej}^{(12)} U_{\mu j}^{*(22)} \pm U_{ej}^{(22)} U_{\mu j}^{(12)}) M_W/M_j \tag{92}$$

Comparing the above expressions with the analogous ones for 0ν $\beta\beta$-decay (equs. (80) - (84)), we notice their unfavourable dependence on the neutrino mass. Thus for neutrinos which are very light (~ 1 eV) or very heavy ($>M_W=80$ GeV) the muon number violating processes, even though they are the analog for the leptonic sector of the well known strangeness violating hadronic decays, are less likely to be seen than 0ν $\beta\beta$-decay even though the latter violates lepton number! For a recent review see ref. 12.

4. CONCLUSION

Even though the neutrino is by now an old particle it still remains elusive and mysterious. Does it have a non-zero mass? Is it a Majorana or a Dirac Particle? Does the right-handed neutrino exist? Is the left-handed weak interaction theory the result of an exact symmetry or only a low energy approximation? Are the neutrinos produced in weak interactions non-stationary?

If the neutrinos produced in weak interactions are non-stationary, i.e. if they do not have a definite mass but they are linear combinations of mass eigenstates, different experiments measure different mass combinations. Decay experiments can, in principle, measure both the mass of an eigenstate and its probability of occurence in a given decay. Neutrino oscillations measure the differences $\delta m^2 = m_k^2 - m_j^2$. For two flavor oscillations one can disentangle the mass from the mixing angle dependence by changing L. For three oscillation lengths the situation is more complex. In any case here the smallness of δm^2 can be compensated by the possibility of large L/E_ν ratio. The latter cannot increase at will, of course, but many such possibilities already exist. Thus if the neutrinos are mixed and non-degenerate the Bugey results [26] should soon be confirmed by other experiments.

Muon number violating neutrino mediated processes are expected to have too small branching ratios to be observed. Not so much because such process are by themselves very slow, but because the muon does not live long enough to tell us about them.

0ν $\beta\beta$-decay had better be observed before unavoidable backgrounds send its signature to oblivion i.e. at the 10^{24} y level of the next generation experiments (this is already a hard goal; one decay in 100 years out of 1 gr of material!). Such an observation will be a great discovery. Not only because it will answer Majorana's question about the nature of the neutrino but because it may reveal that in gauge theories there must be more than two mass scales.

We know that the experiments in the search of the neutrino identity are very fundamental but extremely hard and costly. We hope that they will continue unhindered.

REFERENCES

1. Review of Particle Properties, Rev. Mod. Phys. <u>56</u>, No 2, Part II (1984) 51

2. F.T. Avignone et al, Phys, Rev. Lett. <u>54</u> (1985) 2309; F.T. Avignone Private Communication; E. Belloti et al, Lett. Nuov. Cim. <u>33</u> (1982) 273

3. A. Baderscher et al, Nucl. Phys. <u>377A</u> (1982) 406

4. W. Bertl et al, (Sindrum Collaboration), Search for the decay $\mu^+ \to e^+e^+e^-$, PR-85-06, SIN (1985) (to appear in Nuclear Physics)

5. W. Kinnison et al, Phys. Rev. <u>D25</u> (1982) 2846

6. G. Azuelos et al, Phys. Rev. Lett. <u>51</u> (1983) 164

7. D.A. Bryman et al, Phys. Rev. Lett. <u>55</u> (1985) 465

8. S. Weinberg, Phys. Rev. Lett. <u>19</u> (1967) 264

9. A. Salam in Elementary Particle Theory:Relativistic Groups and Analyticity (Nobel Symposium No. 8), edited by N. Svartholm (Almqvist and Wiksells, Stockholm 1967) p. 367

10. S.L. Glashow, Nucl. Phys. <u>22</u> (1961) 579

11. P. Langacker, Phys. Rep. $\underline{72}$ (1981) 185

12. J.D. Vergados, The neutrino Mass and Family, Lepton and Baryon Non-conservation in Gauge Theories (to appear in Physics Reports)

13. E. Witten, Phys. Rev. Lett. $\underline{91B}$ (1980) 81
    G. Branco and A. Masiero, Phys. Lett. $\underline{97B}$ (1980) 95
    K. Tamvakis and J.D. Vergados, Phys. Lett. $\underline{155B}$ (1985) 373

14. R.G. Robertson, Measurements of Neutrino Mass, Cretan Int. Meeting on Sabatonic Physics on "Current Problems in Nuclear Physics" 23-29 June 85, Crete, Greece

15. R. Abela et al, Phys. Lett. $\underline{146B}$ (1984) 431
    M. Daum et al, Phys. Rev. $\underline{D20}$ (1979) 2692
    H.B. Anderhub et al, Phys. Lett. $\underline{114B}$ (1982) 76

16. C.A. Blocker et al, Phys. Lett. $\underline{109B}$ (1982) 119

17. A. Mateuzzi et al, Phys. Rev. Lett. $\underline{52}$ (1984) 1869

18. K.E. Bergkvist, Nucl. Phys. $\underline{B39}$ (1971) 317

19. E.T. Tretyakov et al, Proc. Int. Neutrino Conf., Aachen (1976) p. 663

20. V.A. Lubinov et al, Phys. Lett. $\underline{94B}$, (1980) 266

21. J.J. Simpson, Phys. Rev. Lett. $\underline{54}$ (1985) 1891

22. F. Calaprice, Heavy neutrinos and beta spectrum of ^{35}S, ref. 14

23. F. Reines, H.W. Sobel and E. Pasiero, Phys. Rev. Lett. $\underline{45}$ (1980) 1307; Proc. Summer Inst. on Part. Phys. SLAC (1980) p. 441

24. F. Boehm et al, Phys. Lett. $\underline{97B}$ (1980) 310;
    K. Kwon et al, Phys. Rev. $\underline{24D}$ (1981) 1097

25. J.L. Vuilleunter et al, Phys. Lett. $\underline{114B}$ 298

26. J.F. Cavaignac et al, Phys. Lett. $\underline{148B}$ (1984) 387

27. F. Boehm, Neutrino Oscillation Experiments, ref. 14

28. F.W. Bullock, R.C.E. Devenish, Reports on Progress in Physics $\underline{46}$ (1983) 1029

29. D.S. Ayres et al, Phys. Rev. $\underline{D29}$ (1984) 902;
    T. Gaisser and T. Stanev, Phys. Rev. $\underline{D30}$ (1984) 985

30. S. M. Bilenki and B. Pontecorvo, Phys. Rep. $\underline{41}$ (1978) 225

31. W.C. Haxton and G. J. Stephenson, Double β-decay, progress in Particle and Nuclear Physics, Vol. 12 (1985) 409

32. T. Kirsten, H. Richter and E. Jessberger, Phys. Rev. Lett. $\underline{50}$ (1983)

33. T. Kirsten, Geochemical Methods of Double beta Decay, ref. 14

HEAVY AND LIGHT DIRAC NEUTRINOS AND A TECHNIPHOTON

Bob HOLDOM *

Department of Physics, University of Toronto, Toronto, Ontario Canada M5S 1A7

In this talk I would like to present a minimal model of neutrino masses within a technicolor framework. The model will involve a nontrivial family structure among four families. The four flavors of left-handed Weyl neutrinos pair up and obtain two Dirac masses. It is attractive to associate the lighter Dirac neutrino, which is a combination of the e and μ neutrinos, with the ITEP $\sim$ 30 eV neutrino [1]. The heavier neutrino is predicted to be $\sim m_\tau/m_e$ times heavier and thus lies in the KeV to MeV range. Now there is normally several problems associated with having a neutrino in this mass range. So I want to show how these problems can be avoided quite naturally in a technicolor context, thus making the existence of heavy neutrino more plausible.

Recently Simpson [2] claimed evidence for a 17 KeV neutrino in the form of a kink in the β-decay spectrum of tritium. But this has not been confirmed by other experiments. Simpson's value for the mixing angle squared seems to be at least a factor of ten discrepant with results for the ^{35}S β-decay spectrum [3]. And also theoretically, as I will show, Simpson's mixing angle seems to large. But experiments will soon probe mixing angles of interest for a heavy neutrino. In this talk I will take 17 KeV as a good typical value for a heavy neutrino mass.

So what does technicolor say about neutrino masses? The simplest technicolor models have families of technifermions with all the standard $SU(3) \times SU(2) \times U(1)$ quantum numbers except that technifermions also carry technicolor. Thus there are technineutrinos and then neutrinos can gain a mass in the same that any quark or lepton gains a mass in a technicolor theory. That is, there must be sideways gauge interactions which cause transitions between ordinary fermions and technifermions. These interactions are broken at mass scales above the weak scale. They can generate effective terms of the form $(g_S/M_S)^2 \, NN\bar{\nu}\nu$ and then the presence of a technineutrino condensate $<NN> \approx m_N^3$ implies

$$m_\nu \approx m_N^3 (g_S/M_S)^2 \ . \tag{1}$$

Thus the question is why are the neutrino masses small compared to other fermion masses? In the simple models I am discussing the same sideways physics is responsible for neutrino masses and all other quark and lepton masses. Thus the g_S/M_S factor is not the cause of small neutrino masses and our question translates into the question of why is the technineutrino lighter than other technifermions. Now it is possible to discuss various mechanisms to

---

* Research supported in part by the Natural Sciences and Engineering Council of Canada.

produce splittings among technifermions, but I will not mention these here. For this talk a small technifermion mass is a given. For a 17 KeV neutrino mass we are forced to consider m_N as low as a few GeV. This would give the right sort of hierarchy between the heavy neutrino and the heaviest quark, which gains a mass from a $\sim 300 \, GeV$ techniquark.

I need to explain the other half of my title. A techniphoton will refer to an unbroken $U(1)$ gauge symmetry of the technifermions only. (The nonabelian technicolor group is also a symmetry of the technifermions only.) All ordinary fermions have zero techniphoton charge. But nonrenormalizable couplings to ordinary matter will be induced through loops.

In this talk I would like to argue that there are two connections between the existence of a techniphoton and the existence of a heavy neutrino. One has to do with constraints on the neutrino mass matrix that a heavy neutrino implies for compatibility with existing neutrino data. This will be discussed latter. The other connection arises as follows.

A 17 KeV neutrino in the context of the standard model is cosmologically stable and it would contribute far too much energy density to the universe. So there must be new physics to allow a faster decay mode. Thus neutrino couplings to a techniphoton are of interest. The diagram in Fig. (1) turns out to give the largest effect, where the jagged line is a sideways boson. This is to be compared to the diagram responsible for the neutrino masses in Fig. (2). The important point is that diagram (1) can induce off-diagonal couplings between light and heavy neutrinos, since there in no reason that diagram (1) and (2) are flavor diagonal in the same basis. We see this explicitly if we note the following details.

1) There is a sum over sideways bosons of different masses in these diagrams.

2) The dynamical mass for N falls off at high momentum and thus the loop in diagram (1) is dominated by momentum of order m_N.

3) This is not so for diagram (2) which develops an additional dependence on the sideways boson masses. This is a log dependence if the N mass falls off like $\sim 1/p^2$.

4) Different sideways boson mass eigenstates need not couple technineutrinos to orthogonal neutrino states, since in general there is mass mixing among gauge bosons with different couplings.

Figure 1

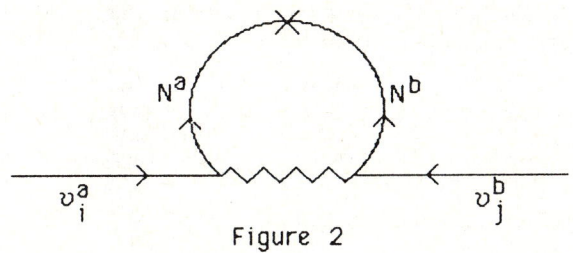

Figure 2

We now estimate the decay rate for a heavy neutrino ν_2 to decay into a light neutrino ν_1 plus a techniphoton γ^*. We consider diagram (1) with the lightest sideways boson (mass M_S and coupling g_S) and introduce an angle θ' to account for suppression of the off-diagonal coupling. This angle is not directly related to the mixing angle as measured in beta decay. The decay rate can be written as

$$\Gamma(\nu_2 \to \nu_1 + \gamma^*) = (m_2/8\pi)F^2 \qquad (2)$$

where F is given by [4]

$$F = (\theta' g^* m_2 m_N / 16\pi^2)(g_S/M_S)^2 \times 0(1) \ . \qquad (3)$$

Sideways generators are normalized like $W^{\pm}$, the N is taken to have a unit γ^* charge, and two technicolors are assumed. But m_2 is also a function of g_S/M_S from (1) and we can eliminate g_S/M_S. Then

$$\Gamma(\nu_2 \to \nu_1 + \gamma^*) \approx \frac{\theta'^2 \alpha^* m_2^5}{512\pi^4 m_N^4} \qquad (4)$$

If we take $m_N = 2$ GeV and $\theta'^2 \alpha^* = 10^{-3}$ then the lifetime of a 17 KeV neutrino would be ~ 10 years. This avoids a cosmological disaster.

Now I would like to describe the second connection between the existence of a heavy neutrino and the existence of a techniphoton. This is related to the observation that a heavy neutrino is not consistent with other data from neutrinoless double beta decay [5] and neutrino oscillation [6] experiments unless the neutrino mass matrix is very constrained. This in turn suggests a global symmetry. One which is particularly attractive [7,8,9,10] is $L_e + L_\tau - L_\mu(-L_\lambda)$ where L_i is an individual lepton number and λ is a possible fourth lepton flavor. If this is a symmetry of the whole lepton sector then it clearly forbids neutrinoless double beta decay as well as $\nu_e \leftrightarrow \nu_\mu$ oscillations. On the other hand it does allow $\nu_e \leftrightarrow \nu_\tau$ and $\nu_\mu \leftrightarrow \nu_\mu$ oscillations, which are not as yet so constrained by experiment. And this symmetry will also permit the heavy neutrino to decay into the light neutrino as described above.

So I want to argue that a techniphoton is connected with the origin of such a symmetry. Remember that we are considering the simple models where families of technifermions have the standard $SU(3) \times SU(2) \times U(1)$ quantum numbers. With just one technifamily it is difficult to make technifermions feel a new $U(1)$ gauge field without running into anomaly problems. But with two families it is easy. There are a number of possible sets of quantum numbers which could be chosen for the first technifamily as long as all quantum numbers have reversed signs for the second technifamily. To make things definite let me choose $T_a - T_b$ where a and b label to two technifamilies. That is, the difference in the numbers of a and b technifermions is conserved and this symmetry is gauged.

Now two technifamilies naturally leads us to consider four ordinary families and a pairing of these families with respect to the sideways physics. Let us look at this structure in the lepton sector. We assume that the following are all multiplets under the same sideways group.

$$
\begin{array}{ll}
(N_a, \nu_\tau, \nu_e)_L & (N_b, \nu_\lambda, \nu_\mu)_L \\
(E_a, \tau, e)_L & (E_b, \lambda, \mu)_L \\
(E_a, \tau, e)_R & (E_b, \lambda, \mu)_R
\end{array}
\qquad (5)
$$

When we consider the possible technilepton condensates we find that the two $U(1)$ gauge symmetries, $T_a - T_b$ and ordinary charge, only allow $<N_a N_b>$, $<\bar{E}_a E_a>$, and $<\bar{E}_b E_b>$. These condensates and the sideways interactions implied by (5) determine the lepton mass matrices. But it is not difficult to see that the global symmetry $L_e + L_\tau - L_\mu - L_\lambda$ remains unbroken. Thus the connection between this global symmetry and the $T_a - T_b$ techniphoton symmetry becomes apparent.

The global symmetry has rather obvious implications for the neutrino masses. The Majorana mass matrix in the weak eigenstate basis $(\nu_e, \nu_\tau, \nu_\mu, \nu_\lambda)$ has to take the form

$$
M = \left(\begin{array}{c|c} 0 & A \\ \hline A^T & 0 \end{array} \right) \qquad (6)
$$

where A is some 2×2 matrix. A unitary transformation is defined by $M = U^T M_D U$ where

$$
2\mathbf{M_D} = \begin{pmatrix} 0 & 0 & m_1 & 0 \\ 0 & 0 & 0 & m_2 \\ m_1 & 0 & 0 & 0 \\ 0 & m_2 & 0 & 0 \end{pmatrix} \text{ and } U = \left(\begin{array}{c|c} P & 0 \\ \hline 0 & Q \end{array} \right). \qquad (7)
$$

Thus two Dirac neutrinos, $\psi_i = (\psi_{Li}, \psi_{Ri})(i = 1, 2)$, have mass m_i where $\psi_{Li} = \Sigma_j P_{ij} \nu_j^a$ and where ψ_{Ri} is the CP conjugate of $\Sigma_j Q_{ij} \nu_j^b$. Here $(\nu_1^a, \nu_2^a, \nu_1^b, \nu_2^b) = (\nu_e, \nu_\tau, \nu_\mu, \nu_\lambda)$. We find that the two Dirac neutrinos have the same $L_e + L_\tau - L_\mu - L_\lambda$ charge, and this allows the heavy neutrino to decay as described.

From the fact that the multiplets in (5) are all multiplets under the same sideways group we note the relation $m_2/m_1 \approx m_\tau/m_e$. We can also deduce that the matrix A in (6) is symmetric and likewise for the charged mass matrix. This implies that in the absence of unnatural cancellations the mixing angle in the charged current should be no more than $\approx (\sqrt{m_1/m_2} + \sqrt{m_e/m_\tau})$. The mixing angle claimed by Simpson exceeds this.

Now we can return to the question of other possible effects of a techniphoton. The dominant coupling of a techniphoton to ordinary matter takes the form

$$(g^*/32\pi^2)(m_\psi/M_T^2)F^{*\mu\nu}\bar{\psi}\sigma_{\mu\nu}\psi \tag{8}$$

M_T is the appropriate technifermion mass. Of most concern is techniphoton emission from stars. γ^* can be compared to a weakly interacting Goldstone boson of the familon variety [11] which has a neutrino coupling to electron coupling ratio of m_2/m_e. In our case (8) implies the ratio is $(m_2 m_{Ea}/m_e m_N)^2$. But we have $(m_{Ea}/m_N)^3 = m_\tau/m_2$ since the same sideways boson couples E_a to τ and N to ν_2. We then find that our neutrino to electron coupling ratio is ~ 75 times larger than for a familon. Thus techniphoton emission from stars is less of a problem than it is for familons.

With regards to searching for the techniphoton it is worth mentioning that $\mu \to e + \gamma^*$ is suppressed to the extent that $L_e + L_\tau - L_\mu - L_\lambda$ is a good symmetry.

Thus far we have only considered the lepton sector. But how do we reconcile a global lepton summetry within a complete theory of quarks and leptons? The corresponding symmetry in the quark sector, $q_{f1} + q_{f3} - q_{f2} - q_{f4}$, is badly broken since otherwise there could be no Cabibbo mixing. In particular, the quark mass generating terms $\bar{Q}Q\bar{q}q$ must not respect such a symmetry. This is odd since the global lepton symmetry seems to associating the first family with the third and the second with the fourth. Some effect of this might be expected to show up in the quark sector.

This raises a question as to whether the $\bar{Q}Q\bar{q}q$ terms can have <u>any</u> approximate global symmetry of this sort. This is of interest since the $\bar{q}q\bar{q}q$ terms could also be constrained by such a symmetry. In particular, a symmetry which treated the s and d quarks differently could end up suppressing $\Delta S = 2$ effects. This would help with the flavor changing neutral current problem of technicolor.

I claim that the $\bar{Q}Q\bar{q}q$ terms can have an approximate symmetry of this type, although I will not give it here. The transformation does not act on the right-handed charge 2/3 quarks, and in this way it allows for Cabibbo mixing. The first and third families have one set up of quantum numbers and the second and fourth families have the opposite set. $\Delta S = 2$ ends up suppressed. In fact I presented [12] a technicolor model a few years ago which was largely motivated by the desire to incorporate this type of approximate symmetry. The outcome was larger than standard model $D^0 - \bar{D}^0$ mixing, close to the experimental level.

But this same model realizes what I have been talking about in the lepton sector. It combines the families into sideways multiplets as I have indicated and it has an exact $T_a - T_b$ symmetry which can be gauged. The global $L_e + L_\tau - L_\mu - L_\lambda$ symmetry follows as described.

But it does not remain an exact symmetry, and the model allows you to study the breaking of this symmetry and other family symmetries. The breaking can be traced to the large right-handed neutrino Majorana masses. These effects feed into the quark sector since the model has couplings between right-handed neutrinos and right-handed charge 2/3 quarks. This indicates why any remaining symmetry does not involve the right-handed charge 2/3 quarks. The right-handed neutrino effects have a much more difficult time feeding into left-handed neutrino mass matrix and there are only very small corrections to the mass matrix presented above.

Just to give some idea of what this model looks like I will give the gauge group and fermion content before any symmetry breaking.

$$SU(4) \times SU(4) \times SU(2) \times SU(2) \times U(1) \times U(1)$$
$$(4, \quad 4, \quad 2, \quad 1, \quad 1, \quad 0)_L$$
$$(4, \quad 4, \quad 1, \quad 2, \quad 0, \quad 1)_R$$
$$(4, \quad 4, \quad 2, \quad 1, \quad -1, \quad 0)_L$$
$$(4, \quad 4, \quad 1, \quad 2, \quad 0, \quad -1)_R$$

References

1) V. Lubinov et. al., Phys. Lett. 94B (1980) 266;
   V. Lubinov, Proceedings of the XXII International Conference on High Energy Physics, Leipzig (July 1984).
2) J. J. Simpson, Phys. Rev. Lett. 54 (1985) 1891.
3) T. Altzitzoglou et. al., Princeton preprint 1985;
   T. Ohi et. al., University of Tokyo preprint 1985;
   ITEP preprint 1985.
4) B.W. Lee and R. Shrock, Phys. Rev. D16 (1977) 1444.
5) P. Langacker, Proceedings of the XXII International Conference on High Energy Physics, Leipzig (July 1984).
6) U. Dore, Proceedings of the XXII International Conference on High Energy Physics, Leipzig (July 1984).
7) O. Shanker, Nucl. Phys. B250 (1985) 351.
8) M.J. Dugan, G.B. Gelmini, H. Georgi, and L.J. Hall, Phys. Rev. Lett. 54 (1985) 2302;
   S.L. Glashow and A. Manohar, Phys Rev. Lett. 54 (1985) 2306.
9) Y.B. Zeldovich, DAN SSSR 86 (1952) 505;
   E.J. Konopinsky and H. Mahmoud, Phys. Rev. 92 (1953) 1045.
10) S.M. Bilenky and B. Pontecorvo, Phys. Lett. 102B (1981) 32;
    S.T. Petcov, Phys. Lett. 110B (1982) 245.
11) D.B. Reiss, Phys. Lett. 115B (1982) 217;
    F. Wilczek, Phys. Rev. Lett. 49 (1982) 1549;
    G.B. Gelmini, S. Nussinov and T. Yanagida, Nucl. Phys. B219 (1983) 31.
12) B. Holdom, Phys. Lett. 143B (1984) 227.

FERMION MASSES IN GUTs WITH INTERMEDIATE SCALES [*]

C. PANAGIOTAKOPOULOS

The Rockefeller University, New York, NY 10021, USA

The relation $m_t/m_c \simeq m_b/m_s$, with $m_s \neq m_\mu$, which predicts a top quark mass close to 40GeV is consistently derived in the context of SO(10) and E_6 grand-unified theories with appropriate intermediate scales.

In the present talk I am going to describe work done in collaboration with Q. Shafi and C. Wetterich[1] on fermion mass relations in a class of grand-unified theories (GUTs) with intermediate scales.

There seems to exist some experimental evidence[2] that the top (t) quark is found at a mass somewhere between 30 and 50 GeV. Most earlier attempts at predicting its mass (for an incomplete list of references see [1]) have yielded lower values. One naturally wonders whether it is possible to construct GUTs "predicting" <u>consistently</u> a t-quark mass within the above range. In the following I will try to argue that this is possible in SO(10) or E_6 GUTs with "suitable" intermediate scales. In such models the relation

$$m_t/m_c \simeq m_b/m_s \qquad (1)$$

holds, without having at the same time $m_s = m_\mu$ (in an obvious notation). From (1), using $m_s(1Gev)=0.15GeV$, $m_c(2m_c)=1.2GeV$ and $m_b(2m_b)=4.5GeV$, one obtains $m_t(m_t)=38Gev$.

In the SO(10) model quarks and leptons of one family belong to a <u>16</u> dimensional spinor representation. Since 16x16=<u>10</u>+<u>126</u>+<u>120</u> one uses <u>10</u>, <u>126</u> and <u>120</u> of Higgs to give masses to the fermions of the <u>16</u>. One real <u>10</u> is not enough (at the tree level at least) because it gives equal mass to every body. (Remember that there is only one Yukawa couping per generation and the 2 weak doublets in a real <u>10</u> get equal vacuum expectation values (vev's)). Therefore it seems that one needs at least a complex <u>10</u>. In practice one uses more than that (i.e. 126 and/or 120).

---

[*] Work supported in part by Department of Energy under contract number DE-AC02-81ER40033B.000.

In general there are two Yukawa couplings for every SO(10)-real, complex Higgs representation, unless an appropriate global symmetry is imposed. Let us assume that such a symmetry has been imposed allowing us to hope for predictions.

The Yukawa couplings can be written as

$$L_{yuk} = f_{ij} 16_i 16_j 10 + h_{ij} 16_i 16_j \overline{126} + k_{ij} 16_i 16_j 120 \qquad (2)$$

(i,j are generation indices, f and h are symmetric matrices, and k is an antisymmetric matrix). There are 8 color singlet $SU(2)_L$ doublets in 10, 126 and 120, which in a $SU(4)_C \times SU(2)_L \times SU(2)_R$ notation are (1,2,2) in 10 and 120 and (15,2,2) in 126 and 120. The mass matrices M_U, M_D, and M_L for charge 2/3 quarks, charge -1/3 quarks, and the charged leptons, respectively, are given at M_X(GUT scale) by

$$M_U = \tfrac{1}{2}\sqrt{2}\, f\langle 10\rangle^u + \tfrac{1}{6}\sqrt{6}\, h\, \langle 126\rangle^u + \tfrac{1}{2}\sqrt{2}k\langle 120\rangle^u_{1,2,2} + \tfrac{1}{6}\sqrt{6}k\langle 120\rangle^u_{15,2,2}$$

$$M_D = \tfrac{1}{2}\sqrt{2}f\langle 10\rangle^d + \tfrac{1}{6}\sqrt{6}h\langle 126\rangle^d + \tfrac{1}{2}\sqrt{2}k\langle 120\rangle^d_{1,2,2} + \tfrac{1}{6}\sqrt{6}k\langle 120\rangle^d_{15,2,2} \qquad (3)$$

$$M_L = \tfrac{1}{2}\sqrt{2}f\langle 10\rangle^d - \tfrac{1}{2}\sqrt{6}h\langle 126\rangle^d + \tfrac{1}{2}\sqrt{2}k\langle 120\rangle^d_{1,2,2} - \tfrac{1}{2}\sqrt{6}k\langle 120\rangle^d_{15,2,2}$$

where the superscripts u (for up) and d (for down) refer to the scalar components that couple to the up-type and the down-type quarks (and leptons). We will see later how one can obtain some predictions from (3), in spite of the large number of free parameters, in models in which the order of magnitude of certain vev's is controlled by ratios of symmetry-breaking scales[3].

Let us consider the limit in which only the Higgs 10 contributes to the fermion mass matrices. At M_X we have $M_U^o \propto M_D^o = M_L^o$ implying $m_t^o/m_c^o = m_b^o/m_s^o$, $m_\tau^o = m_b^o$, $m_\mu^o = m_s^o$, $m_e^o = m_d^o$ and the vanishing of all mixing angles. These relations predict, among other wrong things, that $m_t < 20$ GeV due to the fact that m_s is large. Can one modify the wrong relations $m_\mu = m_s$, $m_e = m_d$ while keeping $m_t/m_c = m_b/m_s$ and $m_b \simeq m_\tau$?

In SO(10) models which do not break via SU(5) one usually uses a 126 Higgs with a large vev in the ($\overline{10}$,1,3) direction. This Higgs breaks B-L and $SU(2)_R$ and gives the ν_R a large Majorana mass. Even if one plans to use only a 10 Higgs for the weak scale, this is not possible anymore. When $SU(4)_C$ breaks at M_C the color singlets in (15,2,2) of 126 mix with the (1,2,2) of 10 and acquire an induced vev. The induced vev of the doublets in 126 is suppressed by powers of M_C/M_X relative to $\langle 10\rangle$ and if $M_C \ll M_X$

$m_\mu \neq m_s$ but $m_\tau = m_b$. In general $m_t/m_c = m_b/m_s$ does not hold anymore. However, if we can arrange through symmetries, that $<126>^u \propto <10>^u$ and $<126>^d \propto <10>^d$ with

$$\frac{<126>^u}{<126>^d} = \frac{<10>^u}{<10>^d} \qquad (4)$$

then also the prediction for the m_t is kept. In general there are small corrections to (4) from subleading effects, which can help create mixing angles without changing substantially the m_t prediction.

One can assume that there is a heavy 120 present as well. Analogous arguments apply to the color singlets in (15,2,2) of the 120 (the (1,2,2) doublets in 120 are not different from the (1,2,2) doublets of 10). They can be also used to create mixing.

To show the implementation of the above considerations I am going to discuss now a specific SO(10) model to which a global Peccei-Quinn [U(1)$_{PQ}$] symmetry is appended. There are two intermediate scales, M_C and $M_{PQ} \simeq M_R$, at which the SU(4)$_C$ and SU(2)$_R \times$U(1)$_{B-L} \times$U(1)$_{PQ}$, respectively, are broken. The symmetry breaking pattern and the Higgs representations responsible for the various breakings are as follows:

$$SO(10) \times U(1)_{PQ} \xrightarrow[M_X]{210(1,1,1)_0} SU(4)_C \times SU(2)_L \times SU(2)_R \times U(1)_{PQ}$$

$$\xrightarrow[M_C]{45_C(15,1,1)_0} SU(3)_C \times SU(2)_L \times SU(2)_R \times U(1)_{B-L} \times U(1)_{PQ} \qquad (5)$$

$$\xrightarrow[M_{PQ} \simeq M_R]{45_{PQ}(15,1,1)_4, 126(\overline{10},1,3)_2} SU(3)_C \times SU(2)_L \times U(1) \xrightarrow[M_W]{10(1,2,2)_{-2}} SU(3)_C \times U(1)_{em}.$$

Only the primary vev's of the Higgs fields are given in (5). the subscripts attached to the SU(4)$_C \times$SU(2)$_L \times$SU(2)$_R$ representations are the U(1)$_{PQ}$ charges. With this chain and $\alpha_s(M_W) \simeq 0.1$, $\sin^2\theta_W(M_W) \simeq 0.23$ one could have $M_X/M_C \simeq 50$, $M_C/M_{PQ} \simeq 40$, and $M_{PQ}/M_W \simeq 10^{10}$, allowing for an acceptable proton lifetime. We assume that the parameters of the scalar potential are such that a suitable linear combination of 10^u and 10^d acquires vev of order M_W. Let us now examine how the doublets in 126 acquire an induced vev. Consider the coupling

$$210 \times 45_C \times 126 \times 10 \qquad (6)$$

Given that the (15,2,2) component in the 126 has mass of order M_X it follows that the induced vev's are

$$\langle 126 \rangle^u \simeq M_c/M_X \langle 10 \rangle^u$$
$$\langle 126 \rangle^d \simeq M_c/M_X \langle 10 \rangle^d \qquad (7)$$

and also (4) holds. Note that for relation (4) to hold it is essential that both $\langle 210 \rangle$ and $\langle 45_c \rangle$ do not break $SU(2)_R \times U(1)_{PQ}$. Corrections to (4) arise from couplings like $210 \times 45_{PQ} \times 126 \times 10^*$ which lead to

$$\langle 126 \rangle^u \simeq M_{PQ}/M_X \langle 10 \rangle^d$$
$$\langle 126 \rangle^d \simeq M_{PQ}/M_X \langle 10 \rangle^u \qquad (8)$$

The fact that $M_{PQ} \ll M_c$ guarantees that those effects are subdominant and (4) holds to a good approximation.

Analogous things happen with the (15,2,2) doublets in 120. They acquire induced vev's which maintain the relation

$$\frac{\langle 120 \rangle^u_{15,2,2}}{\langle 120 \rangle^d_{15,2,2}} = \frac{\langle 10 \rangle^u}{\langle 10 \rangle^d} \qquad (9)$$

from couplings like $210 \times 45_c \times 120 \times 10^*$ (or $45_c \times 120 \times 10^*$) while subleading effects (suppressed by power(s) of M_{PQ}/M_c) and breaking (9) come from couplings like $45_{PQ} \times 120 \times 10$.

The result is that relation (1) holds to a good approximation like the relation $m_\tau = m_b$, but $m_\mu \neq m_s$ and $m_e \neq m_d$. Acceptable mixing can be induced, due to the subleading effects, which is used to correct the m_u/m_d ratio. One also has to a good approximation $m_d = m_s \sin^2 \theta_c$ (θ_c being the Cabibbo angle).

These considerations can be extended to E_6 broken to $SO(10) \times U(1)$ with the (local) $U(1)$ playing the role that the Peccei-Quinn symmetry played in the $SO(10)$ model.

REFERENCES

1) C. Panagiotakopoulos, Q. Shafi, and C. Wetterich, Phys. Rev. Lett. 55 (1985) 787.

2) G. Arnison et.al., Phys. Lett. 145B, (1984) 493.

3) G. Lazarides, Q. Shafi, and C. Wetterich, Nucl. Phys. B 181 (1981) 287.

QUANTUM COSMOLOGY

Alexander VILENKIN

Physics Department, Tufts University, Medford, Massachusetts 02155, USA

I would like to discuss some recent work on quantum cosmology which can be thought of as an attempt to determine the initial conditions for the universe. We are used to thinking that inflation erases all traces of the initial conditions. This, however, is not quite true. For example, the topology of the universe does not change in the course of classical evolution, and thus is determined by the initial conditions. (This issue is particularly important for Kaluza-Klein theories). Besides, at present we have four different types of inflationary scenarios: (i) "Standard" new inflation[1] which is driven by a false vacuum energy; (ii) Chaotic inflation[2] where the energy is supplied by a very large value of a Higgs field, which, however, does not correspond to an extremum of the effective potential; (iii) Starobinsky scenario[3] in which inflation is driven by quantum corrections to vacuum Einstein's equations and (iv) Kaluza-Klein inflation[4], where the role of the "inflation" is played by the radius of the extra dimensions. All four scenarios require different initial conditions, and so we have to face the problem of what determines the initial conditions for the universe.

I will mostly discuss the approach to quantum cosmology I suggested a few years ago[5] in which the universe is created by quantum tunneling from "nothing". By "nothing" I mean a state with no classical space and time. I will mention only briefly the alternative approach suggested by Hawking.[6]

Quantum cosmology is based on quantum gravity, so let me summarize what we have to remember from the classic works of Wheeler and DeWitt[7]. The formalism of quantum gravity is somewhat different for the cases of open and closed universes. We shall only consider the closed case, since only in that case the universe as a whole has a non-zero probability to tunnel through a barrier. The wave function of the universe is defined on the space of all possible 3-geometries and all matter field configurations, which is called superspace (no relation to sypersymmetry). So we can write $\psi(^{(3)}g,\phi)$, where ϕ stands for matter fields. This wave function is independent of the time coordinate t, and so it satisfies a zero-energy Schrodinger equation,

$$\mathcal{H}\,\psi(^{(3)}g,\phi) = 0 \qquad (1)$$

where $\mathcal{H}$ is the Hamiltonian. This equation is called the Wheeler-De Witt equation.

One can wonder how anything can ever happen in a world described by a time-independent wave function. The standard answer to this[7] is that the time coordinate t is an arbitrary label and physics should be independent of it. A physically meaningful time can be defined internally, by using the geometrical or matter variables.

Finally, we shall take a probablistic interpretation of ψ, so that the probability of a certain 3-geometry and matter field configuration is proportional to $|\psi(^{(3)}g,\phi)|^2$.

To illustrate what is meant by creation of universes from nothing, consider a simple model,

$$S = \int (R/16\pi G - \rho_v) \sqrt{-g}\ d^4x, \qquad (2)$$

where the cosmological term is thought of as arising from a false vacuum energy density, ρ_v. We shall assume that the universe is homogeneous and isotropic, so that it is described by a closed Robertson-Walker metric,

$$ds^2 = dt^2 - a^2(t)(dr^2 + \sin^2 r\, d\Omega^2). \qquad (3)$$

Then the classical Einstein's equations reduce to just one equation for the scale factor a:

$$\dot{a}^2 + 1 = H_v^2 a^2, \qquad (4)$$

where $H_v = (8\pi G \rho_v/3)^{1/2}$. This is a very simple model, but it is an important one, since it plays a central role in the inflationary universe scenario. The solution of Eq. (4) is the de Sitter space,

$$a(t) = H_v^{-1} \cosh(H_v t). \qquad (5)$$

The universe contracts at $t<0$, bounces at $t=0$ and re-expands at positive t. This behavior is similar to that of a particle bouncing off a potential barrier. We know that in quantum mechanics particles can tunnel through potential barriers. This suggests the possibility that the universe could tunnel through the classically forbidden range, $0 < a < H_v^{-1}$, directly to the bounce point and start expanding from there.

To describe the tunneling mathematically, we write the Wheeler-DeWitt equation corresponding to the model (2), (3):

Quantum cosmology

$$\left\{ \frac{\partial^2}{\partial a^2} - U(a) \right\} \psi(a) = 0 \tag{6}$$

where

$$U(a) = \left(\frac{3\pi}{2G}\right)^2 a^2 (1-H_v^2 a^2). \tag{7}$$

(I ignore the factor-ordering ambiguities which do not affect the semiclassical wave function). "Nothing" (a=0) is separated from the classically allowed range ($a > H_v^{-1}$) by a barrier. Eq. (6) is easily solved in the WKB approximation:

$$\psi_\pm^{(1)} \propto \exp\left\{\pm i \int_{H_v^{-1}}^{a} P da'\right\} \tag{8}$$

for $a > H_v^{-1}$ and

$$\psi_\pm^{(2)} \propto \exp\left\{\pm \int_{a}^{H_v^{-1}} |P| da'\right\} \tag{9}$$

for $0 < a < H_v^{-1}$. Here, $P(a) = \sqrt{U(a)}$. Any linear combination of ψ_+ and ψ_- solves Eq. (6). To determine the solution completely, one has to impose the boundary conditions. The tunneling solution corresponds to the choice of an "outgoing" wave for $a > H_v^{-1}$: $\psi(a > H_v^{-1}) = \psi_-^{(1)}(a)$. The under-barrier solution is then determined using the WKB connection formulae at the turning point, $a = H_v^{-1}$. The tunneling probability is proportional to [8,9]

$$\exp\left\{-2\int_{0}^{H_v^{-1}} |P| da\right\} = \exp(-3/8G^2 \rho_v) \tag{10}$$

In a more general case there can be several false vacua to tunnel to. For example, suppose we have a grand unified model with a Higgs field ϕ and effective potential $V(\phi)$,

$$S = \int \{R/16\pi G + \frac{1}{2}(\partial_\mu \phi)^2 - V(\phi)\}\sqrt{-g}\, d^4x. \tag{11}$$

If the effective potential is bounded from above, then Eq. (10) suggests that the tunneling is most probable to the highest maximum of $V(\phi)$. Expanding $V(\phi)$ near that maximum,

$$V(\phi) = \rho_v - \frac{1}{2}\mu^2 \phi^2, \tag{12}$$

and keeping only the homogeneous mode of ϕ ($\vec{\nabla}\phi = 0$), we write the corresponding Wheeler-De Witt equation as

$$\{(\partial^2/\partial \tilde{a}^2) - \tilde{a}^{-2}(\partial^2/\partial \tilde{\phi}^2) - \lambda^{-2}\tilde{a}^2(1-\tilde{a}^2 + \frac{\mu^2}{H_v^2}\tilde{\phi}^2\tilde{a}^2)\} \psi(\tilde{a},\tilde{\phi}) = 0 \qquad (13)$$

Here, $\tilde{a} = H_v a$, $\tilde{\phi} = (4\pi G/3)^{1/2}\phi$, $\lambda = 16G^2\rho_v/9$, and I have redefined ϕ so that $\phi = 0$ at the maximum of V. To impose the boundary conditions, we require that ψ is finite everywhere and that it is a purely outgoing wave at $a \to \infty$. The first of these conditions implies

$$\frac{\partial \psi}{\partial \tilde{\phi}}(0,\tilde{\phi}) = 0 \qquad (14)$$

[The analysis here is similar to that in Sec. 5 of Ref. 10, and the reader is referred to that paper for more details].

The WKB approximation corresponds to an expansion in powers of λ; it is valid only if $\rho_v \ll G^{-2}$. To the leading order we have $\psi = e^{-S}$ and

$$(\partial S/\partial \tilde{a})^2 - \tilde{a}^{-2}(\partial S/\partial \tilde{\phi})^2 = \lambda^{-2}\tilde{a}^2(1-\tilde{a}^2 + \frac{\mu^2}{H_v^2}\tilde{\phi}^2\tilde{a}^2) \qquad (15)$$

As it will be clear in a moment, an interesting cosmological model can be obtained only if $\mu \ll H_v$. Then, neglecting terms $\sim (\mu/H_v)^4$, we obtain for $\tilde{a} < 1$:

$$S = \frac{1}{3\lambda} (1+\mu^2\tilde{\phi}^2/H_v^2)[1-(1-\tilde{a}^2 + \mu^2\tilde{\phi}^2\tilde{a}^2/H_v^2)^{3/2}]. \qquad (16)$$

We see that ψ is substantially different from zero only within a tube around the tunneling path ($\phi = 0$); the thickness of the tube determines the fluctuations of ϕ. At $a = 0$ the fluctuations of ϕ are unbounded, but as we approach the nucleation point, $a = H_v^{-1}$, the tube narrows, and the probability distribution for ϕ at nucleation is

$$\mathcal{P}(\phi) \propto \exp\left(\frac{-4\pi^2}{3}\frac{\mu^2\phi^2}{H_v^4}\right). \qquad (17)$$

After nucleation, the universe evolves along the lines of inflationary scenario, as discussed in Ref. 5. The universe expands exponentially as in Eq. (5), while the field ψ "rolls down" according to

$$\phi \sim (H_v^2/\mu) \exp(\mu^2 t/3H_v). \qquad (18)$$

It is clear that sufficient inflation can be obtained only if $\mu \ll H_v$.

If the effective potential $V(\phi)$ is not bounded from above, then a similar analysis shows that the probability (φ) grows towards large ϕ. For values of ϕ with $V(\phi) > G^{-2}$, the exponential suppression of tunneling disappears and the semiclassical approximation breaks down. Linde[8] interprets this as indicating that universes nucleate with initial conditions corresponding to chaotic inflation. Note, however, that for $V(\phi) > G^{-2}$ one can no longer neglect quantum corrections to the classical Einstein's equations. In fact, an analysis of a model with quantum corrections included[10] seems to indicate that the universe is most likely to nucleate with initial conditions corresponding to the Starobinsky inflationary scenario.

A detailed review of the Starobinsky model can be found in Ref. 10. The model includes corrections to vacuum Einstein's equations due to massless, conformally invariant quantum fields (This is a reasonable approximation for an asymptotically free GUT at high energies); all fields have zero expectation values and there are no real particle excitations. To a good approximation, the model is described by the action

$$S = \frac{1}{16\pi G} \int (R + \frac{R^2}{6M^2} + \frac{R^2}{R_0} 2 \ln \frac{R}{R_0}) \sqrt{-g}\, d^4x \tag{19}$$

Here the last term is related to the conformal anomaly. $R_0 \equiv 12 H_0^2$ is determined in terms of the numbers of fields with different spins in the model. For the minimal SU(5) model, $H_0 = 0.7 G^{-1/2}$; for bigger models H_0 is smaller. M is a free parameter, but it can be shown[11] that the density fluctuations predicted by the model are sufficiently small only if $M < 10^{14}$ GeV.

Starobinsky has shown[3] that the model has an inflationary solution,

$$a(t) = H_0^{-1} \cosh H_0 t. \tag{20}$$

This solution is unstable and gets later replaced by a solution in which the scale factor has an oscillatory component; the oscillations are damped by particle production, the particles thermalize and the universe enters a radiation-dominated phase. The duration of the inflationary phase is determined by the initial curvature fluctuation, $\delta = |\delta R/R_0|$:

$$H_0 \Delta t \sim 3(H_0/M)^2 \ln(4/\delta). \tag{21}$$

I now turn to the quantization of the Starobinsky model. The Langrangian (19) describes a higher-derivative theory of gravity. For such theories the number of gravitational degrees of freedom in superspace is doubled. (This corresponds to rewriting a fourth-order differential equation as a system of two second-order equations). In particular, in the Robertson-Walker case we now need 2 variables, which can be chosen, e.g., as a and R. It turns out that a more convenient choice of variables is

$$q = H_0 a(R/R_0)^{1/2}, \qquad x = \frac{1}{2} \ln (R/R_0)$$

Note that near the Starobinsky solution ($R=R_0$), q is proportional to the scale factor a, and x describes the curvature fluctuation.

The Wheeler-De Witt equation for the model (19) is

$$\{\frac{\partial^2}{\partial q^2} - \frac{1}{q^2} \frac{\partial^2}{\partial x^2} - U(q,x)\} \psi(q,x) = 0. \qquad (22)$$

The exact form of $U(q,x)$ can be found in Ref. 10; its expansion near $x = 0$ is

$$U(q,x) = \lambda_0^{-2}[1-q^2 + (M^2/H_0^2)x^2 q^2], \qquad (23)$$

where $\lambda_0 = GM^2/6\pi$. We note that (22), (23) is of the same form as Eq.(13), and so all results obtained there apply with a suitable change of parameters. Like before, the probability distribution is concentrated in a tube around the tunneling path, $R = R_0$ (or $x = 0$). At the moment of nucleation, the probability distribution for x is

$$\mathcal{P}(x) \propto \exp(-4\pi x^2/GH_0^2). \qquad (24)$$

The tunneling probability is proportional to $\exp(-4\pi/GM^2)$.

If one adds to Eq.(19) a false vacuum energy term with $\rho_v \ll G^{-2}$, then the model has two de Sitter solutions: one describing the Starobinsky phase (slightly modified by the presence of ρ_v) and the other corresponding to a false-vacuum-dominated phase in which the quantum corrections are small. It can be shown[10] that the tunneling to the Starobinsky phase has a higher probability. Thus, quantum-cosmological analysis seems to point in the direction of the Starobinsky model. (However, to make a comparison with chaotic inflation, one has to consider models involving both quantum corrections and scalar fields.)

Hawking[6] has suggested an alternative approach to quantum cosmology, in which the wave function is given by a path integral over compact Euclidean geometries. It is not straightforward to translate this prescription into a boundary condition in superspace, but this can be done in some simple cases. The main difference between the two approaches is that Hawking's wave function is always real, and thus gives a time-symmetric picture of the universe. This problem was addressed in a recent paper by Hawking[12], where he suggested that the arrow of time is reversed in a contracting universe. Such an assumption is not needed in the quantum tunneling approach.

To summarize, it appears that quantum cosmology can be useful in determining the initial conditions for the universe. In particular, it may help to decide between various versions of inflation and, most importantly, between different compactification schemes (or different ground states) in Kaluza-Klein and other higher-dimensional theories.[13]

This work was supported by the National Science Foundation and by contributions from General Electric Co., Dennison Manufacturing Co. and Massachusetts Electric Co.

REFERENCES

1) A. H. Guth, Phys. Rev. D23, (1981) 347. For a review see A. D. Linde, Rep. Prog. Phys. 47, (1984) 925.

2) A. D. Linde, Phys. Lett. 129B, (1983) 177.

3) A. A. Starobinsky, Phys. Lett. 91B, (1980) 99.

4) Q. Shafi and C. Wetterich, Phys. Lett. 129B, (1983) 387; E. W. Kolb, this volume.

5) A. Vilenkin, Phys. Lett. 117B, (1982) 25; Phys. Rev. D27, (1983) 2848. This early treatment contains an error which was then corrected in Refs. 8,9.

6) J. B. Hartle and S. W. Hawking, Phys. Rev. D28, (1983) 2960; S. W. Hawking, Nucl. Phys. B239, (1984) 257.

7) J. A. Wheeler, in Battelle Rencontres, ed. by C. De Witt and J. A. Wheeler (Benjamin, NY, 1968); B. S. De Witt, Phys. Rev. 160, (1967) 1113.

8) A. D. Linde, Lett. Nuovo Cim. 39, (1984) 401.

9) A. Vilenkin, Phys. Rev. D30, (1984) 509; Nucl. Phys. B252, (1985) 141.

10) A. Vilenkin, Harvard preprint HUTP-85 / A017.

11) A. A. Starobinsky, Pis'ma Astron. Zh. 9, (1983) 579.

12) S. W. Hawking, The arrow of time in quantum cosmology, DAMTP preprint, 1985.

13) For example, in the case of a D-dimensional theory, one can look for instantons of the form $S_N \times S_{D-N}$. The most probable tunneling path corresponds to the instanton of the smallest action. Hopefully, it is $S_4 \times S_{D-4}$, in which case S_4 should be interpreted as a Euclidean continuation of the de Sitter space.

THE DYNAMICS OF NEW INFLATION

Robert BRANDENBERGER*

Institute for Theoretical Physics, University of California, Santa Barbara, CA 93106, USA

Cosmological phase transitions are examined using a new approach based on the dynamical analysis of the equations of motion of quantum fields rather than on static effective potential considerations. In many models the universe enters a period of exponential expansion required for an inflationary cosmology. Analytical methods show that this will be the case if the interaction rate due to quantum field nonlinearities is small compared to the expansion rate of the universe. We derive a heuristic criterion for the maximal value of the coupling constant for which we expect inflation. The prediction is in good agreement with numerical results.

1. INTRODUCTION

Inflation,[1] in particular in the form of new inflation,[2] presents a solution of several important cosmological problems of the standard big bang model and has hence attracted a lot of attention both in physics and astrophysics.[3]

Inflation can help to explain why the universe today on large scales appears homogeneous, isotropic and flat. Inflation also solves problems introduced to cosmology by modern grand unified theories. In the standard big bang model the expected energy density in magnetic monopoles would by far exceed the critical energy density ρ_c. Inflation after monopole production will exponentially suppress the monopole number density. Similarly it will decrease the abundance of other topological defects in grand unified field theories.

More importantly, however, inflation not only explains the observed large scale homogeneity of the universe, but also provides a mechanism which in a causal way generates the smaller scale inhomogeneities which are observed today as stars, galaxies and clusters of galaxies.[3]

---

*Address beginning October 1, 1985: DAMTP, University of Cambridge, Silver Street, Cambridge CB39EW, United Kingdom.

Recently serious objections against inflationary universe models have been raised by Mazenko, Unruh and Wald.[4] They claim that the models are based on wrong physics, more specifically on unjustified use of effective potential considerations. They assert that in many models previously considered as candidates for new inflation the universe does not enter an inflationary period.

In this talk I will discuss a new approach to phase transitions in cosmological models which does not use the concept of an effective potential. The method is based on considering the evolution of classical scalar field configurations in an expanding universe.

We conclude that there exists a large class of models for which inflation is realized. In an analytical analysis,[5] performed in collaboration with A. Albrecht, we considered two particular families of models. For Coleman-Weinberg models,[6] models in which the scalar field has zero mass in the unbroken phase and in which the dominant interactions of the scalar field are with gauge fields, we conclude that, provided the gauge coupling constant g is small, the universe will enter a period of inflation starting from a hot Friedmann-Robertson-Walker (FRW) phase. The condition is

$$g < \frac{\sigma}{m_{\rm pl}}. \tag{1}$$

σ is the expectation value of the scalar field ϕ in the broken phase, $m_{\rm pl}$ is the Planck mass. The second class of models consists of $\lambda \phi^4$ scalar field models with a double well potential and which interact only very weakly with other fields. For these models we conclude that new inflation in realized if

$$\sigma > m_{\rm pl}. \tag{2}$$

The analytical analysis is based on a perturbative Green function method which bounds the forces disfavoring inflation from above. Hence inflation may occur even when the constraints (1) and (2) are not satisfied. In a collaboration with A. Albrecht and R. Matzner[7] we investigated the evolution of a scalar field in an expanding FRW universe numerically. For $\lambda \phi^4$ models with a double well potential we derive a condition for $\lambda_{\max}$, the maximal value of the coupling constant for which the universe enters an inflationary period:

$$\lambda_{\max}(\sigma, N) = CN\left(\frac{\sigma}{m_{\text{pl}}}\right)^2 \tag{3}$$

N is the number of particle species in thermal equilibrium above the critical temperature, C is a constant of order unity. Our prediction is confirmed numerically.

In Section 2 we present a brief review of the new inflationary universe. We then summarize the recent objections against the standard approach to inflation, objections emphasized by Mazenko, Unruh, and Wald. In Section 4 we explain the main ideas of our analytical approach. In the final section we derive a criterion for the range of parameters for which new inflation is realized and compare it with our numerical analysis.

2. THE NEW INFLATIONARY UNIVERSE

The main idea in inflationary universe models[1] is to describe matter in terms of a quantum field theory which undergoes a symmetry breaking phase transition at a high temperature. The metric is taken to be a spatially flat Robertson-Walker metric

$$ds^2 = -dt^2 + a(t)^2 d\mathbf{x}^2. \tag{4}$$

We will illustrate the basic mechanism[2] in the case of a double well $\lambda\phi^4$ model. The Lagrangian is

$$\mathcal{L}(\phi(\mathbf{x},t)) = \frac{1}{2}d_\mu\phi d^\mu\phi - \frac{1}{4}\left(\phi^2 - \sigma^2\right)^2 \tag{5}$$

$\phi(\mathbf{x},t)$ is a scalar field, σ the scale of symmetry breaking.

In statistical mechanics the ground state of a system is determined by minimizing its free energy. In quantum field theory the free energy density of a homogeneous state is equal to the effective potential V_{eff}. In the $\lambda\phi^4$ model the effective potential, computed to lowest order in quantum corrections, is just the tree-level potential of Eq. (5). It is sketched in Figure 1.

At high temperatures T the effective potential acquires a temperature dependent mass term.

$$V_{\text{eff}}^{(1)}(T,\phi) = V(\phi) + CT^2\phi^2 \tag{6}$$

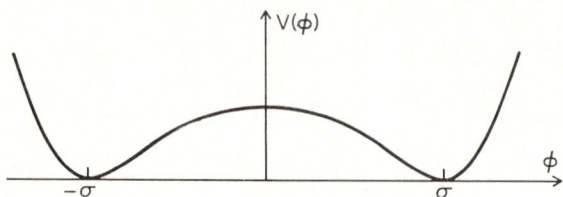

Figure 1: The double well $\lambda\phi^4$ potential.

where C is a constant of the order λ. At high temperatures $\phi = 0$ is the absolute minimum of $V_{\text{eff}}^{(1)}(T, \phi)$. The critical temperature T_c, defined as the temperature below which $\phi = 0$ ceases to be the absolute minimum of the effective potential, is of the order σ.

The conventional scenario of new inflation is as follows: For $T \gg T_c$ the scalar field $\phi(\mathbf{x}, t)$ is constrained to be close to the origin by the finite temperature effective potential. For $T < T_c$, $\phi(\mathbf{x}, t) \simeq 0$ still holds due to the finite temperature potential barrier. The tunneling probability is negligible.[8] When T drops below the Hawking temperature[9] $T_H = H/2\pi$, $\phi(\mathbf{x}) = 0$ becomes unstable to quantum fluctuations. The scalar field, more precisely the expectation value of the operator $\Phi^2(x)$ in the quantum state $|\psi\rangle$

$$\phi_0(t) = \langle\psi|\Phi^2(x)|\psi\rangle^{\frac{1}{2}} \tag{7}$$

will start to increase and gradually approaches its equilibrium value in the phase of broken symmetry. Initially $\phi_0(t)$ evolves slowly and one speaks of the slow rolling period.

The cosmological implications of the standard scenario of new inflation follow from considering the energy-momentum tensor $T_\mu{}^\nu$ of the universe. Both the scalar field and radiation contribute

$$T_\mu{}^\nu = T_\mu{}^\nu(\phi) + T_\mu{}^\nu(\text{rad.}) \tag{8}$$

In a homogeneous state

$$T_\mu{}^\nu = \text{diag}(-\rho, p, p, p). \tag{9}$$

The contributions of $\phi(\mathbf{x}, t)$ to ρ and p are

$$\rho_\phi = V(\phi) + \frac{1}{2}\dot\phi^2 + \frac{1}{2}a^{-2}(t)(\nabla\phi)^2$$
$$p_\phi = -V(\phi) + \frac{1}{2}\dot\phi^2 - \frac{1}{6}a^{-2}(t)(\nabla\phi)^2 \tag{10}$$

The universe starts out in a hot radiation dominated FRW period. The equation of state is $p = \frac{1}{3}\rho$. For $T > T_c$ the scalar field contribution to $T_\mu{}^\nu$ is negligible. As soon as T falls below T_c, the scalar field contribution $T_\mu{}^\nu(\phi)$ begins to dominate $T_\mu{}^\nu$. Since the scalar field configuration is homogeneous and localized at $\phi = 0$, the potential energy term is the only one which contributes to ρ_ϕ and p_ϕ. The equation of state is $p = -\rho$ and $\dot\rho = 0$. Hence by the FRW equation

$$\left(\frac{\dot a}{a}\right)^2 = \frac{8\pi G}{3}\rho \tag{11}$$

the scale factor increases exponentially

$$a(t) = e^{tH}$$
$$H^2 = \frac{8\pi}{3}GV(0). \tag{12}$$

Inflation persists throughout the slow rolling phase, since the equation of state remains $p \simeq -\rho$. We demand that the period τ of inflation be greater than $60H^{-1}$ in order to solve the horizon and flatness problems.

3. OBJECTIONS AGAINST THE CONVENTIONAL PICTURE OF NEW INFLATION

Mazenko, Unruh and Wald[4] have raised two main objections against the standard picture of new inflation presented in the previous section. The first point is a mathematical point and turns out to be unimportant for physical considerations. It does, however, indicate that one must be very careful when basing physical arguments on concepts like the effective potential. The second objection will be the crucial one.

Point one: The effective potential $V_{\text{eff}}(\phi)$ is a convex function of ϕ. This result is well known both in quantum field theory and in statistical mechanics. The mathematical proof is simple. Any function defined as a Legendre transform is convex.[10] For homogeneous configurations, the effective potential is the density of the effective action $\Gamma(\phi)$. On the other hand, the effective action is the Legendre transform of the generating functional $W(J)$ for connected Feynman graphs, the logarithm of the partition function $Z(J)$. Hence $V_{\text{eff}}(\phi)$ is a convex function. This is true not only on the classical level, but to all orders in perturbation theory.

It is not hard to understand the physical reason why $V_{\text{eff}}(\phi)$ is convex. Starting point is the variational definition of the effective potential:

$$V_{\text{eff}}(\phi) = \min \langle s|H|s\rangle$$
$$|s\rangle$$
$$\langle s|s\rangle = 1$$
$$\langle s|\phi(\mathbf{x})|s\rangle = \phi. \tag{13}$$

Consider as an example the double well $\lambda\phi^4$ potential of Fig. 1. The states $|+\rangle$ and $|-\rangle$ are defined by

$$\phi(\mathbf{x})|+\rangle = \sigma|+\rangle \;\forall \mathbf{x}$$
$$\phi(\mathbf{x})|-\rangle = -\sigma|-\rangle \;\forall \mathbf{x} \tag{14}$$

For each value ϕ_0 between $-\sigma$ and σ we now construct a state with vanishing expectation value of H and with expectation value of $\Phi(\mathbf{x})$ equal to ϕ_0. Pick $0 \leq \alpha \leq 1$ such that $2\alpha - 1 = \phi_0/\sigma$ and define the state $|\alpha\rangle$ by

$$|\alpha\rangle = \alpha^{\frac{1}{2}}|+\rangle + (1-\alpha)^{\frac{1}{2}}|-\rangle \tag{15}$$

Then

$$\langle \alpha|\Phi(\mathbf{x})|\alpha\rangle = \phi_0 \tag{16}$$

and

$$\langle \alpha|H|\alpha\rangle = \langle \alpha|V|\alpha\rangle = 0 \tag{17}$$

Since all three terms in H are positive definite operators, it is impossible to construct a state with negative expectation value of H. Hence

$$V_{\text{eff}}(\phi) = 0 \quad -\sigma < \phi < \sigma \tag{18}$$

In particular, the effective potential is nonanalytic. The construction used above is the Maxwell construction.

The effective potential we constructed in Section 2 is obviously not convex. We now argue that nevertheless for metastability analyses it, and not the full $V_{\text{eff}}(\phi)$, is the appropriate function to consider. The important point is that the wave functionals of the states $|\alpha\rangle$ used in the construction of $V_{\text{eff}}(\phi)$, have support only at $\phi = \pm\sigma$. $V_{\text{eff}}(\phi_0)$ is hence insensitive to the shape of the potential and thus to the microphysical forces near ϕ_0. The evolution of a wave functional which is initially localized at ϕ_0 depends on the microphysical forces and not on $V_{\text{eff}}(\phi_0)$. It is possible to construct an object with more significance for the evolution of an initially localized wave functional by restricting the variation in Eq. (13) to states whose wave functional is Gaussian about ϕ_0.

$$V_{\text{eff}}^{(G)}(\phi_0) = \min\langle s|H|s\rangle$$

$$\langle s|s\rangle = 1$$

$$\langle s|\Phi(\mathbf{x})|s\rangle = \phi_0$$

$$|\psi(\phi(\mathbf{x}))|^2 \text{ Gaussian about } \phi_0 \qquad (19)$$

Since the one loop approximation to $V_{\text{eff}}(\phi_0)$ is obtained by taking the Gaussian approximation about ϕ_0 in the functional integral, $V_{\text{eff}}^{(G)}(\phi_0)$ is precisely the one loop effective potential. If we have the freedom to choose an initial homogeneous and localized wave functional, its evolution will be determined approximately by $V_{\text{eff}}^{(1)}(\phi)$.

Getting to the main criticism of Mazenko, Unruh and Wald, in cosmology we do not have the freedom to specify an initial homogeneous localized wave function. The universe starts out at a high temperature and thus there will be large fluctuations in scalar field configurations which contribute to the functional integral. Due to the large inhomogeneities it is unjustified to use $V_{\text{eff}}(\phi)$ or $V_{\text{eff}}^{(1)}(\phi)$.

The authors of Ref. (4) consider a thermal classical initial scalar field configuration at $T = \sigma$. $\phi(\mathbf{x})$ will be approximately uniformly distributed in $[-\sigma, \sigma]$. In Ref. (4) it is then claimed that as the universe expands coming out of the hot FRW phase, $\phi(\mathbf{x}, t)$ will relax to $-\sigma$ at all points in space where intially $\phi(\mathbf{x}) < 0$. Similarly $\phi(\mathbf{x},t)$ will relax to $+\sigma$ wherever initially $\phi(\mathbf{x}) > 0$. Hence at $T < T_c$ spatial domains $\phi(\mathbf{x}) = \pm\sigma$ will form. Since $V(\phi) = 0$ inside the domains, the equation of state will not be $p = -\rho$. Hence there will be no inflation.

We[5,7] agree with the objections against using $V_{\text{eff}}(\phi)$, at least in the case of weakly coupled $\lambda\phi^4$ models. We also agree that eventually the scalar field configuration will consist of domains of $\phi(\mathbf{x}) = \pm\sigma$. On the other hand, the dynamical evolution before domain formation is nontrivial. We will show that in many models the expansion of the universe leads to sufficient Hubble damping of the field configuration to give rise to a long intermediate phase of inflation. In the semi-classical approach we would see a contraction of the wave functional.

4. AN ANALYTICAL ANALYSIS

This section is based on work done in collaboration with A. Albrecht.[5] We assume a thermal initial state at the critical temperature $T = T_c$. W consider classical field configurations which are not Boltzmann suppresed (this statement will be made more precise shortly). We then analyze the dynamical evolution of the scalar field in a radiation dominated FRW background. We propose to determine whether the equation of state of the scalar field is $p \simeq -\rho$ at the time $T_\mu{}^\nu(\phi)$ begins to dominate over $T_\mu{}^\nu$ (rad.). The dynamical equation of motion is

$$\ddot{\phi} + 3\frac{\dot{a}}{a}\dot{\phi} - a^{-2}\nabla^2\phi = -V'(\phi) \tag{20}$$

There are two force terms, the nonlinear force $V'(\phi)$ and the Hubble damping force $3\dot{a}a^{-1}\dot{\phi}$. Associated with the two forces are two time scales, the Hubble expansion time H^{-1} and a quantum field interaction time Γ^{-1}.

The nonlinear force $V'(\phi)$ tends to produce spatial domains. To analyze the effect of the Hubble term $3\dot{a}a^{-1}\dot{\phi}$ we will replace $V'(\phi)$ by $\frac{1}{6}R\phi$, R being the Ricci scalar. Since $R \sim H^2$, the mass term we added is very small compared to other terms in the equation. The new equation of motion

$$\ddot{\phi} + 3\frac{\dot{a}}{a}\dot{\phi} - a^{-2}\nabla^2\phi = \frac{1}{6}R\phi \tag{21}$$

is the equation of a conformally coupled free scalar field. By conformal transformation, every solution $\phi(\mathbf{x},t)$ of Eq. (21) can be written as

$$\phi(\mathbf{x},t) = a^{-1}(t)\hat{\phi}(\mathbf{x},\tau) \tag{22}$$

where τ is conformal time given by

$$dt^2 = a^2 d\tau^2 \tag{23}$$

and $\hat{\phi}(\mathbf{x},\tau)$ is a solution of the Klein-Gordon equation in flat space-time. Consider for example an initial plane wave $\phi(\mathbf{x},0) = \hat{\phi}(\mathbf{x},0))$. In flat space-time the solution $\hat{\phi}(\mathbf{x},\tau)$ will be an oscillating wave with constant amplitude. Thus in an expanding universe $\phi(\mathbf{x},t)$ will be a damped oscillation in conformal time. Thus the expansion of the universe has two effects. First, the scalar field amplitude at all points in space decreases

$$\phi(\mathbf{x},t) \rightarrow 0 \tag{24}$$

(strictly speaking $\phi(\mathbf{x},t)$ relaxes towards the initial spatial average). Second, the spatial gradient term $(\nabla\phi)^2$ is redshifted. Both Hubble damping and Hubble redshifting lead to a scalar field configuration which is homogeneous and localized at $\phi = 0$ and which yields inflation.

The main idea of the calculation is to bound the deviation from the free field evolution in the presence of the nonlinear force terms. We conclude that if $\Gamma \ll H$ then the free field evolution dominates for a time period τ sufficiently long to lead to an intermediate period of inflation. In particular, if $\tau > 60H^{-1}$ the inflationary period is sufficiently long to solve the cosmological problems mentioned in the introduction. H depends only on the background expansion of the universe, Γ on the other hand on the particle physics model. Thus the condition $\Gamma \ll H$ will be model dependent.

Our analysis consists of the following steps: We first fix the initial scalar field configuration. The free field evolution is then given by Eq. (22), so we need only bound the deviation from the free field evolution.

The initial scalar field configuration is assumed to be thermal at $T = T_c$. Given a cutoff volume V in comoving coordinates we Fourier expand $\hat{\phi}(\mathbf{x}, \tau)$

$$\hat{\phi}(\mathbf{x}, \tau) = V^{-\frac{1}{2}} \sum_{\mathbf{k}} e^{-i\mathbf{k}\mathbf{x}} q_{\mathbf{k}}(\tau) \tag{25}$$

For a free field all $q_{\mathbf{k}}$ are harmonic oscillator modes. The initial conditions are specified by giving the amplitudes $|q_{\mathbf{k}}|$ and the phases $\alpha_{\mathbf{k}}$

$$q_{\mathbf{k}} = |q_{\mathbf{k}}| e^{i\alpha_{\mathbf{k}}} \tag{26}$$

We assume random phases $\alpha_{\mathbf{k}}$. The amplitudes are determined by requiring each oscillator to have thermal energy T. We also require the amplitudes to be quantized

$$\frac{1}{2}(k^2 + m^2)|q_{\mathbf{k}}|^2 = n_{\mathbf{k}}(k^2 + m^2)^{\frac{1}{2}} \tag{27}$$

Thus $|q_{\mathbf{k}}|$ is given by Eq. (27) for the largest integer $n_{\mathbf{k}}$ for which

$$n_{\mathbf{k}}(k^2 + m^2)^{\frac{1}{2}} \leq T. \tag{28}$$

In particular, $|q_{\mathbf{k}}| = 0$ for $|\mathbf{k}| > T$.

We use a perturbative Green function method to estimate the deviation from the free field theory evolution model the equations of motion in Fourier space are (in Landau gauge)

$$\ddot{q}_{\mathbf{k}} + k^2 q_{\mathbf{k}} = g^2 V^{-1} \sum_{\mathbf{k}', \mathbf{k}''} q'_{\mathbf{k}} A_\mu(\mathbf{k}'') A^\mu(\mathbf{k} - \mathbf{k}' - \mathbf{k}'') \equiv I_{\mathbf{k}}(\tau). \tag{29}$$

The interaction term stems from the $D_\mu \phi D^\mu \phi$ term in the Lagrangian. Since the scalar self coupling constant λ is of the order g^4, scalar field self-interactions are subdominant and have been omitted from Eq. (29). The solution $q_{\mathbf{k}}(\tau)$ is decomposed as

$$q_{\mathbf{k}}(\tau) = q_{\mathbf{k}}^{(0)}(\tau) + q_{\mathbf{k}}^{I}(\tau). \tag{30}$$

$q_{\mathbf{k}}^{(0)}(\tau)$ is the solution of the unperturbed equation with the given initial conditions, $q_{\mathbf{k}}^{I}(t)$ is the effect of the interaction forces. In our perturbative analysis we evaluate $I_{\mathbf{k}}(\tau)$ using $q_{\mathbf{k}}^{(0)}(\tau)$. By the Green function formula we then have

$$q_{\mathbf{k}}^I(\tau) = q_{\mathbf{k}}^{(1)}(\tau) \int_{\tau_0}^{\tau} d\tau' \epsilon(\tau') I_{\mathbf{k}}(\tau') q_{\mathbf{k}}^{(2)}(\tau') - q_{\mathbf{k}}^{(2)}(\tau) \int_{\tau_0}^{\tau} d\tau' \epsilon(\tau') I_{\mathbf{k}}(\tau') q_{\mathbf{k}}^{(1)}(\tau') \qquad (31)$$

$q_{\mathbf{k}}^{(1)}(\tau) = \sin k\tau$ and $q_{\mathbf{k}}^{(2)}(\tau) = \cos k\tau$ are the solutions of the homogeneous equation, $\epsilon(\tau)$ is the Wronskian

$$\epsilon(\tau) = \left[q_{\mathbf{k}}^{(1)}(\tau)\dot{q}_{\mathbf{k}}^{(2)}(\tau) - q_{\mathbf{k}}^{(2)}(\tau)\dot{q}_{\mathbf{k}}^{(1)}(\tau)\right]^{-1} = k^{-1}. \qquad (32)$$

The random phase approximation is used to estimate the magnitude of $I_{\mathbf{k}}(\tau)$. Since modes with $k = T$ dominate the phase space, the magnitude is given by the amplitude of a single term for $k \sim k' \sim k'' \sim T$ multiplied by the number of terms in a single sum in Eq. (39), *i.e.*,

$$I_{\mathbf{k}}(\tau) \sim g^2 V^{-1} V \left(\frac{T_c}{2\pi}\right)^3 T_c^{-3/2} \sim g^2 T_c^{3/2}. \qquad (33)$$

Hence

$$q_{\mathbf{k}}^I(\tau) \sim g^2 \tau T_c^{\frac{1}{2}}. \qquad (34)$$

The free field evolution will be a good approximation as long as

$$q_{\mathbf{k}}^I(\tau) < q_{\mathbf{k}}^{(0)}(\tau) \qquad (35)$$

(on the right hand side we naturally mean the amplitude of $q_{\mathbf{k}}^{(0)}(\tau)$). The condition will be true for $\tau < \tau_c$. By Eqs. (34) and (35), τ_c, is given by

$$\tau_c \sim g^{-2} T_c^{-1}. \qquad (36)$$

Demanding $\tau_c > H^{-1}$ yields the constraint

$$g^2 < \frac{H}{T_c} \qquad (37)$$

or equivalently, using $T_c \sim g\sigma$

$$g < \frac{\sigma}{m_{\rm pl}}. \qquad (38)$$

Our analytical analysis shows that if the above condition is satisfied there will be an intermediate period of inflation of sufficient length to solve the flatness and horizon problems.

The analysis in the case of the $\lambda\phi^4$ model is similar. In anology to Eq. (37), we get the constraint

$$\lambda < \left(\frac{H}{\sigma}\right)^2 \tag{39}$$

but since H^2 depends linearly on λ the coupling constant drops out and the condition becomes

$$\sigma > m_{\text{pl}}, \tag{40}$$

a condition which is satisfied in many supersymmetry[11] and supergravity[12] models in which inflation is generated by a scalar field which couples only very weakly to all other fields.

Since the perturbative Green function method estimates the maximal effect of the nonlinear force terms, the conditions (38) and (40) are conservative estimates for the parameter space region for which inflation is realized. If the conditions are not met it does not mean there will be no inflation. An important goal of a numerical analysis is to determine the precise parameter space region for which we get inflation. $\lambda_{\max}(\sigma)$ will denote the maximal value of the coupling constant which gives inflation.

Another motivation for a numerical analysis is to justify the appoximations made above, in particular the random phase approximation, the ad hoc condition for τ_c and the perturbative approach.

5. NUMERICAL ANALYSIS

The numerical work reported on in this section was performed in collaboration with A. Albrecht and R. Matzner.[7] We consider the evolution of a single scalar field $\phi(\mathbf{x}, t)$ beginning at $T = \sigma$ in a hot FRW background metric $a(t) \sim t^{\frac{1}{2}}$. We evolve the system until $T_\mu{}^\nu(\phi) > T_\mu{}^\nu$ (rad.) at which point we determine the equation of state. If $p_\phi \simeq -\rho_\phi$ the model will enter an inflationary period.

First we determine a qualitative criterion for $\lambda_{\max}(\sigma)$. Given an initial plane wave configuration

$$\phi(\mathbf{x}, t_0) = A \sin kz \tag{41}$$

there are three forces which will influence the time evolution. The tension force T (stemming from spatial gradient terms in the Klein-Gordon equation) will set up an oscillation. The nonlinear force F_N is always in direction of the closest of the minima of $V(\phi)$, and the Hubble damping force D points towards $\phi = 0$ (strictly speaking towards the spatial average of $\phi(\mathbf{x}, t_0)$). If $D > F_N$ then the scalar field will begin to evolve towards a configuration which gives inflation. Thus the criterion for $\lambda_{\max}(\sigma)$ is

$$D > F_N \tag{42}$$

In the case of the $\lambda\phi^4$ model of Eq. (5) the condition becomes [see Ref. (7)]

$$H^2 \sigma > \lambda \sigma^3 \tag{43}$$

Using the Hubble parameter in the FRW era

$$H^2 = \frac{8\pi}{3} G\rho = \frac{8\pi^3}{90} GNT^4 \tag{44}$$

where N is the number of particle species in thermal equilibrium at T, (43) becomes

$$\lambda < \frac{8\pi^3}{90} N \left(\frac{\sigma}{m_{\text{pl}}}\right)^2. \tag{45}$$

In the case of a scalar field with potential

$$V(\phi) = \lambda \phi^4 \left(\ell n \frac{\phi^2}{\sigma^2} - \frac{1}{2}\right) \tag{46}$$

(a toy model for a Coleman-Weinberg theory) the condition is identical up to a constant factor slightly larger than 1 multiplying the right hand side of (45), coming from the logarithmic suppression of F_N.,

The two crucial predictions from (45), the linear increase of $\lambda_{\max}(\sigma)$ with σ^2 and with N, are confirmed in our numerical analysis.

We use different sets of initial conditions. Thermal initial conditions in the sense of Section 4 are

$$A = k = \sigma \tag{47}$$

condition (45) has been derived for these initial values. In weakly coupled $\lambda\phi^4$ models it may however be unjustified to assume thermal equilibrium at $T = \sigma$. Instead, we should

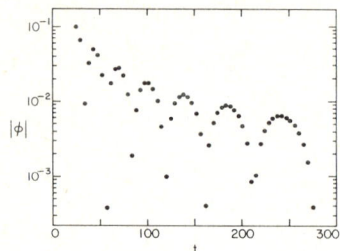

Figure 2: Value of the scalar field as a function of time at the position of the maximum of the initial standing wave ($\lambda \phi^4$ model, $\sigma = 10^{-1}$, $\lambda = 10^{-2}$, $N = 1$, $A = k = \sigma$).

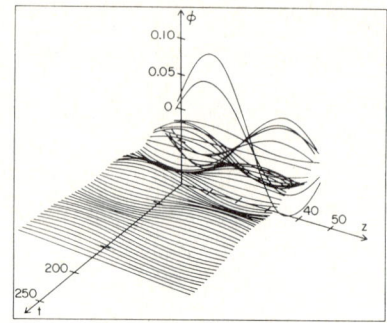

Figure 3: A 3D plot showing the value of the scalar field (vertical axis) as a function of space (right axis) and time (left axis) for the run considered in Fig. 2.

impose classical equipartition, equality of potential and kinetic energy for each mode. This leads to modified initial data

$$k = \lambda^{1/4}\sigma, \; A = \lambda^{-1/4}\sigma. \tag{48}$$

In Figures 2 and 3 the numerical results for a sample run are shown. The parameters are $\sigma = 10^{-1}$, $\lambda = 10^{-2}$ in the $\lambda\phi^4$ model of Eq. (5). The initial conditions are $A = k = \sigma$ and N was chosen to be 1. Figure 2 presents the values of the scalar field as a function of time at the position z of the maximum of the initial configuration. The scales are in terms of natural units $G = 1$. Figure 3 shows the value of the scalar field ϕ (vertical axis) as a function of space (right axis) and time (left axis). The envelope of the curve in Figure 2 corresponds to the decrease in the amplitude of oscillation discussed above. We verified that the equation of state becomes inflationary. In Figure 3 we can visually follow the decrease in amplitude of oscillation.

Our results confirm the two crucial predictions from Eq. (45). In Figure 4 we plot the numerical results for $\lambda_{\max}(\sigma)$. Checks indicate runs for which inflation is realized, crosses those for which it is not. The results are for the $\lambda\phi^4$ double well potential model and assume $k = A = \sigma$ and $N = 1$. The dashed line indicates the result for $\lambda_{\max}(\sigma)$ for the "Coleman-Weinberg" model of Eq. (46). For both models $\lambda_{\max}(\sigma)$ scales approximately as σ^2.

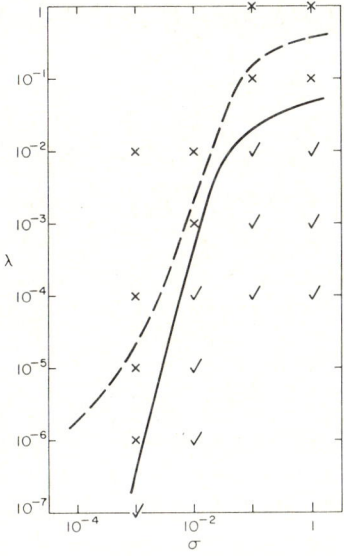

Figure 4: λ_{max} as a function of σ for plane wave initial conditions $A = k = \sigma$, $N = 1$, in the $\lambda\Phi^4$ model. Checks mark runs which yield an inflationary equation of state, x marks runs which fail to give inflation. The dashed line indicates $\lambda_{max}(\sigma)$ for the "Coleman-Weinberg" model of Eq. (46).

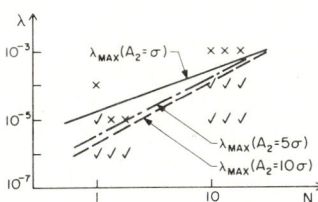

Figure 5: Dependence of λ_{max} on A_2 (Coleman-Weinberg model, $\sigma = 10^{-2}$). The first column gives the results for $A_2 = \sigma$, the second for $A_2 = 5\sigma$ and the third for $A_2 = 10\sigma$.

Figure 6: N dependence of λ_{max} (Coleman-Weinberg model, $\sigma = 10^{-2}$, Hubble radius scale fluctuation with amplitude $A_2 = \sigma$ included).

In the case of the toy model of Eq. (46) we also considered the effect of adding a long wavelength fluctuation. The initial scalar field configuration is given by the wavenumbers $k_1 = \lambda^{1/4}\sigma$ and $k_2 = H$ and the amplitudes $A_1 = \lambda^{-1/4}\sigma$ and A_2 of the two plane waves. We considered $\sigma = 10^{-2}$ and analyzed how λ_{max} depends on A_2 and N. Since a change in A_2 corrsponds to a small change in the total amplitude the effect of varying A_2 is small (Figure 5). On the other hand, the linear dependence of $\lambda_{max}(\sigma)$ on N is verified. The results are shown in Figures 5 and 6.

6. CONCLUSIONS

Phase transitions in the early universe have been studied based on investigating the dynamical equations of motion and without using concepts such as the effective potential. Our analysis applies in particular to the case in which the phase transition is generated by a nonvanishing expectation value of a weakly coupled scalar field. We analytically derived sufficient conditions for the universe to enter an inflationary period. We numerically verified a formula which gives the parameter space region for which inflation is realized and which was derived using rather heuristic arguments.

ACKNOWLEDGEMENTS

I thank A. Albrecht and R. Matzner for a very enjoyable collaboration. I also wish to thank the organizers of the "Particles and the Universe" symposium, in particular G. Lazarides, for their special efforts to make the conference interesting and pleasant.

This research was supported in part by the National Science Foundation under Grant No. PHY82-17853, supplemented by funds from the National Aeronautics and Space Administration, at the University of California at Santa Barbara.

REFERENCES

1. A. Guth, *Phys. Rev.* **D23** (1981) 347; K. Sato, *Not. Roy. Astron. Soc.*, **195** (1981) 467; A. Guth and S.-H. Tye, *Phys. Rev. Lett.* **44** (1980) 631.

2. A. Linde, *Phys. Lett.* **108B** (1982) 389; A. Albrecht and P. Steinhardt, *Phys. Rev. Lett.* **48** (1982) 1220; S. Hawking and I. Moss, *Phys. Lett.* **110B** (1982) 35.

3. For recent reviews see e.g. A. Linde, *Rep. Prog. Phys.*, **42** (1984) 389; R. Brandenberger, *Rev. Mod. Phys.*, **57** (1985) 1.

4. G. Mazenko, W. Unruh, and R. Wald, *Phys. Rev.* **D31** (1985) 273.

5. A. Albrecht and R. Brandenberger, *Phys. Rev.* **D31** (1985) 1225.

6. S. Coleman and E. Weinberg, *Phys. Rev.* **D7** (1973) 1888.

7. A. Albrecht, R. Brandenberger and R. Matzner, *Phys. Rev.* **D32** (1985) 1280.

8. A. Linde, *Nucl. Phys.* **216B** (1983) 421.

9. G. Gibbons and S. Hawking, *Phys. Rev.* **D15** (1977) 2738; R. Brandenberger and R. Kahn, *Phys. Lett.* **119B** (1982) 75.

10. See e.g. V. Arnold, *Mathematical Methods of Classical Mechanics,* English ed. (Springer, New York, 1978).

11. J. Ellis, D. Nanopoulos, K. Olive and K. Tamvakis, *Nucl. Phys.* **221B** (1983) 524.

12. A. Linde, *Phys. Lett.* **132B** (1983) 137; A. Linde, *JETP Lett.* **37** (1983) 724; D. Nanopoulos, K. Olive, M. Srednicki and K. Tamvakis, *Phys. Lett.* **123B** (1983) 41, and *Phys. Lett.* **124B** (1983) 171; B. Ovrut and P. Steinhardt, *Phys. Lett.* **133B** (1983) 161; R. Holman, P. Ramond and C. Ross, *Phys. Lett.* **137B** (1984) 343.

GEOMETRIC EFFECTS IN COSMOLOGICAL PHASE TRANSITIONS

B.L. HU

Department os Physics and Astronomy, University of Maryland, College Park, Maryland 20742, U.S.A.

We discuss how the geometry can affect the symmetry behavior in spacetimes with some compact dimensions. We study the infrared behavior of the de Sitter universe and discuss the implications of this finite-size effect on the inflationary cosmology.

The new inflationary scenario[1] based on the radiatively-induced Coleman-Weinberg (CW) effective potential of massless fields has been quite successful in explaining some outstanding issues of cosmology. However, the original results were drawn from a framework not entirely compatible with the prevailing physical conditions of the early universe - e.g. the assumption of flat-space techniques in the computation of effective potentials and thermodynamic equilibrium conditions for cosmology. To ascertain the viability of these attractive proposals one should analyze in detail all of the field theoretical, curved-spacetime and statistical aspects of these problems.[2,3] Two groups of inter-related questions should be re-examined: (1) <u>Was there inflation?</u> - Does there exist a metastable state? How long did the universe stay in this state? (2) <u>Was there a phase transition?</u> - What was its nature? How complete was it? Earlier Brandenberger has reported on some work addressing to the first question, partially in response to issues raised by Mazenko <u>et al</u>[3]. Our concern[4-10] has been mainly with the gravitational aspects of these problems: How do effects of curvature, field coupling, geometry and dynamics of spacetime affect inflation and phase transitions? For example: Will curvature and anisotropy originally present prevent inflation? Our studies on the Einstein[4] and the Taub universe[5] show that they depend on the deformation and the type of coupling: Conformally coupled fields usually give rise to second order transitions while minimally coupled fields can admit first order transitions but with features quite different from flat-space results. As another example, given that a metastable state can exist and the universe can supercool for a sufficiently long time, how does the geometry of the background spacetime affect the phase transition? The success of the new inflation depends crucially on the universe being able to "roll-over" to the true vacuum slowly and this in turn is due to the extended flatness of the effective potential $V(\hat{\phi})$ near the symmetric vacuum $\hat{\phi} = 0$ ($\hat{\phi}$ being the order parameter), i.e., $V''(\hat{\phi}=0) = 0$ [Eq. (15)]. Our studies on phase transitions in the Einstein and the de Sitter universes[5,6] have identified an important effect due to the compactness (or finiteness) in any spatial dimension of the background geometry. Known as the finite size effect in condensed matter and surface physics, it can alter the behavior of the effective potential near the metastable vacuum and incapacitate new inflation near the Planck energy. These features manifest more distinctly for massless minimal fields. As the gravity-modified mass [Eq. (1)] $M^2 = m^2 + (1-\xi)R/6$ vanishes for massless minimal fields, it is closest to the spirit of the CW mechanism in curved space. Unfortunately this is also the case which has not been properly addressed, despite the large number of previous work on de Sitter space phase transitions.[11] In this talk I want to describe to you the nature of the finite size effect in curved space, the infrared behavior in de Sitter universe and its implication on the inflationary cosmology. The work reported here was done recently in collaboration with D. J. O'Connor[6,7].

Before presentating the result of our calculation, let me describe to you first in qualitative terms how the geometric effect enters. Consider for simplicity a $\lambda\Phi^4$ theory. After a background field splitting $\Phi = \hat{\phi} + \phi$ where $\hat{\phi}$ is the background field, the fluctuation field ϕ obeys the equation $(\Box + m^2)\phi = 0$, in which the generalized mass

$$m^2 = m^2 + (1-\xi)R/6 + \frac{1}{2}\lambda\hat{\phi}^2 + \frac{\hbar\lambda}{2}(\langle\phi^2\rangle_0 + \langle\phi^2\rangle_T) \tag{1}$$

has acquired terms due to classical curvature R and quantum or thermal fluctuations.[2] (Here, $\xi = 0$ denotes conformal and 1 denotes minimal coupling), $\langle \rangle_0$ denotes vacuum expectation value and $\langle \rangle_T$ denotes thermal averaging.) Phase transition would occur at $m^2 = 0$ corresponding to the case of infinite correlation length. In any spacetime with some compact (or finite) dimension so that the invariant operator $A \equiv \Box + m_1^2$ where $m_1^2 = M^2 + \lambda\hat{\phi}/2$ possesses a discrete spectrum (or band structure), it is the lowest mode (or band) which affects most strongly the infrared behavior of the system. The 1-loop effective potential is given by $V^{(1)} = \frac{\hbar}{2\Omega} \text{Tr} \ln A$, where Ω is the spacetime volume. For a D-sphere of radius a (D=3 is Einstein universe and D=4 is the de Sitter universe) the eigenvalues λ_ℓ of A are $\ell(\ell+D-1)/a^2 + m_1^2$ with degeneracy $d_\ell = \frac{(\ell+D-2)!}{\ell!(D-1)!}(2\ell+D-1)$. The lowest mode contribution to $V^{(1)}$ is given by

$$V^{(1)}_{IR} = \frac{1}{2\Omega} \ln m_1^2 \mu^{-2} \tag{2}$$

where μ is the renormalization mass (usually set at $\mu^{-1} = a$, the only scale length in the system). For massless minimal fields near $\hat{\phi} = 0$, the zero mode gives rise to an infrared divergence in the 1-loop effective potential, but is rendered finite as higher loop contributions are included. Here already we see that the form of the effective potential near $\hat{\phi} = 0$ is not flat, but for S^4 de Sitter universe depends on $\ln|\hat{\phi}|$ and becomes zero only as $a \to \infty$. The functional form varies with the geometry. Indeed, it depends on the number and the size of the compact or finite dimensions. For the Einstein universe, $V^{(1)} \sim |\hat{\phi}|$ near $\hat{\phi} = 0$, and in general for spacetimes with topology $R^d \times S^{D-d}$, $V^{(1)} \sim |\hat{\phi}|^d$ near $\hat{\phi} = 0$. This is because for spacetimes with non-compact dimensions, the zero-mode waves in the open directions can grow to far greater lengths (up to ∞ at critical point) than those in the compact directions and thus dominate the infrared behavior of the system. As such the phase transition of the system behaves effectively like a lower d-dimensional system. The concept of effective dimensions introduced by us earlier in the discussion of the symmetry behavior of the Einstein universe[5] is also applicable here for the de Sitter universe or for cosmological spacetimes with compact spatial sections. As such, some powerful theorems (Mermin-Wagner-Coleman) governing phase transitions in two or less dimensions can be invoked for describing the symmetry behavior of these spaces[5]

As the infrared behavior of the system is dictated by the open dimensions, the system has an effectively reduced IR dimension. Some results are already familiar: Take D=4. For d=4, the theory is just flat space. For d=3, this corresponds to a finite temperature field theory with a periodic imaginary time. For D=5 and d=4, this becomes the Kaluza-Klein theory, or the modern generalization thereof (e.g. D=11, d=4 with S^7 internal space). Superstring theories contain finite size effects at two levels: Firstly as strings they are of finite extent (D=2,d=1) which lends themselves to field theory reduction. Secondly the field theory of strings involve a supergravity Kaluza-Klein compactification (D=10,d=4 on a Calabi-Yau internal space C^6). These higher dimensional geometric theories appear to an observer in four-dimensional spacetime as possessing an infinite ladder of massive particles, with the most important contribution at low energy coming from the zero modes (massless particle). For D=4, d=2 one deals with the class of generalized Einstein-Rosen metrics. For cosmological theories the most interesting cases correspond to d=1, which includes all cosmologies with closed spatial sections (e.g. Robertson-Walker, mixmaster), and d=0, which is the de Sitter universe in the Euclideanized form.

For all the above spacetimes, the effect discussed here is analogous to the finite size effect in surface physics.[12] The distinction between phase transitions in flat and curved spacetimes (with some compact dimensions) to a large extent parallels the distinction between critical behavior in the bulk and in surfaces (the bulk versus the slab or cylinder geometry correspond to R^3 versus $S^2 \times R^1$ and $S^1 \times R^2$ geometries) in condensed matter physics. Viewing our problem in this light gives a unified and simple way of understanding their symmetry behavior.

The logarithimic dependance of $V^{(1)}$ on $\hat{\phi}$ in de Sitter universe [Eq. (3)] was derived from a 1-loop calculation in a $\lambda\phi^4$ theory. To address the more realistic situations of the inflationary cosmology, one needs to consider gauge theories or scalar quantum electrodynamics (SQED) and include higher loop contributions. A number of authors[11] have calculated the effective potential for SQED in de Sitter space before and some have considered higher-loop contributions. But the zero-mode contribution in the massless minimal field was not properly treated and the physical significance of the finite size effect not adequately recognized. For SQED we found that the dominant 1-loop infrared behavior remains logarithmic and comes from the scalar field (like a two-component field) but not from the gauge field. However, it is well-known that 1-loop results are generally unreliable in the study of infrared behavior. To account for the higher-loop contributions one can consider an N-component scalar field theory and use an eigenmode expansion to express the functional integral. In spaces with some finite dimension, the lowest mode contribution to the functional integral reduces it to an ordinary N-dimensional integral, which enables one to account for all IR dominant higher-loop contributions analytically.

The wave operator A^{ab} for the fluctuating field ϕ^a (a,b = 1 to N) is

$$A^{ab} = (\Box + m_1^2)\, \hat{\phi}^a \hat{\phi}^b / \hat{\phi}^2 + (\Box + m_2^2)(\delta^{ab} - \hat{\phi}^a \hat{\phi}^b / \hat{\phi}^2)$$

where $m_1^2 = M^2 + \lambda\hat{\phi}/2$, $m_2^2 = M^2 + \lambda\hat{\phi}/6$.

For compact spacetimes whose invariant operators A^{ab} admit discrete eigenvalues (or bands) λ_n^{ab} with eigenfunctions ϕ_n, i.e., $A^{ab}\phi_n(x) = \lambda_n^{ab}\phi_n(x)$, one can express the measure $[d\phi]$ of the functional integral in terms of the coefficients c_n^a in the Fourier decomposition of the fields ϕ^a, i.e., $\phi^a(x) = \Sigma_n\, c_n^a\, \phi_n(x)$. The successive terms of the action can then be expressed in terms of products of c_n^a and λ_n^{ab}. The lowest mode contribution to the effective action Γ is given by

$$e^{-\Gamma} = \int dc_0^a \exp\{-\tfrac{1}{2} c_0^a \lambda_0^{ab} c_0^b - \tfrac{\lambda}{6} \hat{\phi}^a f_1\, c_0^a c_0^b c_0^b - \tfrac{\lambda}{4!} f_2\, c_0^a c_0^a c_0^b c_0^b\} \quad (3)$$

where $f_i = \int d^4x\, \sqrt{g}\, \phi_0^{i+2}(x)$. This becomes a zero-dimensional theory with two coupling constants λf_1 and λf_2. The functional integral reduces to an N-dimensional ordinary integral over the fluctuation field ϕ^a and can be performed exactly for the case $N=1$ or perturbatively for other N by standard techniques.[14] The result obtained in this way is non-perturbative and contains contribution from all loops.

Alternatively one can consider only those diagrams with leading large N contribution (the superdaisy diagrams). Borrowing results from formal considerations before[15], we obtain the following effective potential for an N-component massless minimally coupled scalar field in de Sitter space ($\Omega = 8\pi^2 a^4/3$) near $\hat{\phi} = 0$:

$$V_{eff}(\hat{\phi}) = \frac{\alpha_{IR}(\lambda)}{a^4} + \frac{\beta}{2} \frac{\lambda^{1/2}}{a^2} \hat{\phi}^2 + \frac{\gamma}{4!} \hat{\phi}^4 \tag{4}$$

where

$$\alpha_{IR}(\lambda) = -\frac{3\hbar N}{32\pi^2}[1 + \ln(\frac{16\pi^2 a^4 \mu^4}{\hbar \lambda N})], \quad \beta = \frac{(\hbar N)^{1/2}}{4\pi} \text{ and } \gamma = \frac{\lambda}{2}.$$

We see that the $\ln|\hat{\phi}|$ behavior near $\hat{\phi} = 0$ obtained earlier from 1-loop calculation is now modified. The curvature of the potential well $V(\hat{\phi})$ at $\hat{\phi} = 0$ gives the effective (induced) mass (or inverse correlation length) of the system:

$$m_{eff}^2 = 2 \left.\frac{\partial V_{eff}}{\partial \hat{\phi}^2}\right|_{\hat{\phi} = 0} = \frac{\beta \lambda^{1/2}}{a^2} . \tag{5}$$

It depends on the size of the system and vanishes as $a \to \infty$.

For scalar QED the infrared behavior remains dominated by the scalar contribution given by (4) with N=2. We find the effective potential for small $s = e\hat{\phi} a$ to be

$$V_{eff}(\hat{\phi}) = V_{eff}(0) + \frac{3}{64\pi^2 a^4}\{4g^{1/2}s^2 - \tau s^4 + \frac{2}{3}[5-4\zeta_R(3)]s^6 + ...\} \tag{6}$$

where

$$V_{eff}(0) = \alpha_{IR}(\lambda)/a^4 + 3M^4/128\pi^2, \quad g \equiv 8\pi^2\lambda/9e^4 \text{ and } \tau \equiv \ln(M_v^2 a^2) + g/2 + 2\gamma_E - 3,$$

M_v is the gauge particle mass in flat space and γ_E is the Euler constant. The coupling constant ξ has been renormalilzed to correspond to minimal coupling in de Sitter space, for consistency with the scalar infrared calculation. Unlike in $\lambda\phi^4$ theory, the quartic term can become negative when $M_v^2 a^2 > \eta_1$ where

$$\eta_1 = \exp(3-2\gamma_E - g/2) \approx 6.33 \exp(-g/2) \tag{7}$$

In a de Sitter universe $a^2 \approx 16\pi M_p^2/M_v^4$, M_p being the Planck mass. This condition will be satisfied provided $M_v^4/M_p^4 \ll 1$, as in Grand Unification (GUT) transitions at $M_v = M_x \sim 10^{14}$GeV. $V(\hat{\phi})$ assumes a maximum of $V_{eff}(0) + 3g/16\pi^2 a^4 \tau$ at $e^2\hat{\phi}_{max}^2 a^2 \approx -2g^{1/2}/\tau$. For $\hat{\phi} > \hat{\phi}_{max}$, $V(\hat{\phi})$ starts the gradual decline down the long plateau characteristic of the CW potential. Finite size effect will have little influence on $V(\hat{\phi})$ for $\hat{\phi} \gg \hat{\phi}_{max}$, where it assumes the usual form.[11] Using the 1-loop potential in the limit $s^2 \gg 1$, one sees that the global minimum of $V(\hat{\phi})$ occurs at $e^2 \hat{\phi}_{min}^2 \simeq M_V^2$. This is the true vacuum where the phase transition ends. The depth of the potential at this global minimum (or the height of the plateau) is defined by the characteristic scale M_V of the theory. The relative height of the potential wells at $\hat{\phi} = 0$ and at $\hat{\phi} = \hat{\phi}_{min}$ is determined by the ratio $(M_V/M_p)^4$. At $M_V^2 a^2 \simeq \eta_2$ where

$$\eta_2 \equiv \exp\{3 - 2\gamma_E + 4[\tfrac{2}{3}(5-4\zeta_R(3))]^{1/2} g^{1/4} - g/2\} \tag{8}$$

the second minimum at $\hat{\phi}_{min}$ which exists for $M_V^2 a^2 > \eta_1$ becomes degenerate with the minimum at $\hat{\phi} = 0$. For $M_V^2 a^2 < \eta_2$ as in Planck scale processes, the potential well at $\hat{\phi} = 0$ is the only one surviving as there is no second minimum at $\hat{\phi}_{min}$. This means that there is primordial (Planck scale) inflation but no phase transition. As the energy is lowered ($M_V^2 a^2$ increased), the depth of potential well at $\hat{\phi} = 0$ decreases rapidly while the shoulder broadens. Throughout this range of energy which covers certain supersymmetric and Kaluza-Klein scales, there is inflation and limited first order transition by tunneling, but it suffers from the same problems of the old inflation.[1] At the GUT scale, the local well at $\hat{\phi} = 0$ has curvature of the order of 10^9 GeV, which is of the same magnitude as the vacuum or thermal fluctuations (at Hawking temperature). So the same fluctuation which overcomes the thermal barrier near the metastable state can also "creep out" of the potential well in this modified picture (at zero temperature). Thus, we do not expect the present finite size correction to drastically affect the existing GUT picture. However, our study does suggest that finite size effect could possibly rule out all phase transitions at or near the Planck scale. In particular, this excludes any assumption that the universe can stay in a de Sitter state through the Planck epoch, as is sometimes assumed in some inflationary or quantum cosmology scenarios[16], for once the universe enters a de Sitter phase near Planck time it will probably remain trapped in the false vacuum and be unable to exit. The subject of phase transition near the Planck time is rather complicated. Proper consideration should include quantum gravitational effects of particle production, tunneling and critical dynamics in non-equilibrium conditions.[2]

Our analysis above is applied to the S^4 coordinatization of the de Sitter universe. In the $R^1 \times S^3$ coordinatization of de Sitter, the finite size effect of S^3 will be coupled with dynamical effects of expansion. Using quasi-local approximations we found earlier that[9] second order variation of the background curvature and field can induce a term in the effective mass of the system. Whe de Sitter space is regarded as an exponentially expanding Robertson-Walker universe, these dynamic terms are of the same nature as that arising from the curvature terms in the S^4 formulation. With proper handling of the kinetic and boundary terms accounting for the two different coordinatization we speculate that one should get the same result as reported here. General discussion of the finite-size effect in cosmological phase transition and details of the present calculation can be found in Refs. 7 and 8.

REFERENCES

1) A. H. Guth, Phys. Rev. D23, 347 (1981); A. D. Linde, Phys. Lett. 108B 389 (1982); A. Albrecht and P. J. Steinhardt, Phys. Rev. Lett. 48, 120 (1982).

2) B. L. Hu, in *Proceedings of the Inner Space/Outer Space Symposium*, Fermilab, May 1984, edited by E. Kolb et.al. (Univ. of Chicago Press, 1985); also in *Proceedings of the 10th International Conference on General Relativity and Gravition* Padua, July 1983, edited by B. Bertotti, F. DeFelice and A. Pascolini (Consiglio Nazionale Delle Ricerche, Roma, 1983).

3) G. Mazenko, W. G. Unruh and R. M. Wald, Phys. Rev. D31, 273 (1985).

4) D. J. O'Connor, B. L. Hu and T. C. Shen, Phys. Lett. 130B, 31 (1983).

5) T. C. Shen, B. L. Hu and D. J. O'Connor, Phys. Rev. D31, 2401 (1985).

6) B. L. Hu and D. J. O'Connor, Phys. Rev. Lett. (1985).

7) D. J. O'Connor and B. L. Hu, "Symmetry Behavior of the de Sitter Universe: Finite Size Effect and Inflationary Cosmology" (1985).

8) B. L. Hu and D. J. O'Connor, "Finite Size Effect in Cosmological Phase Transtion" (1986).

9) B. L. Hu and D. J. O'Connor, Phys. Rev. D20, 743 (1984).

10) B. L. Hu and D. J. O'Connor, "Dynamical Effect in Cosmological Phase Transition" (1986).

11) G. M. Shore, Ann. Phys. (N.Y.) 128, 376 (1980); S. W. Hawking and I. Moss, Phys. Lett. 110B, 35 (1982); B. Allen, Nucl. Phys. B226, 228 (1983); A. Vilenkin, Nucl. Phys. B226, 504 (1983); L. Ford, Phys. Rev. D31, 704, 710 (1985); B. Ratra, Phys. Rev. D31, 1931 (1985).

12) See. e.g., Barber, in *Phase Transitions and Critical Phenomena*, Vol. 8, edited by C. Domb and J. Lebowitz (Academic Press, 1983).

13) Previous authors (e.g. Allen in Ref. 11) took only the one-loop correction from the gauge field into consideration and by assuming that $\lambda \sim e^4$ (e being the electric charge) ignored the one-loop scalar field contribution. Such acts are justified only for cases where $ζ^2 a^2 = 0$, but this doesnot correspond to the Coleman-Weinberg condition. When one is interested in the theory near $ζ^2 a^2 = 0$, as with problems in inflationary transitions, the scalar lowest mode ought to be included. In fact, one should include all higher loop contributions of this mode.

14) S. Coleman, R. Jackiw and D. Politzer, Phys. Rev. D10, 2491 (1974).

15) J. M. Cornwall, R. Jackiw and E. Tomboulis, Phys. Rev. D10, 2428 (1974).

16) R. Brout, E. Englert and P. Spindel, Phys. Rev. Lett. 43, 417 (1979); A. Vilenkin, Phys. Rev. D30, 509 (1984).

COSMIC STRINGS AND GALAXY FORMATION

T.W.B. KIBBLE

Blackett Laboratory, Imperial College, Prince Consort Road, London SW7, England.

The evolution of a system of strings created at a phase transition early in the history of the universe is reviewed. The two possible end points are a string-dominated universe, which behaves much like a matter- dominated one, and a scaling solution, in which the persistence length of the system of strings scales with the horizon distance. The latter is the basis for a very attractive theory of galaxy formation.

1. INTRODUCTION

Galaxy formation has long been a major cosmological problem. In particular it is not easy to reconcile the inhomogeneity that galaxies represent with the smoothness of the microwave background radiation.

Before the decoupling time, when electrons and protons combined to form neutral hydrogen at a temperature of about 3500K, density perturbations are conventionally classified as isothermal or adiabatic. In an adiabatic perturbation, the temperature and baryon number density fluctuate together ($\delta n_b/n_b \simeq 3\delta T/T$) while in an isothermal perturbation only the baryon number fluctuates ($\delta T/T = 0$).

Isothermal perturbations do not grow in the radiation-dominated phase and would have had to be present in the initial state. Unfortunately, they are incompatible with the idea of baryon number generation, which is one of the great successes of grand unification. If the observed nonzero mean baryon number of our universe was created by baryon-number-violating interactions at very early times, then the local baryon number density is determined completely by the local temperature, and isothermal perturbations are impossible.

The alternative is to suppose that the initial perturbations are adiabatic. On small scales adiabatic perturbations are erased during decoupling by photon diffusion. This occurs on all scales below the Silk mass

$$M_{Silk} \simeq 10^{12}(\Omega h^2)^{-2} M_O.$$

Here Ω is the ratio of the density in the universe to the critical density,

$\Omega = \rho/\rho_{cr}$, and h is the Hubble constant in units of 100 km s^{-1} Mpc^{-1}. Thus adiabatic perturbations yield a scenario in which the large scales condense first. The primary objects are galactic clusters, and galaxies form by fragmentation.

There are several problems with this scenario. It is difficult to reconcile with the observation that the galaxy-galaxy correlation function decreases with distance like $r^{-1.8}$. The time scale is problematic: galaxies cannot form until after the clusters have condensed, at a red-shift of only about 1, which appears to contradict the observation of quasars out to red-shifts of 3.5 or more.

Moreover, there is at least a potential conflict with the observed isotropy of the microwave background radiation. Density perturbations present at the decoupling time should have been accompanied by perturbations of a similar magnitude in the temperature. The observed microwave back- ground temperature has a dipole term, presumably due to the motion of the earth relative to the background. Once this has been subtracted, the remaining fluctuations are very small, of order

$$\frac{\delta T}{T} \simeq 10^{-5}.$$

The required magnitude is very close to this limit, if not already in conflict with it.

Finally, there remains the question of the origin of the adiabatic perturbations. The best answer so far is provided by inflation. The new inflationary universe, particularly in supersymmetric form, does seem capable of yielding perturbations of the right magnitude and with the right "scale-invariant" spectrum. However this does tend to require some fine tuning of parameters.

It is clear that neither adiabatic nor isothermal perturbations yield a wholly satisfactory theory of galaxy formation. It is therefore worthwhile to examine any alternative. That cosmic strings might provide the answer was first suggested by Zel'dovich[1] and by Vilenkin[2].

In this talk I shall first review briefly the conditions for the existence of strings and then discuss recent developments that have made the string theory very attractive.

2. EXISTENCE OF STRINGS

Like monopoles and domain walls, strings can appear as topological defects when a gauge symmetry is spontaneously broken. Let us consider a gauge theory with symmetry group G, which is broken to a subgroup H at a high-temperature phase transition. The manifold of degenerate vacuum states can then be identified with the quotient space G/H, the space of left cosets of H in G. The condition for the existence of strings is that the first homotopy group $\pi_1(M)$ be nontrivial, i.e.[3] that there are non-shrinkable loops in M.

The most familiar strings, such as the flux tubes in superconductors, correspond to the complete breaking of an Abelian group $G = U(1)$. In that case M is a circle, and integer, the winding number, n.

In grand unified theories on the other hand we start with a semisimple group G. If G is connected and simply connected, i.e. $\pi_0(G)$ and $\pi_1(G)$ are trivial, then by a well-known theorem

$$\pi_1(M) = \pi_0(H) = H/H_c,$$

where H_c is the connected component of H. Thus strings exist if and only if H has more than one connected component. They are labelled by the elements of a finite group The simplest example is a model with $G = SO(3)$ and two Higgs fields and

$$V(\underline{\phi}_1, \underline{\phi}_2) = -\tfrac{1}{2}\lambda\eta^2(\underline{\phi}_1{}^2 + \underline{\phi}_2{}^2)$$

$$+ \tfrac{1}{4}\lambda\left[(\underline{\phi}_1{}^2)^2 + (\underline{\phi}_2{}^2)^2\right] + \tfrac{1}{2}\varepsilon(\underline{\phi}_1 \cdot \underline{\phi}_2)^2. \tag{1}$$

T en for $0 < \varepsilon < \lambda$ the minima are at $\underline{\phi}_1{}^2 = \underline{\phi}_2{}^2 = \eta^2$, $\underline{\phi}_1 \cdot \underline{\phi}_2 = 0$. Thus $\underline{\phi}_1$ and $\underline{\phi}_2$ are of fixed length and orthogonal, so G is completely broken. Since $\pi_1(SO(3)) = Z_2$, there are strings characterised by a Z_2 quantum number, which means that strings and antistrings coincide.

Alternatively, we may replace G by its covering group SU(2), in which case $H = \{1, -1\}$. Again we find $\pi_1(M) = \pi_0(H) = Z_2$.

Above the critical temperature T_c, the Higgs field (or fields) has vanishing expectation value. Below

$$\|\langle\phi\rangle\| = \eta \sim T_c.$$

A string is a trapped region where $\langle\phi\rangle$ is constrained to pass through zero. It has a tension, equal to its mass per unit length, of $\mu \sim 2\pi\eta^2$. For example, for η in the range 10^{15} to 10^{16} GeV, the dimensionless parameter $G\mu$ (where G is Newton's constant) lies between 10^{-7} and 10^{-5}. As a typical value we may take $G\mu \simeq 10^{-6}$. The value cannot be much larger than this without running into conflict with observation[4]. It could however be very much smaller.

3. CONFIGURATION OF STRINGS

As the temperature falls below T_c, the Higgs field acquires a non-vanishing expectation value lying somewhere on the manifold M of minima of V. It can choose different points in different regions of space. This leads to trapped regions where $\langle\phi\rangle$ passes through zero, namely strings.

We should expect the initial configuration of strings to be essentially random, or Brownian, with a persistence length L determined by the correlation length at the time, and certainly much smaller than the horizon distance. The length of string in a large volume R^3 is then of order R^3/L^2 (i.e. one segment of length L in each volume L^3), and hence the average string density is

$$\rho_s \simeq \frac{\mu}{L^2}. \qquad (2)$$

In the early stages, the motion of strings is heavily damped. Small kinks in the string tend to straighten out. Sometimes strings cross and exchange partners creating new kinks that straighten out in time. These processes lead to a reduction in the total length and a growth in the length scale: in fact $L \propto t^{5/4}$. This process continues until L is of the same order of magnitude as t. This happens when

$$L \sim t \sim t_p/(G\mu)^2 \qquad (3)$$

(where $t_p = G^{1/2}$ is the Planck time), or equivalently at a temperature

$$T \sim (G\mu)^{1/2} T_c \sim 10^{-3} T_c. \qquad (4)$$

Thereafter, the strings move freely with little damping, and acquire relativistic speeds.

An important role in the subsequent evolution is played by the process of closed-loop formation. When a string intersects itself it can give rise to a closed loop. These loops then decay by gravitational radiation. For a string of length $2\pi\ell$ and mass $M = 2\pi\mu\ell$, the oscillation frequency is $\omega = 2/\ell$. The rate of loss of energy by gravitational radiation is very roughly $\dot{M} \simeq - GM^2\omega^6\ell^4$. Thus we can write[5,6]

$$M = -2\pi\beta G\mu^2, \tag{5}$$

where β is a numerical factor somewhat larger than unity, perhaps around 10.

According to (5), the mass, and hence the length, of the loop decreases linearly with time. The lifetime is thus proportional to size,

$$\tau \simeq \ell/\beta G\mu \sim 10^5 \ell \text{ or } 10^6 \ell.$$

The lifetime could be substantially reduced if loops frequently intersect themselves and so split in two, because this would effectively halve the lifetime. Turok and I have shown[7] that although this does happen for many loops there are also many loops that do not intersect themselves. So we should expect a sizeable fraction to have lifetimes of the order suggested.

4. EVOLUTION OF STRINGS AND CLOSED LOOPS

The probability that a segment of string of length L intersects another segment in a time δt is of order $v\delta t/L$. Dimensional arguments suggest that the total length of closed loops formed in a large volume R^3 in a time δt is

$$\alpha p v L \frac{R^3}{L^3} \frac{\delta t}{L},$$

where α is a constant of order unity (or more probably somewhat less, say 0.1), p is the probability of reconnection when strings cross and v is a typical string velocity.

Let us for the moment assume that we can neglect the possibility of reattachment of the loops to long strings. Then the energy E of the long strings within volume R^3 changes with time according to

$$\frac{\dot{E}}{E} = \frac{\dot{R}}{R}(1-2v^2) - \frac{\alpha p v}{L} \tag{6}$$

The first term here represents the effect of the cosmic expansion, while the second is the loss to closed-loop formation.

If the strings remain 'Brownian' during their evolution, so that $E = \mu R^3/L^2$, then from (6) we can immediately find how the ratio $\gamma = L/t$ of persistence length to expansion time evolves. In the radiation-dominated era ($R \propto t^{1/2}$) we find

$$t\frac{\dot{\gamma}}{\gamma} = -\frac{1-v^2}{2} + \frac{\alpha p v}{2\gamma}, \quad (\gamma = \frac{L}{t}). \tag{7}$$

Note that v is a function of γ. As $\gamma \to 0$, when the persistence length is very small, we have $v \simeq 1/\sqrt{2}$, while for longer persistence lengths, $\gamma \gg 1$, $v \to 0$. It is clear that the right hand side of (7) is a monotonically decreasing function of γ. It tends to $+\infty$ as $\gamma \to 0$ and to $-\frac{1}{4}$ as $\gamma \to \infty$. Hence there is a unique value of γ, say γ_0, at which it vanishes. Since $\dot{\gamma} > 0$ for $\gamma < \gamma_0$ and $\dot{\gamma} < 0$ for $\gamma > \gamma_0$, $\gamma = \gamma_0$ is a stable point. The system of strings will evolve towards a scaling solution in which $L = \gamma_0 t$. From (7) we see that, up to a factor of order unity, $\gamma_0 \sim \alpha p$. From (2) it follows that once the scaling solution has been reached, the ratio of the string density ρ_s, to the radiation density ρ is constant,

$$\frac{\rho_s}{\rho} = \frac{32\pi}{3\gamma_0^2} G\mu \sim 10^{-6}. \tag{8}$$

We must ask whether or not the approximation we have made of neglecting reattachment of strings is a good one.

The probability of survival of a loop of size ℓ in a system of strings with $L = \gamma t$ is roughly $e^{-pv\ell t/L^2}$. In particular if ℓ is of order L it is about $e^{-pv/\gamma}$. When γ is of order unity this probability is appreciable and the approximation should be at least qualitatively correct, but it is clear that it cannot be so when $\gamma \ll 1$. For this reason, one really should allow for the possibility of reattachment. To do this we consider separately the energy E in volume R^3 in the form of long strings (including closed loops above a certain size, say $\ell > kL$ for some small constant k), and the energy $e(\ell)d\ell$ in the form of loops of size between ℓ and $\ell + d\ell$. It is not sensible to take k too large because loops of size $\ell \gg L$ never retain their identities for long.

For these quantities one finds[8] a set of coupled differential equations,

$$\dot{E} = \frac{\dot{R}}{R} E (1-2v^2) - Eka(k) \frac{\dot{L}}{L}$$

$$- E \frac{pv}{L} \int_0^{kL} a(\frac{\ell}{L}) \frac{\ell d\ell}{L^2} + \frac{pv}{L} \int_0^{kL} e(\ell) \frac{\ell d\ell}{L}, \tag{9}$$

and.

$$\dot{e}(\ell) = E \frac{pv\ell}{L^3} a\left(\frac{\ell}{L}\right) - \frac{pv\ell}{L^2} e(\ell). \tag{10}$$

Here the function $a(\ell/L)$ determines the probability of formation of a loop of size ℓ in a system of strings with persistence length L.

The second term on the right hand side of (9) is the contribution of loops that have hitherto been regarded as part of the system of long strings but are now reclassified as small loops, as L increases. (In reference 7 these equations were stated without this term and with $k = \infty$. There are however problems in that case with unphysical effects).

5. EQUILIBRIUM LOOP DISTRIBUTION

In a non-expanding universe ($\dot{R} = 0$), the system of strings would evolve towards an equilibrium distribution, in which the number density of loops of size ℓ is

$$N_{eq}(\ell)d\ell = a\left(\frac{\ell}{L}\right) \frac{d\ell}{2\pi \ell L^3}. \tag{11}$$

It follows that if we can establish the equilibrium distribution of loops, then we can determine the function a.

Vachaspati and Vilenkin[9] have performed a numerical simulation on a lattice of the system of strings generated in the breaking of a U(1) symmetry. They find a scaling low for the size of the loop. As one might expect the typical radius r of a loop of length $2\pi\ell$ is roughly

$$r \sim (\ell L)^{1/2}. \tag{12}$$

The number density of loops on all sizes bigger than L scales as

$$dn \approx \frac{1}{r^3} \frac{dr}{r}.$$

On the assumption that this distribution is the same as the equilibrium distribution, this would give

$$N_{eq}(\ell)d\ell \approx \frac{1}{(\ell L)^{3/2}} \frac{d\ell}{\ell} \tag{13}$$

or

$$a\left(\frac{\ell}{L}\right) \propto \left(\frac{L}{\ell}\right)^{3/2}. \tag{14}$$

(Note that this result was stated incorrectly in reference 7, as $a(x) \propto x^{-3}$). This form of a applies only for $\ell > L$; because of the finite lattice spacing, the simulation cannot reveal the distribution of very small loops. Vachaspati and Vilenkin also find that only about 20% of the total length of string is in the form of closed loops, which implies

$$\int_0^\infty a(x)dx \simeq 0.2.$$

Thus we may expect that for $x > 1$, $a(x)$ is very roughly of the form $0.1x^{-3/2}$.

The Z_2 strings formed in many grand unified theories have rather different properties, and it is not obvious that they should have the same equilibrium distribution. I have therefore recently performed a simulation to test this.

Vachaspati and Vilenkin approximated U(1) by Z_3, choosing randomly at each site one of three orientations $\{0, \pm 2\pi/3\}$. Then around any plaquette one can decide unambiguously whether the angle has changed by 0 or by $\pm 2\pi$; in the latter case a string passes through it.

Similarly, one can approximate SO(3) by the tetrahedral group T of order 12, and choose one of 12 random orientations at each site in such a way that each link corresponds to a rotation through no more than $2\pi/3$. Each element of T has nine acceptable neighbours, including itself, and three unacceptable ones that world require rotations through π. If the latter are excluded one can distinguish unambiguously between rotations through 0 and 2π.

As yet the results are preliminary. They indicate as before that the typical radius is $r \simeq (\ell L)^{1/2}$ and that $n(\ell)$ scales in the same way. The only significant difference is that the total number of loops is less. In this case only about 10% of the total length, rather than 20%, is in the form of loops. The reasons for this discrepancy will be explored in a forthcoming paper.

6. EFFECT OF REATTACHMENT

If we allow for the possibility of reattachment of loops we have to consider the coupled equations (9) and (10) rather than simply (6). Then under suitable assumptions, (7) is replaced by

$$t\frac{\dot{\gamma}}{\gamma} = -\frac{1-v^2}{2} + \frac{1}{2} ka(k) + \frac{1}{2}\frac{pv}{\gamma} \int_0^k x[a(x)-f(x)]dx, \tag{15}$$

where f(x) satisfies

$$xf(x) + \frac{3}{2} f(x) = \frac{pv}{\gamma} x [f(x)-a(x)]. \tag{16}$$

In principle, the right hand side of (15) ought to be independent of the arbitrary choice of the length cutoff kL between what we call long strings and loops. In fact, because of the approximations made, it is not. If $a(x) = Cx^{-3/2}$, then $f(x) = a(x)$, so that the integral in (15) vanishes, while the second term on the right hand side clearly depends on k. The choice of k must be made on a physical basis to match the approximations made: large loops are stretched by the expansion of the universe and should be included in E, while small loops are not.

Unlike (7), the right hand side of (15), say $F(\gamma)$, has a maximum as a function of γ and decreases as $\gamma \to 0$, and is negative at $\gamma = 0$. There are thus two possibilities. Suppose first that the maximum of F is positive. Then the larger of the two roots of $F(\gamma) = 0$, γ_0 say, is a stable point, and provided that the initial value of γ is not too small we may expect to reach a scaling solution.

On the other hand if γ is initially small, or more plausibly if $F(\gamma)$ never reaches positive values, then γ will tend to zero and a string-dominated universe will become inevitable. The idea that the universe could be string-dominated was first suggested by Vilenkin[10], but in the context of a model where the "intercommuting probability" p was very small compared to unity. It is however possible that even with $p \simeq 1$ the universe could be string-dominated, provided it has become so only recently.

It is clear from (9) that if the net effect of creation of loops is small then once the scale size L has become small compared to t, so that $v \simeq 1/2$, then E will be approximately constant, or very slowly decreasing. Thus in a string-dominated universe $\rho \propto R^{-(3+\varepsilon)}$ for some small value of ε. The universe behaves much as though it were matter-dominated. To ensure that the strings come to dominate only at late times, one must choose a very small value of μ, perhaps $G\mu \sim 10^{-15}$ to 10^{-12}, much less than the usual values in grand unified theories. (These are revised estimates from those given in the talk on which this article is based). The features of a string-dominated universe will be discussed in a forthcoming paper.

7. LOOPS AND GALAXY FORMATION

Let us however assume that the universe does approach a scaling solution, rather than string dominance.

Taking account of the finite lifetime of loops we find[8] that the number distribution of small loops is given by

$$n(\ell)d\ell = \frac{C}{2\pi} \frac{d\ell}{(\ell+\beta G\mu t)^{5/2} (\gamma t)^{3/2}}, \qquad (17)$$

where C is a constant, determined by the function $a(x)$, and perhaps of order 0.1.

The density in loops as a fraction of the total density is

$$\frac{\rho_{\text{loops}}}{\rho} = \frac{128\pi}{9} \frac{C}{\gamma^{3/2}\beta^{1/2}} (G\mu)^{1/2} \sim 10^{-3}. \qquad (18)$$

The commonest loops are the smallest still surviving, with masses given in terms of the mass M_{hor} inside the horizon by

$$\frac{M}{M_{\text{hor}}} \sim (G\mu)^2 \sim 10^{-12}. \qquad (19)$$

The spacing between the loops is typically $(G\mu)^{1/2}t \sim 10^{-3}t$, while their size is $G\mu t \sim 10^{-6}t$.

Strings can make their presence felt in several ways. A long straight string exerts no gravitational field[11]. The space around it is flat but cone-shaped, with a deficit angle around the string of $\delta = 8\pi G\mu$. This would lead to double images separated by angles of order $4\pi G\mu \sim 10^{-5}$ radians[12]. From loops of string one can get very interesting time varying structure, as shown by Hogan and Narayan[13]. Because of the deficit angle, a long string moving across the universe will create a wake behind it[14]. These may very well be of cosmological significance. However the main role in generating structure is probably that of the small loops which contribute up to 10^{-3} of the total density. They may act as seeds for galaxy formation. They have several properties that make them attractive candidates for this role.

Firstly, because there is no relevant intrinsic scale, the spectrum of density perturbations produced by loops is approximately "scale-invariant", in the sense that $\delta\rho/\rho$ at the time of horizon-crossing is independent of scale. This is the Harrison-Zel'dovich spectrum. It means that on very large scales the fluctuation in mass inside a given volume goes as

$\delta M/M \propto M^{-2/3}$, while on small scales $\delta M/M \propto M^{-1/3}$. Where this scenario differs essentially from the conventional adiabatic picture is that unlike adiabatic perturbations loops are not erased at the time of decoupling. There is therefore much more structure on small scales, and we do not have to wait for galaxy formation until larger structures have condensed.

Another very important feature is that in the Fourier components $\delta_{\underset{\sim}{k}}$ of $\delta\rho/\rho$ there are non-random phases, arising both from the fact that the loops are intrinsically very small compared to the distances between them and because there are important correlations between loops, particularly between those resulting from fragmentation of larger loops.

Later in the proceedings, Turok will describe his recent work[15], based on numerical simulations of string evolution by Albrecht and Turok[16], which shows that the correlations between loops provide a very natural explanation for the observed correlations between rich clusters.

8. CONCLUSIONS

In any symmetry breaking from a gauge group G to a subgroup H, strings can be formed if $\pi_1(G/H)$ is non-trivial. If the universe underwent such a phase transition early in its history then a system of strings would have been produced. The subsequent evolution could lead to one of two alternative final states, either string domination or a scaling solution.

A string-dominated universe would behave much like a matter-dominated one, with $R \propto t^{2/3}$. This is possible only if string domination occurred rather late, which in turn means a small value of $G\mu$.

In the scaling solution the persistence length of the system of strings scales with the horizon distance, $L = \gamma t$. Loops lose energy slowly by gravitational radiation and have long lifetimes. There is therefore a large population of small surviving loops. Those present at and after decoupling can act as seeds for galaxy formation. The resulting theory has several very attractive features.

REFERENCES

1) Ya. B. Zel'dovich, Mon. Not. Roy. Astr. Soc. 192 (1980) 663.
2) A Vilenkin, Phys. Rev. Lett. 46 (1981) 1169, 1496(E).
3) T.W.B. Kibble, J. Phys. A9 (1976) 1387.
4) C.J. Hogan and M.J. Rees, Nature 311 (1984) 109.
5) N. Turok, University of California at Santa Barbara preprint UCSB TH-3 (1984).

6) T. Vachaspati and A. Vilenkin, Harvard University preprint HUTP-84/A065 (1984).
7) T.W.B. Kibble and N. Turok, Phys. Lett. B116 (1982) 141.
8) T.W.B. Kibble, Nuc. Phys. B252 (1985) 227.
9) T. Vachaspati and A. Vilenkin, Phys. Rev. D30 (1984) 2036.
10) A. Vilenkin, Phys. Rev. Lett. 53 (1984) 1016.
11) A. Vilenkin, Phys. Rev., D23 (1981) 852; W.A. Hiscock, Montana State University preprint.
12) A. Vilenkin, Tufts University preprint TUTP-84-3 (1984); J.R. Gott, Princeton University preprint.
13) C. Hogan and R. Narayan, Caltech preprint GRP-010 (1984).
14) J. Silk and A. Vilenkin, Phys. Rev. Lett. 53 (1984) 1700.
15) N. Turok, Phys. Rev. Lett., to be published (1985).
16) A. Albrecht and N. Turok, Phys. Rev. Lett. 54 (1985) 1868.

RECENT DEVELOPMENTS IN THE COSMIC STRING THEORY OF GALAXY FORMATION

Neil TUROK

Department of Theoretical Physics, Imperial College, London, SW7 2BZ, England

Recent progress in the cosmic string theory of galaxy formation is reviewed, and a remarkable agreement between the correlation function for cosmic string loops and that for Abell clusters is reported.

In this talk I will review some of the recent developments in the cosmic string theory of galaxy formation[1] and report on an apparently remarkable agreement between the theory and the observed correlation function of rich clusters of galaxies[2]. Of course I should begin by acknowledging our debt to the Ancient Greeks who were the first to suggest that galaxies lay on strings (see any picture of the constellations) although theirs had ends and branches!

Over the past year there has been considerable progress in three main areas. First, many of the most interesting grand unified theories and in particular those based on superstring theories are now known to predict cosmic strings[3,4]. Second, as a result of numerical and analytical work, the evolution of a network of cosmic strings is now significantly better understood[5,6]. Lastly, the numerical simulations have allowed a series of detailed calculations of the specific predictions of the string theory to be made, from which it has so far emerged very successfully[7,8,9].

In spontaneously broken theories, strings may form if the space of degenerate vacua V_0 (the space of minima of the higgs potential) is nonsimply connected[10]. In the simplest case, the group $U(1)$ is broken completely and V_0 is just a circle. Here we get the simplest, directional strings (the higgs field can wind round the circle in either direction). Strings can also form if a simply connected group is broken to a subgroup with one or more disconnected components. For example, if the subgroup has one disconnected piece one gets "Z_2" strings, where two identical strings can annihilate. This case occurs in many candidate grand unified theories[3].

In the recently discovered superstring theories examples of both types occur. In the $E_8 \times E_8$ theory, most of the proposed symmetry breaking patterns involve breaking extra U(1)'s, giving directional strings. The SO(32) theory yields strings for slightly more subtle reasons. The universal covering group of SO(32) has centre $Z_2 \times Z_2$ which remains unbroken since all fields are in the adjoint representation, but is generally not part of a continuous symmetry in the unbroken subgroup. Hence we again get nondirectional strings[4].

In these theories, strings would form at a symmetry breaking phase transition in the very early universe. The basic idea, due to Kibble[10], is that as the as the universe cools below some temperature $T_c \sim m$, the symmetry breaking scale, the Higgs field Φ falls into the minimum of the potential V_0. Φ is only correlated over some scale $\xi \sim T_c^{-1} < t$ and the universe is broken up into domains of size $\sim \xi$, with Φ choosing a random position on V_0 in each. To minimise gradient terms in the energy Φ smoothes itself out by choosing the shortest path along V_0 in going from one domain to the next. Edges where Φ winds right around V_0 then become string segments. In these theories strings have no ends, so the segments are all part of closed loops or infinite lengths. One then has a tangled network of loops and lengths with a fixed number of segments of length ξ per volume ξ^3. Simulations of this process were first performed by Vachaspati and Vilenkin, who found that about 80% of the string is in the form of infinite lengths and the remainder in closed loops with a scale invariant size distribution i.e. a fixed number of loops of size $\sim r$ in a volume r^3. Both infinite lengths and closed loops are in the form of Brownian trajectories[11].

A crucial parameter in the cosmic string theory is the dimensionless number $G\mu \sim \pi(m/m_{pl})^2$ where μ is the mass per unit length of the string.

Between the time when they are formed, at a temperature $T \sim (G\mu)^{1/2} m_{pl}$ and the time when $T \sim G\mu m_{pl}$, the strings are heavily damped by collisions with relativistic particles^{10}. This results in the string straightening out – the scale ξ on which the string is straightened out grows as $t^{5/4}$ and $\xi \sim (G\mu)^{1/4} t$ at the time when damping ceases to be important.

The subsequent evolution of this network is a complicated problem, and the early literature on the subject contained several errors. It is obviously crucial to know whether the string density comes to dominate over the radiation, whose density $\rho_{rad} \sim 1/Gt^2$. The earliest discussions claimed that the network would evolve by straightening out so that $\xi \sim t$, and the density in strings $\rho_s \sim \mu t^{-2}$, whatever the string-string interactions were^{1,12}. This behaviour is desirable in the sense that the fractional density perturbation provided by the strings $\rho_s/\rho_{tot} \sim G\mu$ is constant, so the strings neither come to dominate nor disappear as the universe expands. The perturbation they provide is also of the right order of magnitude (for $m \sim 10^{16}$ GeV, a typical GUT scale) to seed the formation of galaxies and clusters of galaxies later on. However simulations by Albrecht and myself have shown that the evolution of the string network does depend crucially on how the strings interact. In particular if the strings simply pass through each other without interacting, the energy in infinite strings remains constant, $\rho_s \propto t^{-3/2}$, and strings rapidly come to dominate.^{5}

Luckily for the string theory, it seems more likely from the simulations of Shellard of "global" strings^{13} that in fact when two strings cross they almost always reconnect the other way. This makes it possible for the infinite lengths of string to slowly chop themselves up into closed loops. Albrecht and myself also simulated this case and found strong evidence that with $\xi_1 \sim t$ initially, as one would expect for "grand unified" strings, and for directional strings the energy density in the string network does indeed fairly rapidly evolve towards the "scaling" behaviour $\rho \sim \mu t^{-2}$.

All our results are consistent with the following simple picture. The network of strings longer than the horizon straightens out on scales of the order of the horizon scale, with a fixed number of lengths crossing each horizon. The self-intersection of these long strings continually produces loops with radii r of order the horizon scale, and more or less straightened out on that scale i.e. with length $1 \sim 2\pi r$. These "parent" loops are very likely to intersect themselves, breaking up into several smaller loops, but this process rapidly terminates, producing a clump of non-self-intersecting loops. These occupy a small and rapidly decreasing fraction of space and the probability of them reconnecting to the network of long strings or colliding with another loop rapidly becomes negligible. Whilst they are produced with considerable paculiar velocities $v \sim 0.1$ or so this rapidly redshifts away and the total distance loops of a given size move relative to a comoving frame is a small fraction of their mean separation. The loops just sit at almost fixed comoving location, oscillating in a periodic way and slowly radiating gravitational radiation at a calculable rate[14,15], until they eventually disappear. Their mass being almost constant, their energy density scales like matter and the smallest loops (those just about to disappear) dominate the total energy density in strings. The positions of the loops are correlated due to the way in which they are produced - "parent" loops producing a clump of "child" loops".

This "scaling" solution may be understood in a fairly simple way. Let us assume that there is always a scale ξ below which strings are straight and above which they are more or less random. For $\xi \gg t$ the strings are conformally stretched by the expansion of the universe and $\rho_s \propto R^{-2} \propto t^{-1}$ in a radiation dominated universe, where R is the scale factor. For $\xi \sim t$ long strings self-intersect rarely so few loops are chopped off and $\rho \sim t^{-3/2}$ as discussed above. This means that the number of string segments ξ per horizon grows. When this number becomes much larger than one, strings frequently collide and reconnect, reducing ξ to avalue less than t.

On the other hand, for $\xi \ll t$, long strings frequently self-intersect and chop off loops. With typical velocities ~ 1, and neglecting reconnection of loops, the time scale for a long string to chop off some fraction of it's length is $\sim \xi$ so as we reduce ξ the rate of loss of string froom the network increases indefinitely. This must result in ξ growing, so for $\xi \sim t$ a balance is struck and the string density $\rho_s \sim \mu t^{-2}$.

This picture contains two assumptions. First, strings are a straight on a scale ξ but not above ξ. This is reasonable and in fact because of the way the strings were formed, ξ cannot grow faster than t (recalling the way the strings were formed, the higgs field cannot align itself on scales larger than t). Secondly, we have ignored reconnection of loops to long strings. This is justified provided that the probability of a loop reconnecting is much smaller than that of one being produced. This is particularly plausible in an expanding universe when $\xi \sim t$ - loops produced of size < t remain constant in size whilst the network of long strings expands with the universe - the probability for a loop to reconnect rapidly becomes negligible. However when $\xi \ll t$ or equivalently for a string network in a non-expanding universe, the situation is not so clear.

As he has discussed at this meeting, Kibble has developed an analytic approach to this problem on the assumption that for a string network in a non-expanding universe an equilibrium distribution of lengths and loops would be reached resembling the initial conditions of the network described above. With this assumption it it possible that for initial conditions $\xi \ll t$ in an expanding universe that reconnecting of loops compensates at least to some extent for chopping off of loops, and the density in strings does indeed come to dominate. Kibble has explored the consequences of this scenario for lighter strings, with $G\mu \sim 10^{-13}$ or so [16]. Whether such an equilibrium exists is not clear at this stage - it may be that the processes of a loop being chopped off from or joining onto a length are not at all symmetrical, and no equilibrium exists at all. After this conference, Albrecht and myself simulated this case too and it is clear from our simulations that if such an equilibrium exists, it is very different from the initial conditions, with a far higher fraction of the string in loops[5]. In my opinion we will have to do more numerical work to build a clearer picture of the details of the processes of chopping off, reconnection of and breaking up of loops before we will be able to construct analytical models with any level of precision.

Let me turn now to the most interesting part of the cosmic string story, the idea that loops chopped off the network as the universe expanded later seeded the formation of galaxies and clusters of galaxies. For simplicity I will only deal with the case of an $\Omega = 1$ cold dark matter dominated universe[7].

Perturbations start to grow around loops at the time t of equal matter and radiation density. At this time the distribution of loops is as in Figure 1. The mean separation of loops of size r is ~ r when they are produced but has now grown to $r(t/r)^{1/2}$. Loops of size r have number density ~ $(rt)^{-3/2}$ and contribute a density $\mu r^{-1/2} t^{-3/2}$ so the smallest loops dominate. Larger loops are more massive but rarer.

Figure 1 Distribution of loops when perturbations start to grow.

Density perturbations grow around loops in a fairly simple way, well approximated by a spherical model on scales much larger than the loop radius and smaller than the mean separation of loops. Larger loops produce larger potential wells, accreting larger masses containing smaller loops as well as matter about them. Thus a hierarchy of objects, smaller ones in bigger ones, is produced which we shall identify with galaxies and clusters of galaxies^2.

Now imagine drawing a sphere of radius R<t and moving it through space. Neglecting correlations between loops, r.m.s. fluctuations in the mass inside R due to loops of radius r are given by

$$\delta M/M = (N)^{1/2} (\mu r)/(4\pi \rho R^3/3) \sim G\mu R^{-3/2} r^{1/4} t^{5/4} \quad (1)$$

with ρ the background density and N the mean number of loops in the sphere, for N >>1. These fluctuations are dominated by the largest loops for which N ~ 1 i.e. r ~ R^2/t. Now imagine looking for regions containing much greater mass excess than this typical fluctuation. These are places where the sphere contains a much larger loop of radius r with R > r >> R^2/t. In fact this selection procedure will 'find' all loops of a mean separation equal to the mean separation of the regions so selected[2].

This selection procedure is in fact precisely the one employed observationally by George Abell to define rich clusters of galaxies over 25 years ago[17]. He defined them as regions containing more than 50 bright galaxies within a radius 1.5h^{-1} Mpc[18]. For comparison, the mean separation of bright galaxies is 5h^{-1} Mpc so Abell clusters are exceptionally dense clusters of galaxies. Their mean separation d_c is known[19] to be 55h^{-1} Mpc. Furthermore, in theories where small scale structure forms first, such as the cosmic string theory, $\delta M/M$ decreases with M and on as large scales as d_c, there has been hardly any gravitationally induced motion of matter since perturbations started to grow. The positions of Abell clusters should thus accurately trace the positions of the loops that produced them.

Now Abell himself noticed that the two dimensional distribution of Abell clusters on the sky is in fact highly non random,[17] but it was not until quite recently that enough redshift data became available for a precise determination to be made of a good statistic for measuring this, the two point correlation function $\xi(r)$ (the excess probability over random of finding two objects at a separation r). This was done in 1983 by Bahcall and Soneira,[19] and Klypin and Kopylov,[20] who found that the two point correlation function for clusters was similar in form, but 18 times larger than that for galaxies. It was also observed to be positive out to distances of more than 100h^{-1} Mpc[19].

This suggested a very clean test of the cosmic string theory[2]. The correlations of loops may be measured in simulations such as those described above. The two point correlation function $\xi(r)$ of loops of mean separation d must in fact be dimensionless function of r/d. This is because all loops are produced in the same way, and at the time loops of a given size are formed, there is only one relevant scale, the horizon scale. At formation therefore, and measuring distances in terms of the horizon scale, all loops show identical correlations. Subsequent expansion of the universe simply stretches all separations by the same amount, so $\xi(r/d)$ does not change. Furthermore, there are no free parameters in $\xi(r/d)$ at all, the dynamics of string being independent of μ, the mass per unit length, and independent of the cosmological parameters h or Ω. $\xi(r/d)$ thus provides a very good test of the cosmic string theory.

Figure 2 shows $\xi(r/d)$ calculated from numerical simulations compared to $\xi(r/d)$ as calculated for Abell clusters by Bahcall and Soneira. As you can see, the agreement is remarkably good. Some simple models that explain the form of $\xi(r/d)$ in the cosmic string theory were presented in ref. 2.

In the simulations 'U(1)' type strings were used and assumed to always reconnect the other way when they crossed. Loops with the separation of Abell clusters today were produced just at the end of the radiation dominated era, so a flat Robertson-Walker background with scale factor $a(t) \propto t^{\frac{1}{2}}$ was used. The effect of peculiar velocities of loops was shown to be minimal (producing the slight flattening in ξ at small r/d) and negligible in terms of accretion of matter[7,8].

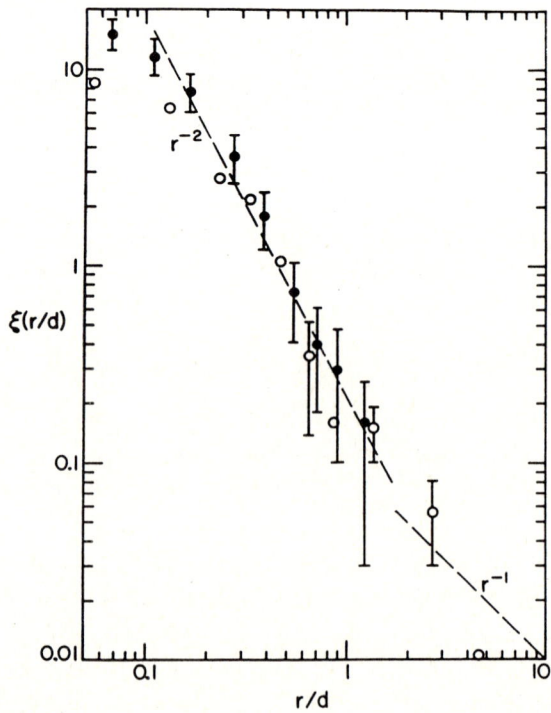

Figure 2 Correlation Function for Abell Clusters (circles) and Cosmic String Loops (dots).

Having identified loops with the separation of Abell clusters with Abell clusters, one can then demand that they be massive enough to have accreted objects with the mass of an Abell cluster by today. This determines μ, the mass per unit length of the string, in a much cleaner and more direct way than previous calculations^7.

Dynamical mass measurements of the masses of Abell clusters suggest the overdensity inside an Abell radius is $\frac{\delta M}{M}$ ~170, slightly larger than that in the spherical model after virialisation (~150). This suggests that Abell clusters formed very recently, and allows one to calculate^7, with Ω=1 and h=0.5

$$G\mu = 2 \times 10^{-6} \qquad (2)$$

from requiring that loops with the separation of Abell clusters virialised an Abell overdensity inside an Abell radius by the present day. A baryon dominated universe with Ω=0.1 has trouble explaining the existence of Abell clusters at all^6.

A completely independent calculation of Gμ can be performed by requiring that loops with the mean separation of galaxies gave rise to objects as massive as galaxies and with the observed galaxy-galaxy correlation function. In fact, since $d_g \sim d_c/10$, the 'bare' correlation function of 'galaxy' loops is roughly one hundredth that for clusters, so a gravitational enhancement of the 'bare' galaxy loop correlation function by a factor of about 5 is needed to fit observations.

In ref. 7 a simple model for the gravitational enhancement of correlations is used to calculate Gμ. One finds $G\mu \sim 4 \times 10^{-6}$ which is quite close to (2), the difference being well within the accuracy of the calculation. Similarly, using the overdensity in a typical galaxy derived from rotation curves one finds $G\mu \sim 4 \times 10^{-6}$ as well. These values are far more imprecise than (2) however, so the most one can legitimately claim is that the cosmic string theory seems at present consistent with galaxy and cluster masses and correlations for $G\mu \approx 2 \times 10^{-6}$, Ω = 1, h = 0.5. These are however considerable successes, which no other theory has so far achieved. More precise calculations will clearly require large N-body simulations.

Finallly, what are the prospects of detecting cosmic strings? Thanks to the numerical simulations, fairly precise statements can now be made. The most direct effect of strings is their gravitational lensing effect - a straight length[21] of string produces double images at a typical angular separation

$$\Delta\theta = 4\pi G\mu = 2\times 10^{-5} \approx 5''$$

using the value for $G\mu$ in (2). Larger or smaller values are obtained as the geometry is varied.

All five of the observed cases of gravitational lensing are within 3" of this value. For only one of them has a candidate object for producing the lensing so far been observed. We may thus have already seen the effects of cosmic strings!

A second effect is the anisotropy induced in the microwave background. A single string moving at velocity $v_\perp$ perpendicular[22] to the line of sight produces a discontinuity

$$\frac{\delta T}{T} = 8\pi G\mu v_\perp$$

which is always less than $\sim 4\times 10^{-5}$ and has rms value $\sim 2\times 10^{-5}$, below present observational limits. These discontinuities are rare : with a beam separation of 5' one sees a discontinuity on average 1/20 of the time.[8]

The Sachs-Wolfe effect due to accreting loops has also been calculated and produces rms fluctuations of similar magnitude[8].

The best limit on the cosmic string theory will soon come from the effects of the gravitational radiation from loops. Despite recent claims, this is not[9] in contradiction with the standard nucleosynthesis scenario. However it would lead to variations in the observed frequency of the millisecond pulsar. The limit on $G\mu$ obtained if the predicted variations are not observed gets rapidly[23] better as the period T (in years) of observation increases :[8]

$$G\mu < 10^{-5} (\alpha/T)^8$$

with $\alpha \sim 2\pi$. This limit may become very stringent in a decade or so.

In conclusion the cosmic string theory has proved to be an eminently testable theory of galaxy formation. It has so far yielded several important predictions in good agreement with observation, and we can look forward to several new tests in the near future.

ACKNOWLEDGEMENT

I would like to thank A.Albrecht, R.Brandenberger and J.Traschen for helpful discussions.

REFERENCES

[1] Ya. B. Zel'dovich, Mon. Not. Roy. Ast. Soc., 192 (1980) 663; A. Vilenkin, Phys. Rev. Lett. 46 (1981) 1169, 1496 (E). For reviews see T.W.B. Kibble, Phys. Rep. 67 (1980) 183; A. Vilenkin, Phys. Rep. 121 (1985) 263.

[2] N. Turok, Phys. Rev. Lett., to be published (1985).

[3] T.W.B. Kibble, G. Lazarides and Q. Shafi, Phys. Lett. 113B (1982) 237.
D. Olive and N. Turok, Phys. Lett. 117B (1982) 193.

[4] E. Witten, "Cosmic Superstrings", Princeton preprint (1985)

[5] A. Albrecht and N. Turok, Phys. Rev. Lett. 54 (1985) 1868; in preparation.

[6] T.W.B. Kibble, talk at this meeting.

[7] N. Turok and R. Brandenberger, "Cosmic Strings and the Formation of Galaxies and Clusters of Galaxies", UCSB/ITP preprint (1985).

[8] R. Brandenberger and N. Turok, "Fluctuations from Cosmic Strings and the Microwave Background", ITP preprint (1985).
J. Traschen, N. Turok and R. Brandenberger, "Microwave Anisotropies from Cosmic Strings", preprint (1985).

[9] R. Brandenberger, A. Albrecht and N. Turok, "Gravitational Radiation from Cosmic Strings and the Microwave Background", preprint (1985).

[10] T.W.B. Kibble, J. Phys. A9 (1976) 1387.

[11] T. Vachaspati and A. Vilenkin, Phys. Rev. D30 (1984) 2036.

[12] A. Vilenkin, Phys. Rev. D24 (1981) 2082.

[13] P. Shellard, unpublished.

[14] N. Turok, Nuc. Phys. B242 (1984) 520.
[15] T. Vachaspati and A. Vilenkin, Phys. Rev. D (1985).
[16] T.W.B. Kibble, Nuc. Phys. B252 (1985) 227; "String Dominated Universe", Imperial College preprint (1985); talk at this meeting.
[17] G.O. Abell, Ap. J. Suppl. 3 (1958) 211, Ap. J. 66 (1961) 607.
[18] M. Davis and J. Huchra, Ap. J. 254 (1982) 437.
[19] N.A. Bahcall and R.M. Soneira, Ap. J. 270 (1983) 20.
[20] A.A. Klypin and A.I. Kopylov, Sov. Astr. Lett. 9 (1983) 41.
[21] A. Vilenkin, Phys. Rev. D23 (1981) 852; Ap. J. Lett. 282 (1984) L51; J.R. Gott, Princeton preprint (1984); C.J. Hogan and R. Narayan, Mon. Not. Roy. Astr. Soc (1985).
[22] N. Kaiser and A. Stebbins, Nature 310 (1984) 391.
[23] C.J. Hogan and M.J. Rees, Nature 311 (1984) 109.

"INVISIBLE" AXION DETECTORS[†*]

P. SIKIVIE

Physics Department, University of Florida, Gainesville, Florida 32611

1. INTRODUCTION

The axion[1,2] was postulated eight years ago to explain why the strong interactions conserve P and CP[3]. The parameter that sets the amount of P and CP violation in QCD is[4]

$$\bar{\theta} = \theta - \arg \det m \tag{1.1}$$

where m is the quark mass matrix and θ is the coefficient of $g^2/32\pi^2 \, G^a_{\mu\nu} \tilde{G}^{a\mu\nu}$ the action density for QCD. Because $G\tilde{G}$ is a 4-divergence, the $\bar{\theta}$ dependence of QCD is due purely to non-perturbative quantum effects. The present upper limit on the neutron electric dipole moment requires[5] $\bar{\theta} \lesssim 10^{-8}$. If the CP violation necessary to explain $K_L \to 2\pi$ is introduced into the standard $SU_L(2) \times U_Y(1) \times SU^c(3)$ model of particle interactions in the manner of Kobayashi and Maskawa[6], then arg det m is an arbitrary (random) angle and there is absolutely no reason why $\bar{\theta} \lesssim 10^{-8}$. Other methods of introducing CP violation into the standard model also suffer from this difficulty which is believed to be quite general[7] and which has been given the name of "strong CP problem".

Peccei and Quinn[1] proposed a simple and elegant solution to this problem. They postulated a $U_{PQ}(1)$ symmetry for the classical action density under which the quark fields rotate chirally and which is spontaneously broken by the vacuum expectation value of a scalar field $\langle \phi \rangle = v e^{i\alpha}$. As a result, arg det m becomes proportional to α and hence

$$\bar{\theta} = \theta - N\alpha \qquad (1.2)$$

becomes a dynamical degree of freedom. In Eq. (1.2), N is a model dependent integer given by

$$N = \sum_f Q_f t_f \qquad (1.3)$$

where the sum is over all colored Dirac fermions f, the Q_f are the Peccei-Quinn charges of these fermions and the t_f are given by $Tr(T_f^\alpha T_f^\beta) = \frac{1}{2} t_f \delta^{\alpha\beta}$ where the T_f^α ($\alpha=1$ to 8) are the $SU^c(3)$ generators for the color representation to which f belongs ($t_3 = t_{\bar{3}} = 1$, $t_6 = t_{\bar{6}} = 5$, etc.). Because of the $U_{PQ}(1)$ symmetry, the value of α is indifferent at the classical level. But the quantum effects (instantons...) which make the physics of QCD $\bar{\theta}$-dependent will lift this degeneracy and align α in a particular direction. One finds that α aligns in such a way that $\bar{\theta} = 0$. The strong CP problem is thus solved.

Weinberg and Wilczek[2] independently pointed out that the Peccei-Quinn solution to the strong CP problem implies the existence of a light pseudoscalar particle, which they called the axion. The axion is the Pseudo-Nambu-Goldstone boson associated with the spontaneous breaking of the $U_{PQ}(1)$ quasi-symmetry, i.e. $a=\alpha v$ where a is the axion field, and α and v are the angle and vacuum expectation value defined above. The mass and the couplings of the axion have been determined (see Section II below). They are all inversely proportional to v. At first, it was thought that the breaking of $U_{PQ}(1)$ occurred at the electroweak scale, i.e. $v \sim 250$ GeV. The corresponding axion was searched for in various laboratory experiments but was not found. Soon, however, it was discovered[8] how to construct axion models with arbitrarily large values of v. These were called "invisible" axion models because for $v \gg 250$ GeV, the axion is so weakly coupled that the event rates in the axion search experiments mentioned above are hopelessly small. For a while, it was thought that the strong CP problem was solved without any presently observable consequences whatsoever. Fortunately, astrophysics[9] and cosmology[10,11] came to the rescue.

Stellar evolution implies that[9,12] $f_a = \frac{v}{N} \geq 3.10^7$ GeV. The constraint arises because stars emit the weakly coupled axions from their whole volume whereas they emit photons only from their surface. Unless $f_a \geq 3.10^7$ GeV, axion emission by stars is too copious to be consistent with our understanding of stellar evolution. Cosmology, on the other hand, implies that[11] $f_a \leq 2.10^{12}$ GeV. This constraint arises because of the relative suddenness with which the axion mass turns on when the universe is about 10^{-6} sec old and the temperature has reached about 1 GeV. The sudden switch-on of the axion mass by QCD instantons produces large numbers of non-relativistic axions. Unless $f_a \leq 2.10^{12}$ GeV, the axion energy density today would exceed the critical energy density for closing the universe.

If f_a is near 10^{12} GeV, there are enough axions to constitute the dark halos of galaxies and galaxy clusters. Axions are in fact an excellent candidate for the halo matter[13]. Not only are axions sufficiently abundant and, for such large values of f_a, so very weakly interacting that an axionic halo would automatically be dark, but axions also have the additional desirable properties of very large phase space density and vanishingly small free-streaming distance. The large phase space density allows them to cluster easily into galactic halos, whereas the absence of free-streaming allows the growth of primordial density perturbations to proceed on all scales. Because of these properties, axions fall into the "cold" dark matter category. Other cold dark matter candidates are photinos, gravitinos, quark nuggets... Various recent studies[14] of the evolution of primordial density perturbations into the large scale structure of our universe have shown that cold dark matter is indeed preferable to "hot" dark matter (e.g. neutrinos).

There is an additional cosmological constraint due to the presence of domain walls in axion models[10]. The domain walls must be gotten rid of before they dominate the energy density of the universe. This can, in fact, be achieved through inflation or by constructing the axion model is such a way that it has a unique vacuum[15].

The main purpose of this talk is to give a brief review of various ideas[16,17,18,19] which have been put forth to detect "invisible" axions, i.e. axions with 3.10^7 GeV $\leq f_a \leq 2.10^{12}$ GeV. These experiments would attempt to detect the axions which constitute the halo of our galaxy or axions which are emitted by our sun; or they would attempt to detect the force mediated by virtual axions. In the next section, we give various relevant axion parameters: mass, couplings, flux ... as functions of f_a. The remaining sections describe the experiments.

2. AXION PROPERTIES

The axion mass is given by[2,20]

$$m_a = \frac{f_\pi m_\pi}{f_a} \frac{\sqrt{m_u m_d}}{m_u + m_d} \approx .6 \; 10^{-5} \; \text{eV} \; \left(\frac{10^{12} \text{GeV}}{f_a}\right) \tag{2.1}$$

where $f_a = \frac{v}{N}$, with v and N as defined in section I. In units where $\hbar = c = 1$, one has

$$.6 \; 10^{-5} \; \text{eV} = 2\pi(1.45 \; \text{GH}_z) = \frac{2\pi}{20.7 \; \text{cm}} \tag{2.2}$$

All the experiments described below exploit the fact that the inverse mass of the "invisible" axion is a macroscopic length scale.

The electromagnetic coupling of the axion is given by[20,21]

$$\mathcal{L}_{a\gamma\gamma} = - g_\gamma \frac{\alpha}{\pi} \frac{a}{f_a} \vec{E}\cdot\vec{B} \tag{2.3}$$

where

$$g_\gamma = \frac{4}{3} - \frac{1}{3} \frac{4m_d + m_u}{m_d + m_u} \approx .36 \; . \tag{2.4}$$

To obtain (2.3), one need only assume that the axion model is grand unified (or grand unifiable) with $\sin^2\theta_W^0 = \frac{3}{8}$ where θ_W^0 is the electroweak angle at the grand unification scale. Note that the electromagnetic coupling strength [Eq. (2.3)] of the axion is given uniquely in terms of the axion mass.

The coupling of the axion to the electron[21] is more model dependent. If we write it as

$$\mathcal{L}_{aee} = -i g_e \frac{m_e}{v} a \bar{e}\gamma_5 e, \qquad (2.5)$$

then usually g_e is of order one. For example, in the Dine-Fischler-Srednicki model[8]

$$g_e = \frac{2v_u^2}{v_u^2 + v_d^2} \qquad (2.6)$$

where both v_u and v_d are vacuum expectation values of order the electroweak scale. In Kim's model[8] however, g_e vanishes at the tree level; loop corrections yield[21] $g_e = O(10^{-3})$.

If the Milky Way halo is composed of axions, the local axion energy density is approximately[22] $.6 \times 10^{-24}$ gr/cm^3. Galactic halo axions would have velocities of order the galactic virial velocity, i.e. 10^{-3} times the speed of light. They would form a highly degenerate Bose gas with average quantum state occupation number of order $10^{26} (\frac{f_a}{10^{12} \text{GeV}})^4$.

If axions exist, the flux of solar axions on earth is approximately

$$\text{solar axion flux} \approx 10^{12}/\text{sec. cm}^2 \left(\frac{10^8 \text{GeV}}{f_a}\right)^2 \qquad (2.7)$$

Solar axions have a broad spectrum of energies of order the temperature in the solar interior. Their average energy is approximatively 1.8 keV.

3. GALACTIC AXION DETECTOR USING A CAVITY[16,18,19,23-25]

Through the coupling (2.3), axions can convert to photons in the presence of an externally applied magnetic (or electric) field. Galactic halo axions have velocities β of order 10^{-3} and hence their energies

$$\varepsilon_a = m_a + \frac{1}{2} m_a \beta^2 \qquad (3.1)$$

have a spread of order $\frac{1}{2} 10^{-6}$ above the axion mass. Consider then a cylindrical electromagnetic cavity of arbitrary cross-sectional shape, permeated by a large static approximately homogeneous longitudinal magnetic field $\vec{B} = B_0 \hat{z}$. When the frequency ω of an appropriate cavity mode equals $m_a[1 + 0(10^{-6})]$ galactic halo axions can convert to quanta of excitation (photons) of that cavity mode. Provided the cavity is much smaller in size than the de Broglie wavelength $\lambda_a = 2\pi(\beta m_a)^{-1} \simeq 2\pi \cdot 10^3 \, m_a^{-1}$ of the galactic halo axions, the coupling of a given cavity mode to the galactic halo axion field a is

$$\frac{\alpha}{\pi} g_\gamma \frac{1}{f_a} a \, B_0 \int_V d^3x \, \vec{E}_\omega(\vec{x},t) \cdot \hat{z} \qquad (3.2)$$

where $\vec{E}_\omega$ is the electric field amplitude for that mode and V is the volume of the cavity. From (3.2) one sees that only $TM_{n\ell 0}$ modes couple. The power on resonance ($\omega_{n\ell} = m_a[1 + 0(10^{-6})]$) from axion $\rightarrow$ photon conversion into the $TM_{n\ell 0}$ mode is[18,19]

$$P_{n\ell} = .8 \, 10^{-19} \text{Watt} \left(\frac{V}{500 \text{ liter}}\right)\left(\frac{B_0}{8 \text{ Tesla}}\right)^2 C_{n\ell} \left(\frac{\rho_a}{10^{-24} \text{ gr/cm}^{-3}}\right)$$

$$\left[\frac{m_a}{2\pi(3\text{GH}_z)}\right] \text{Min}\left(\frac{Q_{n\ell}}{Q_a}, 1\right) \qquad (3.3)$$

where ρ_a is the galactic axion energy density. Eq. (3.3) indicates that the power is proportional to the quality factor $Q_{n\ell}$ for the $T_{n\ell 0}$ mode provided $Q_{n\ell} < Q_a$ where $Q_a = \frac{\varepsilon_a}{\Delta\varepsilon_a} \simeq 2.10^6$ is the "quality factor" of the galactic halo axions, i.e. the ratio of their energy to their energy spread. $C_{n\ell}$ is a mode dependent factor. For a cavity of rectangular cross-sectional shape,

$$C_{n\ell} = \frac{64}{\pi^4} \frac{1}{n^2 \ell^2} \quad \text{for } n \text{ \underline{and} } \ell \text{ odd}$$

$$= 0 \quad \text{otherwise} . \tag{3.4}$$

For a circular cross-section

$$C_{nm} = \frac{4}{(\chi_{0n})^2} \delta_{m0} \tag{3.5}$$

where χ_{0n} in the nth zero of the Bessel function $J_0(x)$.

To detect the power, a hole must be made in the cavity walls through which the electromagnetic radiation can be brought to shine upon a microwave detector. Note, in this regard, that the quality factor $Q_{n\ell}$ which appears in Eq. (3.3) is the \underline{loaded} quality factor Q_L given by

$$\frac{1}{Q_L} = \frac{1}{Q_W} + \frac{1}{Q_h} \tag{3.6}$$

where $\frac{1}{Q_W}$ is the contribution due to absorption into the cavity walls and $\frac{1}{Q_h}$ is the contribution due to the hole. The sensitivity of a microwave detector is usually expressed by its noise temperature T_n. From T_n one can compute how long a measurement integration time t is required to detect a power P_d in a bandwidth B with a given signal to noise ratio s/n by using the formula

$$\frac{s}{n} = \frac{P_d}{T_n} \sqrt{\frac{t}{B}} . \tag{3.7}$$

In particular, we can use Eq. (3.7) to compute how much time a galactic halo axion search takes. If we write Eq. (3.3) in the form $(f = \frac{\omega}{2\pi})$

$$P = P_0 \frac{f}{3 \text{GH}_z} \text{Min}\left(\frac{Q_L}{Q_a}, 1\right) \qquad (3.8)$$

one obtains for the search rate

$$\frac{df}{dt} = \frac{10 \text{ GH}_z}{\text{year}} \left(\frac{P_0}{.4 \; 10^{-19} \text{Watt}}\right)^2 \left(\frac{20^\circ K}{T_n}\right)^2 \left(\frac{f}{3 \text{GH}_z}\right)^2 \times$$

$$\times \begin{cases} Q_W/Q_a & \text{if } Q_W < 3Q_a \\ \frac{27}{4}\left(1 - \frac{Q_a}{Q_W}\right)^2 & \text{if } Q_W > 3Q_a \end{cases} \qquad (3.9)$$

This result was obtained under the following assumptions:
- that the search is carried out with a signal to noise ratio of three,
- that when $Q_L < Q_a$, i.e. when the cavity bandwidth is larger than the axion bandwidth, one does use the possibility of looking at Q_a/Q_L axion bandwidths simultaneously,[26]
- that Q_h has been adjusted so as to minimize the search rate. For $Q_W < 3Q_a$, the optimal Q_h is $\frac{1}{2} Q_W$ (and hence $Q_L = \frac{1}{3} Q_W$) whereas for $Q_W > 3Q_a$ the optimal Q_h is such that $Q_L = Q_a$.

Typical noise temperatures for state of the art microwave receivers are as follows:

T_n noise temperature	f frequency	detector type
< 20°K	.3 to 6 GH$_z$	Ga-As FETs
20 - 100°K	6 - 20 GH$_z$	Ga-As FETs
100 - 200°K	20 - 100 GH$_z$	SIS or Schottky diode mixers.

The cavity must be cooled to decrease the thermal noise. Presumably, it is sufficient to cool the cavity to a physical temperature well below the noise temperature of the receiver; e.g. the cavity may be placed in a liquid helium bath. The cavity must also be tunable. Tuning over a small frequency range ($\lesssim 10\%$) can be achieved by squeezing the cavity walls, and/or by inserting a dielectric rod into the cavity, and/or by arranging to have a variable level of liquid helium in the cavity. Hence, it will be necessary to build many cavities, each one designed to cover a relatively small range of values of the axion mass. Note that the power into a given cavity mode decreases sharply with mode number [see Eqs. (3.3), (3.4) and (3.5)]. It is therefore preferable to use the fundamental TM mode as much as possible. On the other hand, for cavities of rectangular or circular transverse cross-section, the frequency ω of the fundamental TM mode is $\sim \frac{\pi}{d}$ where d is the shortest linear dimension of the transverse cross-section. To explore high values of the axion mass using the fundamental TM mode of such cavities, their volume would have to be correspondingly reduced. One can use instead many small cavities whose fundamental TM frequencies (before coupling) are close to each other and which are then strongly coupled together[23]. This allows one to have simultaneously large volume, high frequency and low mode number. The unloaded quality factor for these "multi-cell" cavities is typically

$$Q_W \simeq .6 \; 10^5 \; (\frac{3 GH_z}{f})^{2/3} \qquad (3.10)$$

In this design, the major part of the losses are in the metal walls that separate the cells from one another. Morris[24] has proposed an alternate design in which the metal walls separating the cells are replaced by half-wavelength dielectric plates of high dielectric constant and low loss-tangent (e.g. alumina, sapphire, certain ceramics...). This improves the quality factor substantially because the losses in the dielectric plates are negligible compared to those in the metal walls. The quality factor can be further improved by adding quarter-wavelength dielectric plates facing the end-cap metal walls. In this design, assuming $\epsilon = 9$ for the dielectric plates, one has

$$Q_w \simeq 5 \; 10^6 \; (\frac{f}{3\text{GH}_z})^{1/3} \; (\frac{A}{50 \text{ cm}}) \qquad (3.11)$$

where A is the transverse size of the cavity. On the basis of Eqs. (3.3), (3.9) and (3.11), it appears that the experiment described in this section is feasible with existing technology.

4. "SPIN COUPLED" AXION DETECTION

Recently, Krauss, Moody, Morris and Wilczek[19] have proposed an interesting variation on the above experiment. These authors pointed out that the two Compton-like scattering graphs $a + e \to \gamma + e$ (e = electron) produce the following effective coupling in the non-relativistic limit

$$\mathcal{L}_{eff} = g_e \frac{e}{2m_e} \frac{a}{v} \vec{S} \cdot \vec{E} \qquad (4.1)$$

where $\vec{S}$ is the number density of aligned electron spins. Comparing Eqs. (2.3) and (4.1), one finds that, for the purpose of converting axions to photons, a density of aligned electron spins is equivalent to a magnetic field

$$\vec{B}_{eq} = \frac{g_e}{g_\gamma} \frac{1}{N} \frac{2\pi^2}{e} \frac{1}{m_e} \vec{S} \qquad (4.2)$$

In certain ferrite materials, $S \simeq 10^{23}$ cm^{-3} and hence $B_{eq} \simeq 230$ Tesla (for $\frac{g_e}{N} \simeq \frac{1}{6}$), which is much larger than the field strengths of present superconducting magnets. However, this idea faces a serious difficulty[19,27]. The ferrite used should not be a conductor, otherwise the photons produced will be absorbed in the ferrite before they can be detected. On the other hand, if the ferrite is an insulator, then the effective $a\vec{S} \cdot \vec{E}$ coupling strength is much smaller than given in Eq. (4.1). Eq. (4.1) was derived assuming the electrons to be free. In insulators, the recoil of an electron hit by an axion is suppressed by atomic forces. This decreases the $a\vec{S} \cdot \vec{E}$ coupling strength by a factor of order $\frac{m_a}{1\text{eV}} \simeq 0(10^{-5})$ as compared to Eq. (4.1).

5. AXION → PHOTON CONVERSION IN AN INHOMOGENEOUS STATIC MAGNETIC FIELD

The cross-section for $a \to \gamma$ conversion in a region of free space ("free" here means that there are no metal walls present) of volume V, permeated by a static inhomogeneous magnetic field $\vec{B}_0(\vec{x})$ is[16,18]

$$\sigma = \frac{1}{16\pi^2 |\vec{\beta}_a|} \left(\frac{\alpha}{\pi} \frac{g_\gamma}{f_a}\right)^2 \int d^3 k_\gamma \delta(E_a - \omega) \left| \int_V d^3 x \, e^{i(\vec{k}_\gamma - \vec{k}_a)\cdot \vec{x}} \, \hat{n} \times \vec{B}_0(\vec{x}) \right|^2 \qquad (5.1)$$

where $(E_a, \vec{k}_a) = E_a(1, \vec{\beta}_a)$ is the 4-momentum of the axion and $(\omega, \vec{k}_\gamma) = \omega(1, \vec{n})$ is the 4-momentum of the photon. Because the magnetic field is static, $E_a = \omega$ and hence $\vec{k}_\gamma \neq \vec{k}_a$. Therefore 3-momentum must be provided for the transition to occur. That is why the cross-section (5.1) is proportional to the power in $\vec{B}_0(\vec{x})$ on the momentum scale $\vec{q} = \vec{k}_\gamma - \vec{k}_a$. On the basis of Eq. (5.1) a galactic halo axion detector has been proposed[16,23] which may be suitable for the exploration of a range of high values of the axion mass $[m_a \gtrsim 2\pi(100 \text{ GH}_z)]$ not accessible to the cavity experiment described in Section 3.

6. A SOLAR AXION DETECTOR

Axion → photon conversion is an inhomogeneous magnetic field can also be used to search for solar axions[16,18]. Since the solar axions are highly relativistic, the momentum transfer q_z is small

$$q_z = (\vec{k}_\gamma - \vec{k}_a) \cdot \vec{n} = E_a - \sqrt{E_a^2 - m_a^2} = \frac{1}{2} \frac{m_a^2}{E_a} = \frac{2\pi}{70 \text{ cm}} \left(\frac{10^8 \text{GeV}}{f_a}\right)^2 \frac{\text{keV}}{E_a} \qquad (6.1)$$

where $\vec{n}$ is the direction from the sun. Resonant conversion occurs when the magnetic field is inhomogeneous on a length scale $d \simeq \frac{2\pi}{q_z}$. Multiplying Eqs. (4.7) and (5.1), one finds the event rate

$$\frac{\#_{x-ray}}{time} = \frac{2}{day} \frac{Vd}{(\text{meter})^4} \left(\frac{B_0}{8 \text{ Tesla}}\right)^2 \left(\frac{10^8 \text{GeV}}{f_a}\right)^4 . \qquad (6.2)$$

Assuming that one event/month can be distinguished from the background, this experiment could improve the stellar evolution constraint[9,12] $f_a \geq 3.10^7$ GeV by approximately an order of magnitude.

7. MACROSCOPIC FORCES MEDIATED BY AXIONS

Moody and Wilczek[17] have made a thorough analysis of the apparent deviations from the $\frac{1}{r^2}$ gravitational force law due to the exchange of virtual axions. The coupling of the axion to a fermion can be written in the general form

$$\mathcal{L}_{aff} = g_f \frac{m_f}{v} a \bar{f}(i\gamma_5 + \theta_f)f \tag{7.1}$$

where g_f and θ_f are model dependent. Because CP is violated in the electroweak interactions, one expects the θ_f to have very small but non-zero values. The 2^d term in Eq. (7.1) produces a coupling $\theta_f g_f \frac{m_f}{v} a f^\dagger f$ of the axion field to the mass density of a macroscopic collection of non-relativistic fermions, whereas the 1st term in Eq. (7.1) produces a coupling $g_f \frac{1}{2v} (\vec{\nabla}a) \cdot f^\dagger \vec{\sigma} f$ of the axion field to the spin density of that macroscopic body. The first type of coupling is called "monopole", the second type "dipole". The axion mediated forces between two macroscopic bodies thus fall into three categories: monopole-monopole, monopole-dipole and dipole-dipole. The (monopole)^2 force is suppressed by two factors of θ_f and is therefore very small. [One may expect θ_f for nucleons to be of order $\bar{\theta}$. Recall then that $\bar{\theta} \leq 10^{-8}$ is required by the experimental upper limit on the neutron electric dipole moment. However, in superweak type theories of CP violation such as the Kobayaski-Maskawa model, one expects $\bar{\theta} = O(10^{-14})$.] The (dipole)^2 force has a very large background from ordinary magnetic forces. This background can be suppressed by the use of superconducting shields but not well enough for the (dipole)^2 force to be detected. It appears that the monopole-dipole force is the one least difficult to detect because it is suppressed by only one factor of θ_f and because it can be modulated. Modulation greatly increases the experimental sensitivity. Moody and Wilczek estimate that future experiments may reach a sensitivity of 10^{-20} cm/sec^2 which would be sufficient to detect the axion mediated monopole-dipole force for $\theta_f \simeq 10^{-14}$ if $f_a \simeq 10^{10}$ GeV and $\theta_f \simeq 10^{-10}$ if $f_a \simeq 10^{12}$ GeV.

ACKNOWLEDGEMENT

This work was supported in part by the DOE under contract No. DE-AS-05-81ER40008.

REFERENCES AND FOOTNOTES

1. R. D. Peccei and H. Quinn, Phys. Rev. Lett. 38 (1977) 1440, and Phys. Rev. D16 (1977) 1791.

2. S. Weinberg, Phys. Rev. Lett. 40 (1978) 223; F. Wilczek, Phys. Rev. Lett. 40 (1978) 279.

3. G. 't Hooft, Phys. Rev. Lett. 37 (1976) 8, and Phys. Rev. D14 (1976) 3432; R. Jackiw and C. Rebbi, Phys. Rev. Lett. 37 (1976) 172; C. G. Callan, R. F. Dashen and D. J. Gross, Phys. Lett. 63B (1976) 334.

4. S. L. Adler, Phys. Rev. 117 (1969) 2426; J. S. Bell and R. Jackiw, Nuovo Cimento A60 (1969) 47.

5. V. Baluni, Phys. Rev. D19 (1979) 2227; R.J. Crewther, P. Di Vecchia, G. Veneziano and E. Witten, Phys. Lett. 88B (1979) 123.

6. M. Kobayashi and K. Maskawa, Progr. Theor. Phys. 49 (1973) 652.

7. See, however: J. Ellis and M. K. Gaillard, Nucl. Phys. B150 (1979) 141; A. Nelson, Phys. Lett. 136B (1984) 387; S. Barr, Phys. Rev. Lett. 53 (1984) 329 and Phys. Rev. D30 (1984) 1805.

8. J. Kim, Phys. Rev. Lett. 43 (1979) 103; M. Shifman, A. Vainshtein and V. Zakharov, Nucl. Phys. B166 (1980) 493; M. Dine, W. Fischler and M. Srednicki, Phys. Lett. 104B (1981) 199.

9. D. Dicus, E. Kolb, V. Teplitz and R. Wagoner, Phys. Rev. D18 (1978) 1829 and Phys. Rev. D22 (1980) 839; M. Fukugita, S. Watamura and M. Yoshimura, Phys. Rev. Lett. 48 (1982) 1522 and Phys. Rev. D26 (1982) 1840; N. Iwamoto, Phys. Rev. Lett. 53 (1984) 1198; L. M. Krauss, J. E. Moody and F. Wilczek, Phys. Lett. 144B (1984) 391; D. Morris, LBL preprint 18690 (1984); D. Dearborn, D. Schramm and G. Steigman, to be published.

10. P. Sikivie, Phys. Rev. Lett. 48 (1982) 1156.

11. L. Abbott and P. Sikivie, Phys. Lett. 120B (1983) 133; J. Preskill, M. Wise and F. Wilczek, Phys. Lett. 120B (1983) 127; M. Dine and W. Fischler, Phys. Lett. 120B (1983) 137.

12. The lower limits upon f_a obtained in the papers of Ref. (9) vary from 3.10^6 GeV to $(3.10^9 \text{ GeV}) \frac{1}{N}$, depending upon what star and what axion producing process in that star are used to set the limit. Some limits are much more model dependent than others. The limit quoted in the text is a relatively model independent one based upon the recent work of Dearborn et al.[9] It correspond to the value of f_a for which the energy loss by red giants, due to axion production through the Primakoff process only, becomes too large to allow the start of helium burning (D. Schramm and G. Steigman, private communication).

13. J. Ipser and P. Sikivie, Phys. Rev. Lett. 50 (1983) 925; F. W. Stecker and Q. Shafi, Phys. Rev. Lett. 50 (1983) 928; M. S. Turner, F. Wilczek and A. Zee, Phys. Lett. 125B (1983) 35; M. Axenides, R. Brandenburger and M. Turner, Phys. Lett. 126B (1983) 178; M. Fukugita and M. Yoshimura, Phys. Lett. 127B (1983) 181.

14. A. L. Melott, J. Einasto, E. Saar, I. Suisalu, A. Klypin and S. F. Shandarin, Phys. Rev. Lett. 51 (1983) 935; C. S. Frenk, S. D. M. White and M. Davis, Astrophys. J. 271 (1983) 417; G. R. Blumenthal, S. M. Faber, J. R. Primack and M. J. Rees, Nature 311 (1984) 517.

15. For a review and list of references, see P. Sikivie, "Axions in Cosmology", Univ. of Florida preprint UFTP-83-6 (1983), published in the Proceedings of the Gif-sur-Yvette Summer School in Particle Physics, Sept. 1982.

16. P. Sikivie, Phys. Rev. Lett. 51 (1983) 1415.

17. J. Moody and F. Wilczek, Phys. Rev. D30 (1984) 130.

18. P. Sikivie, "Detection rates for 'invisible' axion searches", Univ. of Florida preprint UFTP-85-5 (May 1985), to be published in Phys. Rev. D.

19. L. Krauss, J. Moody, F. Wilczek and D. Morris, "Calculations for cosmic axion detection", Univ. of California, Santa Barbara preprint NSF-ITP-85-76 (August 1985).

20. W. A. Bardeen and S.-H. H. Tye, Phys. Lett. 74B (1978) 229.

21. D. Kaplan, Harvard preprint HUTP-85/A014 (1985); M. Srednicki, Univ. of California, Santa Barbara preprint (1985).

22. J. N. Bahcall, Astroph. J. 276 (1984) 169; M. S. Turner, "On the cosmic and local mass density of 'invisible' axions", Fermilab preprint 85/93-A (June 1985).

23. P. Sikivie, N. Sullivan and D. Tanner, "A search for the 'invisible' axion", experimental proposal (April 1984).

24. D. Morris, "An electromagnetic detector for relic axions", LBL preprint LBL-17915 (May 1984).

25. A. C. Melissinos, J. Rogers, W. Wuensch, H. Halama, A. Prodell, P. Thompson and W. B. Fowler, "A search for galactic axions", preprint (Dec. 1984).

26. This was emphasized to me by D. Morris (private communication).

27. John C. Slonczewski, IBM T. J. Watson Research Center preprint (1985).

COSMIC AXION DECAY TO PHOTONS IN THE MICROWAVE AND INFRARED REGIONS

Thomas J. WEILER

Department of Physics and Astronomy, Vanderbilt University, Nashville, TN, 37235, USA[1]

The origin and properties of cosmic axions are reviewed. A possible axion contribution to the cosmic photon gas is analyzed in terms of the axion mass, lifetime and number density. "Erasure" of the axion signature due to early Universe interactions of the decay photons with the electron plasma is discussed. It is concluded that the enhancement of the photon spectrum due to decay of the standard invisible axion is insignificant. A possible photon depletion due to axion formation, and possibly useful bounds on non-standard axions are mentioned.

1. INTRODUCTION

A popular candidate for the dark matter of the Universe is the invisible[1] axion[2]. "Invisible" refers to the fact that, due to its incredibly tiny coupling to matter, the axion evades detection. In the absence of detection, we rely upon cosmological arguments to tell us what the axion is not. Such that stellar interiors not burn (via axion emission) faster than observed, the mass of the standard (i.e. satisfying eq. (5)) axion must be less than[3] 0.1 eV. Such that axions not fill the Universe with more energy density than allowed by observational bounds, the standard axion mass must be greater than[4] 10^{-5} eV (in an inflationary Universe the mass may be smaller[5]).

The primary decay mode of such a light axion is two-photon decay. From the mass bounds on the standard axion one deduces the bound on the energy of the monoenergetic decay photons: 0.1K < E < 1000K, or 10 cm > λ > .001 cm for the wavelength. Thus one may search for a signature of the axion dark matter in the microwave and infrared regions of the cosmic background radiation (CBR) spectrum. Since the extremely weak coupling of axions to matter renders them nearly stable, with a lifetime far exceeding the age of the Universe, it is advantageous to sum axion decays over the age of the Universe. Red-shifting of early decay events then yields photons which today have very long wavelengths; we should seek the electromagnetic remnants of axion decay at all wavelengths longer than about 10 microns. This talk constitutes a progress report of such a search, being conducted in collaboration with Tom Kephart.

---

[1]This work was supported in part by a DOE Outstanding Junior Investigator Award DE-FG05-85ER-40226 and a Vanderbilt URC grant.

2. COSMIC PHOTON INTERACTIONS

Decay photons added to the black body spectrum are subject to several interactions; each tends to smooth out any distortions in the CBR. The two interactions relevant for decay of the standard axion, namely bremmstrahlung and single Compton scattering, are carefully discussed in ref. 6. Here we present the salient points.

2.1. Bremsstrahlung

Bremsstrahlung processes are occurring until electron recombination at $z \sim 1.4 \times 10^3$. Bremsstrahlung processes thermally and chemically equilibrate the low energy end of the spectrum at

$$E/T < 2 \times 10^{-3} \sqrt{z} \; (z + 10^4)^{-1/4} \tag{1}$$

Writing $T = (1 + z)T_o$, and $T_o = 2.7°K$, one deduces that decay photons will hide their axion parentage of the axion mass satisfies

$$m_a \lesssim 1.0 \times 10^{-6} \theta \; z^{3/2} \; (z + 10^4)^{-1/4} \; eV \tag{2}$$

2. Single Compton scattering

Single Compton scattering effects thermal equilibrium for the photon gas, but cannot change photon numbers. A detailed analysis reveals that an excess of photons at $E < 4/3 \; \bar{E}$, where $\bar{E} = 2.7T$ is the black body mean energy, is pushed down the spectrum to $E = 0$ (Bose condensation) by single Compton scattering. Since the axion mass is 2E, we conclude that the decay of axions with

$$m_a < 7.2T = 1.7 \times 10^{-3} eV(1 + z)\theta \tag{3}$$

leaves no signature in the black body spectrum. On the other hand, for decay of an axion with mass above this bound, a chemical potential is induced in the thermal distributin function, givng a signal of the axion parentage. However, single Compton scattering becomes insignificant after $z \sim 6 \times 10^4$. Thus, only decaying axions with a mass exceeding ~100eV will contribute a nonzero chemical potential.

2.3. Recombination

At $z \sim 1.4 \times 10^3$, the electrons combine (recombine is a misnomer) with protons and alpha particles to form a neutral plasma, leaving the photon gas free at all wavelengths. After this point, all distortions are frozen into the CBR.

2.4. Summary of section 2

Scattering of the decay photons by the electron plasma serves to assimilate the photons into the CBR. According to eqs. (2) and (3), the decay photons from an axion with mass less than 100 eV will not be assimilated, and therefore will leave a signature, if the decay occurs after

$$z_{max} = \begin{cases} 6.3 \times 10^4, & \text{for } 1 \text{ eV} < m_a < 100\text{eV} \\ 6.3 \times 10^4 (m_a/\text{eV})^{4/5}, & 0.1\text{eV} < m_a < 1\text{eV} \\ 4.7 \times 10^4 (m_a/\text{eV})^{2/3}, & .005\text{eV} < m_a < 0.1\text{eV} \\ 1.4 \times 10^3, & m_a < .005\text{eV} \end{cases} \quad (4)$$

From the red giant mass bound we are thus led to expect $z_{max} = 1.4 \times 10^3$ to 1.0×10^4.

3. COSMIC AXION EVOLUTION

Peccei and Quinn (PQ) introduced an approximate global U(1) symmetry into the standard theory to eliminate CP violation from the QCD sector[7]. As this U(1) symmetry is broken at some high temperature, $T \sim f_a$, Goldstone's theorem guarantees the existence of an associated, nearly massless boson. This is the axion[2]. The axion mass is given by (N is a small model-dependent integer)

$$m_a f_a = N \frac{\sqrt{m_u m_d}}{m_u + m_d} m_\pi f_\pi \sim m_\pi f_\pi \quad (5)$$

The bulk of the axion quanta are produced at $T \sim$ GeV, when expansion no longer inhibits coherent oscillations of the axion field. From their inception, the axion quanta are nonrelativistic. This is a consequence of their coherent production and their tiny interaction rate: thermalization does not occur so the coherent axions remain Bose-condensed. As the Universe expands, the energy density of radion falls as T^4, while that of the axions falls as T^3. Thus the axions become an increasingly important contribution to the Universe's energy content.

That the axion mass and decay constant f_a in eq. (5) scale with those of the pion is a consequence of the fact that since the PQ and chiral symmetries are both symmetries of the same u and d quark fields, the axion and pion fields mix. More general axions will not result from symmetry among the light quarks and need not obey this relation. However, in this report we shall restrict our attention to the axion obeying eq. (5), which we call the standard axion. It is for this standard axion that the mass bound and lifetime relations hold:

$$10^{-5} \text{ eV} < m_a < .1 \text{ eV} , \quad (6)$$

$$\tau_a \sim (m_\pi/m_a)^5 \tau_\pi \sim 1.2 \times 10^8 \text{ Gyr}(\text{eV}/m_a)^5 . \quad (7)$$

For an axion mass in the range of eq. (6), this lifetime far exceeds the age of the Universe, $\tau_{uni} \sim 15$ Gyr, and one must work hard to find a signature of axion decay!

We may use the lifetime formula (7) to derive some interesting constraints on the axion mass, weaker than, but independent of, the stellar bound. Let Ω_a be the present energy density of axions in units of the critical energy density $\rho_c = 2.0 \times 10^{-29}$ g/cc. It is well known that theoretical calculations for the nucleosynthesis of light elements at $T_N \sim 1$ MeV, or $z_N \sim 10^{10}$, yield the observed abundances, provided there exists at most one light neutrino species beyond the known three. (Further particle species increase the Universe's energy density, and therefore its expansion rate, incorrectly influencing the nucleosynthesis equations.) This same argument limits the axion energy density at the time of nucleosynthesis to the equivalent of one neutrino species: $\rho_a \lesssim \rho_{\nu_i}$ at z_N. Allowing for possible axion decay, and for the fact that matter energy density scales as z^3 whereas radiation scales as z^4, the inequality becomes $\exp(\tau_{uni}/\tau_a)\rho_a \lesssim z_N \rho_{\nu_i}$, or

$$\tau_a \gtrsim \tau_{uni}/\ln(z_N \Omega_{\nu_i}/\Omega_a). \tag{8a}$$

Substituting $z_N \sim 10^{10}$ and $\Omega_{\nu_i} = 7/8(4/11)^{4/3}\Omega_{\gamma_0} = 5.0 \times 10^{-6}\theta^4/h^2$, where the present Hubble parameter is $H_0 = 100h$ km/s/Mpc, one gets

$$\tau_a \gtrsim \tau_{uni}(11 + \ln(\theta^4/h^2\Omega_a))^{-1}. \tag{8b}$$

For an axion dominated Universe $\Omega_a \sim 1$, and we conclude that $\tau_a \gtrsim \tau_{uni}/11$, or equivalently, $m_a \lesssim 40$ eV. As shown in eq. (6), the standard axion satisfies this bound.

4. AXION CONTRIBUTION TO CBR AND IR SPECTRA

4.1. Homogeneous axions

If one assumes that the "dark matter" of our present Universe is dominantly axions, the number density and occupation number of these axions is easily determined. The axion number density is $n_a = \rho_c \Omega_a/m_a = 1.1 \times 10^4 h^2 \Omega_a (\text{eV}/m_a)/\text{cc}$. The favored axion mass range of eq. (6) yields a range for the present axion density of $10^{7\pm2}$/cc.

In terms of ther axion occupation number f_{ax}, $n_a = \int d^3p \, f_{ax}/(2\pi)^3$, so for Bose-condensed axions a good approximation to the occupation number is

$$f_{ax} = \frac{\rho_c \Omega_a}{m_a}(2\pi)^3 \delta^3(\vec{p}) \tag{9}$$

Consider the reversible process: $\text{axion}(p) \longleftrightarrow \gamma(k_1) + \gamma(k_2)$. The zero temperature axions are out of equilibrium with the thermal photons. Hence, detailed balance does not apply and the time average of this reaction favors one direction. The change in photon density $n_\gamma = N_\gamma/V$ is

$$\frac{dn_\gamma}{dt} = \frac{1}{(2\pi)^5}\int\frac{d^3p}{2p^0}\int\frac{d^3k_1}{2k_1^0}\int\frac{d^3k_2}{2k_2^0}\,\delta^4(p - k_1 - k_2)|M|^2[f_{ax}(1 + f_{\gamma_1})(1 + f_{\gamma_2}) - (1 + f_{ax})f_{\gamma_1}f_{\gamma_2}]\,. \tag{10}$$

$|M|^2$ is the squared and spin summed matarix element for a $\longleftrightarrow \gamma\gamma$, related to the axion width in vacuo by $\Gamma(a \longleftrightarrow \gamma\gamma) = |M|^2/32\pi\, m_a$.

The f's are the occupation numbers for axions and photons. Since the relic photons are in thermal equilibrium at temperature T, $f_{\gamma_i}(E_i) = (\exp(E_i/T)-1)^{-1}$. The factors $1 + f$ express the fact that in addition to the spontaneous emission process, there is also stimulated emission induced by the relic axion and photon backgrounds.

Expansion of the Universe is incorporated into the process by:
i) scaling energies and the photon temperature by the red shift factor $(1+z)$,
ii) scaling number densities by $(1 + z)^3$, and
iii) replacing dt with $dz/(H_0(1 + z)^2\sqrt{1 + z\,\Omega_m + z(2 + z)\Omega_r}\,)$.
Here $\Omega_{m(r)}$ is the present energy density of matter (radiation) in the Universe as a fraction of the closure value ρ_c. Empirically, $\Omega_r \sim 2.3 \times 10^{-5}\,\theta^4/h^2$, and $0.02 < \Omega_m < 2.0$. In the expanding Universe, eq (10) becomes

$$\frac{d\rho}{dx} = \frac{3H_0\,\Gamma(a \to \gamma\gamma)}{4\pi G}\,\frac{1}{y^2}\left[\Omega_a f(x,y,\Omega_i) - \frac{T_0^4}{\pi^2\rho_c}\,\frac{x^{5/2}}{e^x - 1}\,g(y^2/x)\right] \tag{11}$$

with

$$f(x,y,\Omega_i) = \frac{x^2}{\sqrt{x^2(1 - \Omega_0) + \Omega_m xy + \Omega_r y^2}}$$

$$\left[\frac{1}{e^x - 1} + \frac{1}{2}\right]\Theta(y - x)\,\Theta(x - y/(1 + z_{max})) \tag{12}$$

and

$$g(v) = \frac{-v^{3/2}}{2}\int_{(1+z_{max})^{-2}}^{1}d\xi\,\frac{\ln(1 - e^{-\xi v})}{\sqrt{1 - \Omega_0 + \Omega_m/\sqrt{\xi} + \Omega_r/\xi}}\,; \tag{13}$$

ρ is the photon energy density, Ω_0 is the sum $\Omega_m + \Omega_r$, G is the gravitational constant, $x \equiv E_\gamma/T_0$ is the present photon energy scaled by the present black body temperature, and $y = m_a/2T_0 = (2.2/\theta) \times 10^3(m_a/eV)$. $(1 + z)$ has been replaced with y/x. The bracketed factor in eq. (12), reveals that photons with x below 1 arise from stimulated axion decay, while those with x above 1 arise from spontaneous axion decay. The back reaction, $\gamma\gamma \to$ axion, is suppressed by the factor $T_0^4/(\pi^2\rho_c) = 3.5 \times 10^{-6}\,\theta^4/h^2$. However, numerical studies of $g(v)$ seem to indicate a peak value of $\sim z_{max}$ at $v \sim z_{max}^2$ for $(\Omega_r/\Omega_m \gtrsim 1/z_{max})$. Thus, the back reaction may actually lead to a <u>depletion</u> in the CBR. The back reaction is still being analyzed and will not be reported on here. In the following we neglect it.

As a fraction of the unadulterated black body spectrum $(d\rho/dx)_{BB} = T_0^4 x^3 \pi^{-2}(e^x-1)^{-1}$, the axion contribution is

$$r(x) \equiv \frac{(d\rho/dx)_{a \to \gamma\gamma}}{(d\rho/dx)_{BB}} = 1.0 \times 10^{-10}\Omega_a \frac{[e^x + 1](m_a/eV)^{5/2}}{x^{3/2}\sqrt{\Omega_m + (y/x)\Omega_r + (x/y)(1-\Omega_0)}} \frac{h}{\theta^{3/2}} \quad (14)$$

in the region of support $y(1 + z_{max})^{-1} \le x \le y$. $z_{max}(m_a)$ is given in eq. (4). We have used eq. (7). For a black body spectrum, $\langle x \rangle = 2.7$ and the spectral peak occurs at $x = 2.8$. CBR measurements span a range in x from .007 (λ = 75cm) to 5.4 (λ = 0.1cm). At wavelengths longer than 100cm, radiation from our galaxy makes CBR measurements impossible. At wavelengths shorter than .04cm, emissions from interstellar dust dominate the CBR at $.04 \gtrsim \lambda \gtrsim .01$cm, while interplanetary dust emissions are important for $\lambda \lesssim 10^{-3}$cm. There is an observational window available between 10 and 100 microns, which we will discuss shortly. $r(x)$ increases on either side of $x = 1.76$, but even at the measured extremes $x = .007$ and $x = 5.4$ the ratio is only 1.7×10^{-7} and 1.8×10^{-9}, respectively, times $\Omega_a(m_a/eV)^{5/2}$. Such an enhancement is not detectable for $m_a < 100$ eV.

Away from the black body spectrum we have the simple spontaneous emission result (again assuming eq. (7)): for $\xi \equiv 2E\gamma/m_a$ in the interval $[(1 + z_{max})^{-1}, 1]$,

$$\frac{d\rho}{dE\gamma}\bigg|_{a \to \gamma\gamma} = 1.7h \times 10^{-3}\,\Omega_a(m_a/eV)^4 \frac{\xi^{3/2}}{\sqrt{\Omega_m + \Omega_r/\xi + \xi(1-\Omega_0)}} \text{ cm}^{-3}. \quad (15)$$

M. G. Hauser et al.[8] have analyzed recent IRAS satellite data, and report a possible anomalous photon flux at 100μm. Their flux estimate is 2.4 ± 1 MJy/sr, equivalent to $d\rho/dE = 9.7 \pm 4$cm^{-3} at $E = .012$ eV. The contribution of eq. (15) at this energy is $6.3h\Omega_a(m_a/eV)^{5/2} \times 10^{-6}$/cc, and we see that the anomalous datum is accommodated by a 250 eV axion, yielding $\xi \sim 10^{-4} \sim (1+z_{max})^{-1}$. However, such a large mass violates the stellar bound by three orders of magnitude, and the nucleosynthesis bound by a factor of 6, making the standard axion an unlikely candidate for the source of the new flux.

4.2. Gravitationally clustered axions

Since the axions are nonrelativistic matter, it is expected that they are gravitationally clumped along with the luminous matter. The most red-shifted quasars have a z ~ 3.7, so it is commonly believed that the clustering occurred (abruptly, due to the nonlinear growth stage of density perturbations) at z ~ 4. If so, axions have been in the clustered phase for ninety per cent of the age of the Universe. The light transit distance during this long time scale is essentially the horizon size, meaning that the clustering is effectively averaged out, returning us to our prior formulae. However, there is an additional local contribution due to the enhanced axion density in the cluster containing our planet earth. The local axion density is increased by the ratio of total volume in the Universe to cluster volume, i.e. by an additional factor as large as ~10^6. We may well be surrounded by an ambient axion density of more than a trillion per cc! When we relize that the total CBR number density is just 400 θ^3/cc, the possibility for distortion of the CBR by axion decay in the local cluster demands scrutiny!

The price to be paid for the gain in axion density is that the range of integration back in time may not exceed the light transit time from the edge of the axion cluster at distance d. Explicitly, the integration range is Δt = .0033(d/Mpc) Gyr. For such a small time period, cosmic expansion is irrelevant, and simple formulae result: For decay stimulated by the CBR, the relative enhancement at $x = m_a/2T_0$ is

$$r(x) = 1.6 \cdot 10^{-22} \, h^2 \theta \Omega_a^{(cl)} (d/Mpc) x^2 (e^x + 1)/\Delta x \quad , \tag{16}$$

where Δx is the larger of (i) the experimental resolution in x, and (ii) the inherent spread βx, due to the virialized velocities[9] ($\beta = \langle v \rangle \sim 10^{-3}$) of the clustered axions: For spontaneous emission, the contribution at $E_\gamma = m_a/2$ is

$$\frac{d\rho}{dE_\gamma} = 3.0 \cdot 10^{-7} h^2 \Omega_a^{(cl)} (d/Mpc)(m_a/eV)^5 \, \Delta E^{-1} \text{ eV/cc} \quad , \tag{17}$$

where ΔE is the larger of (i) the experimental resolution in E, and (ii) the inherent energy spread, $\beta m_a/2$, in the axion contribution.

Even putting in the numbers d = 30 Mpc (appropriate for clustering on the scale of the Virgo supercluster), and $\Omega^{(cl)} = 10^6$, one finds that: (i) $r(x)$ is many orders of magnitude below unity in the x range of interest, leaving little hope for discerning an axion signature in the CBR; (ii) while clustered axions of mass .025eV can yield an excess of photons at 100 microns, the yield is 10 orders of magnitude too small to explain the observed[8] galactic excess.

5. SUMMARY

Apparently the standard axion does not leave a signature in its decay photons. To be seen an axion must violate the relation expressed in eq. (5). Axions arising from the breaking of global symmetries among other than light quark fields will not in general obey eq. (5). Axions of this ilk, e.g. techniaxions, may in fact leave a signature in the microwave sky. A comparison of eq. (11) with the measured microwave data then presents a bounded region for allowable mass, lifetime and density of the general axion. Explorations along this line are still in progress, as is the study of a possible depletion in the CBR due to the back reaction, $\gamma\gamma \to$ standard axion.[10]

REFERENCES

1) J. Kim, Phys. Rev. Lett. 43 (1979); M. Shifman, A. Vainstein and V. Zakharov, Nucl. Phys. B166 (1980) 493; M. Dine, W. Fischler and M. Srednicki, Phys. Lett. 104B (1981) 199.

2) S. Weinberg, Phys. Rev. Lett. 40 (1978) 223; F. Wilczek, Phys. Rev. Lett. 40 (1978) 279.

3) D. Dicus, E. Kolb, V. teplitz and R. Wagoner, Phys. Rev. D18 (1978) 1829 and D22 (1980) 839; M. Fukugita, S. Watamura and M. Yoshimura, Phys. Rev. Lett. 48 (1982) 1522 and Phys. Rev. D26 (1982) 1840; N. Iwamoto, Phys. Rev. Lett. 53 (1984) 1198.

4) J. Preskill, M. Wise, and F. Wilczek, Phys. Lett. 120B (1983) 127; L. Abbott and P. Sikivie, Phys. Lett. 120B (1983) 133; M. Dine and W. Fischler, Phys. Lett. 120B (1983) 137; see, however, T. DeGrand, T. Kephart and T. Weiler, Vanderbilt Preprint VAND-TH85-1.

5) F. Wilczek, Proc. Erice Summer School (1983); S.-Y. Pi, Phys. Rev. Lett. 52 (1984) 1725.

6) J. Silk and A. Stebbins, Ap. J. 269 (1983) 1, and references therein.

7) R. Peccei and H. Quinn, Phys. Rev. Lett. 38 (1977) 1440.

8) M. G. Hauser et al., Ap. J. 278 (1984) L15; F. J. Low et al., Ap. J. 278 (1984) L19.

9) P. Sikivie, Phys. Rev. Lett. 51 (1983) 1415.

10) T. Kephart and T. Weiler, in preparation.

SUPERCONDUCTING STRINGS AND MEMBRANES

G.LAZARIDES

Physics Division, School of Technology, University of Thessaloniki, Thessaloniki, Greece.

The behaviour of the heterotic string as a superconducting axionic string attached to an axionic domain wall is discussed. It is also shown that superconducting domain walls (membranes) readily appear in realistic grand unified models.

1. SUPERCONDUCTING AXIONIC STRINGS

1.1. Introduction

Currently, the most promising superstring theory[1] is the $E_8 \times E_8$ heterotic string theory[2]. It is, therefore, interesting to understand the physical nature of the heterotic string. In particular, the important questions are whether the heterotic string can appear in the real universe as a physical object and what are its physical properties. To answer these questions, the following preliminary considerations are required: i) superconducting strings[3], and ii) axionic strings[4,5] and domain walls[4,6]. The main conclusion[7] is that the heterotic string is a superconducting axionic string attached to an axionic domain wall.

1.2. Superconducting strings

Consider a group G (local or global) broken to a subgroup H at a mass scale M by the vacuum expectation value (VEV) $<\varphi>$ of a scalar field φ, i.e.,

$$G \xrightarrow{<\varphi>/M} H \ . \qquad (1)$$

Topologically stable strings then arise if the first homotopy group of the vacuum manifold $\pi_1(G/H)$ is non-trivial. For example, if $G=U(1)$ and $H=\{1\}$, $\pi_1(G/H)=\pi_1(U(1))=Z$ (the set of integers) and stable strings appear. The minimal string can be taken along the z axis. In this case, the field φ depends only on x and y and vanishes for $x=y=0$. As we go once around the string, the phase of φ changes by 2π. If U(1) is a local gauge group there exists U(1) magnetic field along the string with total flux $2\pi/g$ (g is the gauge coupling). The thickness of the string is $\sim M^{-1}$ and its mass per unit length $\sim M^2/\alpha$ ($\alpha=g^2/4\pi$).

The string can become superconducting if, for example, there are charged fermions trapped in lognitudinal zero modes along the string[3]. Jackiw and Rossi[8] have shown that such modes can arise in models of interest.

Consider a toy model[3] based on the group $U(1)_Q \times U(1)_R$, where $U(1)_Q$ is the usual gauge group of electromagnetism and the extra $U(1)_R$ can be either local or global. $U(1)_R$ can be broken to $\{1\}$ by the VEV of a scalar field φ with charges $Q=0$ and $R=1$. This model contains topologically stable strings. Now introduce left-handed fermions ψ_L and χ_L with charges $Q_\psi=q$, $R_\psi=r$ and $Q_\chi=-q$, $R_\chi=-r-1$ respectively. These fermions obtain their masses through coupling to φ. The fermion part of the Lagrangian can be written as

$$L_f = \bar\psi\, i\, \not{D}\psi + \bar\chi\, i\, \not{D}\chi - \lambda\,(\varepsilon^{\alpha\beta}\psi_{\alpha L}\chi_{\beta L}\varphi + h.c.) \quad . \tag{2}$$

One can then consider the Dirac equations for the ψ and χ fields coupled to the minimal string along the z axis and look for solutions which are independent of z and t. According to an index theorem[9] there exists precisely one solution

$$(\psi,\chi) = (\beta(x,y),\, \tilde\beta(x,y)), \tag{3}$$

which is normalizable, i.e., $\int dxdy(|\psi|^2+|\chi|^2) < \infty$. This solution is an eigenstate of $i\gamma^1\gamma^2$ with eigenvalue $+1$. We can now look for 4 dimensional solutions of the form

$$\psi=\alpha(z,t)\beta(x,y), \quad \chi=\alpha(z,t)\tilde\beta(x,y) \quad . \tag{4}$$

The Dirac equation becomes

$$\left(\frac{\partial}{\partial t} - \frac{\partial}{\partial z}\right)\alpha(z,t)=0 \quad , \tag{5}$$

which gives

$$\alpha(z,t) = f(z+t) \quad .$$

Thus the fermions are trapped in transverse zero modes that travel in the $-z$ direction with the velocity of light and are called left-movers (L-movers). For an one dimensional observer living on the string, these are massless left-handed chiral fermions ($\gamma^0\gamma^3\alpha(z,t)=-\alpha$). By the same token, one can show that left-handed fermions which obtain masses by coupling to φ^* instead of φ get trapped in transverse zero modes travelling in the $+z$ direction (R-movers).

The effective field theory describing charged fermions trapped in transverse zero modes on a string, in the simplest anomaly free case[3] with one L-mover and one R-mover with the same electric charge q, is given by the action

$$I = -\frac{1}{4}\int d^4x F_{\mu\nu}F^{\mu\nu} + \int dzdt\bar\psi(z,t)i\gamma^i D_i\psi(z,t) \quad . \tag{6}$$

Here $\psi(z,t)$ is a two dimensional Dirac field and $i=0,3$. By using the technique of bosonization,

$$\bar\psi\gamma^i\psi = \frac{1}{\sqrt\pi}\varepsilon^{ij}\partial_j\phi(z,t), \tag{7}$$

eq.(6) becomes

$$I = -\frac{1}{4}\int d^4x F_{\mu\nu}F^{\mu\nu} + \int dzdt(\frac{1}{2}(\frac{\partial\phi}{\partial t})^2 - \frac{1}{2}(\frac{\partial\phi}{\partial z})^2 - \frac{q}{\sqrt{\pi}}\phi E). \qquad (8)$$

Here $E = \epsilon^{ij}\partial_i A_j$ is the component of the electric field tangent to the string. From eq.(8) and for a z-independent E field, one obtains

$$\frac{d^2\phi}{dt^2} + \frac{q}{\sqrt{\pi}}E = 0 \quad . \qquad (9)$$

This equation written in terms of the electromagnetic current $J = -q\bar{\psi}\gamma^3\psi = -(q/\sqrt{\pi})(\partial\phi/\partial t)$ becomes

$$\frac{dJ}{dt} = \frac{q^2}{\pi}E \quad . \qquad (10)$$

For a time independent E, one obtains

$$J = \frac{q^2}{\pi}Et + J_0 \quad , \qquad (11)$$

in contrast to the case of a wire of finite conductivity σ, where $J = \sigma E$. This shows that charged fermions trapped in transverse zero modes on a string make the latter superconducting[3].

1.3. Axionic strings

CP violation in strong interactions is parametrized by the effective $\bar{\theta}$-angle

$$\bar{\theta} = \theta - \text{argdet}\, m \quad , \qquad (12)$$

where θ is the QCD θ-angle and m the fermion mass matrix. From considerations of the electric dipole moment of the neutron, one obtains the bound $\bar{\theta} < 10^{-9}$ on an otherwise arbitrary angular parameter of the theory. To solve this strong CP problem, Peccei and Quinn[10] introduced a global $U(1)_{PQ}$ symmetry with QCD anomalies. Due to QCD instanton effects the parameter $\bar{\theta}$ becomes now a dynamical variable (the axion field) with a potential energy $\sim 1-\cos\bar{\theta}$ which is minimized for $<\bar{\theta}> = 0$. $U(1)_{PQ}$ is spontaneously broken at a scale f_a ($10^9 \text{GeV} \lesssim f_a \lesssim 10^{12} \text{GeV}$). This leads to the production of topologically stable strings[4]. As we go around such an axionic string, $\bar{\theta}$ varies by 2π. At a cosmic temperature of the order of $\Lambda_{QCD} \sim 200\text{MeV}$, the QCD instantons come into play and these strings become boundaries of axionic domain walls that separate vacua with $\bar{\theta} = 0$ and $\bar{\theta} = 2\pi$[4,6].

The axionic strings are always superconducting[5]. In the case of no fermions other than the known quarks and leptons, the PQ - mechanism requires two Weinberg - Salam doublet Higgs fields with different PQ-charges. This is so because otherwise the $U(1)_{PQ}$ symmetry has no QCD anomalies. It follows that at least one of these Higgs fields changes phase by an integral multiple of 2π as we go around the string and the fermions that couple to it can be trapped in transverse zero modes. In the case of extra superheavy fermions in vector-like representations of $SU(3)_c \times SU(2)_L \times U(1)_Y$, the PQ-charges of the electroweak Higgs doublets could be identified. The QCD anomalies must then be introduced through the extra fermions. This requires that some of the $SU(3)_c \times SU(2)_L \times U(1)_Y$ singlet Higgs fields which couple to the superheavy fermions have non-zero PQ-charges and therefore changing phases around the string. The zero modes are then obtained from the corresponding superheavy fermions.

The superconducting axionic strings exhibit the following surprising phenomenon. Due to the QQ(PQ) triangular anomaly inherent in axion models, the equation

$$\sum_{\text{L-movers}} q_i^2 = \sum_{\text{R-movers}} q_i^2 \qquad (13)$$

in reference 3 is not satisfied (q_i's are electric charges). This implies[5] that a constant electric field E along the string produces a constant rate of change not only for the supercurrent but also for the electric charge density on the string,

$$\frac{d^2 Q}{dt dl} \propto E \quad . \qquad (14)$$

Thus, charge conservation is "seemingly" violated in axionic strings[5]. This paradox is resolved by noting that the axion field around the string is necessarily varying. An electric field in the presence of a gradient in the axion field induces a normal (non-superconducting) "Hall" current which, in the present case, flows radially inwards towards the string[11]. This is the source of the varying charge on the string.

1.4. The heterotic string

We are now ready to study the physical properties of the $E_8 \times E_8$ heterotic string. The $E_8 \times E_8$ gauge group comes from charged modes propagating on the string. These modes play the role of the fermion transverse zero modes in the above discussion and the heterotic string is most definitely superconducting[7]. Moreover, the fermionic $E_8 \times E_8$ charge carriers move on the heterotic string in only one direction and the anomaly cancellation condition(13) is obviously not satisfied. This indicates that the heterotic string must be an axionic string[7]. We will now show that this is actually true. Consider the antisymmetric tensor field[12] $B_{\mu\nu}$ ($\mu,\nu=0,1,2,3$) which plays a crusial role in anomaly cancellation. The dual $Y^\mu = (1/6)\varepsilon^{\mu\nu\lambda\rho} H_{\nu\lambda\rho}$ of the invariant field strength $H_{\nu\lambda\rho}$ of $B_{\mu\nu}$ obeys the following field equation in the vacuum,

$$\partial_\mu Y_\nu - \partial_\nu Y_\mu = 0 \quad . \qquad (15)$$

Thus, we can find a scalar field φ such that $Y_\mu = \partial_\mu \varphi$. Using the Bianchi identity one can show[13] that φ has the standard coupling of an axion field.

Now consider a heterotic string along the z axis. The variation of φ around the string is given by

$$\Delta\varphi = \int_C \partial_\mu \varphi dx^\mu = \int_C Y_\mu dx^\mu \quad , \quad (16)$$

where the curve C around the string can be taken on the xy plane. This implies that

$$\Delta\varphi = \int_S d\sigma^{\mu\nu}(\partial_\mu Y_\nu - \partial_\nu Y_\mu) = \int_S dxdy\, \partial^\mu H_{\mu 30} \quad , \quad (17)$$

where S is the part of the xy plane bounded by C. Using the fact that $\partial^\mu H_{\mu\nu\lambda} \propto V_{\nu\lambda}$ (the vertex operator) and $V_{30} \propto \delta(x)\delta(y)$ we obtain from eq.(17) that $\Delta\varphi \neq 0$. This shows[7] that the heterotic string is in fact a (superconducting) axionic string. Proper normalization of the axion field φ also shows[7] that there is a single axionic domain wall attached to the heterotic string for cosmic temperatures below the QCD mass scale. Thus, (closed) heterotic strings are "confined" by the axionic walls attached to them[7]. Equating the energy in the string to that in the wall we conclude that the typical "confining" scale for these objects is about 10 Kilometers. The presence of macroscopic axionic walls bounded by heterotic strings is, however, unlikely because they disappear quickly radiating their energy gravitationally.

2. SUPERCONDUCTING MEMBRANES

Recently, anomaly free supergravity theories coupled to Yang-Mills superfields in six spacetime dimensions have been constructed[14]. These theories could be the low energy approximation of six dimensional membrane theories[15]. Motivated by this speculative possibility we will now discuss the superconductivity of some non topologically stable domain walls (membranes) that arise in some spontaneously broken gauge theories. (Topologically stable walls are cosmologically unacceptable [16]).

Let us consider a grand unified model[17,18] based on the gauge group Spin(10). The Lie algebra of this group is generated by $T_{ij} \equiv \sigma_{ij}/2 = [\Gamma^i, \Gamma^j]/4i$ ($1 \leq i,j \leq 10$), where Γ^i (i=1,...,10) are the Dirac matrices in ten dimensions. A 54-plet of Higgs fields φ_{54} can break Spin(10) as follows:

$$\text{Spin}(10) \xrightarrow{\langle\varphi_{54}\rangle} H \quad . \quad (18)$$

The subgroup H consists of two connected components, $H_0 = \text{Spin}(6) \otimes \text{Spin}(4) \simeq SU(4) \otimes [SU(2)_L \times SU(2)_R]$ and $C \cdot H_0$, where

$$C = (i\sigma_{23})(i\sigma_{67})(i\sigma_{89})$$

$$= \exp(i\pi\sigma_{23}/2)\exp(i\pi\sigma_{67}/2)\exp(i\pi\sigma_{89}/2)$$

$$= \Gamma_2\Gamma_3\Gamma_6\Gamma_7\Gamma_8\Gamma_9 \in \mathrm{Spin}(10) \tag{19}$$

is the charge conjugation operator. The Cartan subalgebra of Spin(6) changes sign under C. In particular,

$$C : \tfrac{1}{2}(B-L) \equiv C\tfrac{1}{6}(\sigma_{12}+\sigma_{34}+\sigma_{56})C^{-1} \longrightarrow -\tfrac{1}{2}(B-L). \tag{20}$$

Also,

$$C : T_{3L} \equiv \tfrac{1}{4}(\sigma_{910}+\sigma_{78}) \longrightarrow -T_{3R},$$

$$T_{3R} \equiv \tfrac{1}{4}(\sigma_{910}-\sigma_{78}) \longrightarrow -T_{3L},$$

$$Q \equiv T_{3L}+T_{3R}+\tfrac{B-L}{2} \longrightarrow -Q. \tag{21}$$

The breaking of Spin(10) to H produces[17] Z_2 strings, since

$$\pi_1(\mathrm{Spin}(10)/H) = \pi_0(H) = Z_2. \tag{22}$$

The charge conjugation operator C breaks during the subsequent breaking of H to $SU(3)_c \times U(1)_{em}$ and the strings become boundaries of domain walls of thickness M_W^{-1} [17]. Going around the string from one side of the wall to the other, one performs gauge transformations that interpolate between 1 and C. The Cartan subalgebra of SU(4) changes sign, $T_{3L} \longleftrightarrow -T_{3R}$ and $Q \longrightarrow -Q$. The electroweak breaking is achieved by the VEV of a complex Higgs 10-plet of Spin(10), $\langle\varphi_{10}\rangle$. This must be a vector in the 7-8-9-10 space on which Spin(4) acts, since colour must remain unbroken. Furthermore, the requirement of unbroken electric charge Q, implies that

$$\langle\varphi_{10}\rangle = \varphi_7\bar{u}_7 + \varphi_8\bar{u}_8, \tag{23}$$

where $\bar{u}_7$, $\bar{u}_8$ are the unit vectors in the 7-8 plane and φ_7, φ_8 are complex coefficients. One can then show[18] that $\varphi_d \equiv \varphi_7 + i\varphi_8$ couples to the down quarks and charged leptons and $\varphi_u \equiv \varphi_7 - i\varphi_8$ couples to the up quarks. Going around the string from one side of the wall to the other we perform rotations by π on the 2-3, 6-7 and 8-9 planes. Consequently, $\langle\varphi\rangle \equiv \langle\varphi_u\rangle$ or $\langle\varphi_d\rangle$ goes to $-\langle\varphi\rangle$. Thus, as we go through the wall, φ follows a kink-like configuration which connects $-\langle\varphi\rangle$ to $+\langle\varphi\rangle$.

Now assuming an infinite wall along the xz-plane, we can write the Dirac equations for transverse zero modes,

$$-i\sigma^2 \frac{d\psi(y)}{dy} = g\varphi^*(y)\chi(y) ,$$

$$i\sigma^2 \frac{d\chi(y)}{dy} = g\varphi(y)\psi(y) . \qquad (24)$$

Here $\varphi(y)=\varphi_u(y)$ or $\varphi_d(y)$ is in the kink-like configuration, g is a real Yukawa coupling, ψ is u_L, d_L or e_L, and χ is u_R, d_R or e_R. These equations are solved by

$$\chi(y) = i\sigma^2 \psi(y),$$

$$\psi(y) = c \exp\left(-g \int_0^y \varphi(y')dy'\right) . \qquad (25)$$

The transverse zero modes can be boosted along any direction of the wall. For a boost along the z-axis, for instance,

$$\chi = i\sigma^2 \psi , \quad \psi=\eta(z,t)\exp\left(-g \int_0^y \varphi(y')dy'\right) , \qquad (26)$$

where $\eta(z,t)$ satisfies the following equation,

$$(\partial_0 - \sigma^3 \partial_z) \eta (z,t) = 0 . \qquad (27)$$

Thus, $\eta=\eta(x^0+z)$ for $\sigma^3\eta=\eta$ (left-moving mode) and $\eta=\eta(x^0-z)$ for $\sigma^3\eta=-\eta$ (right-moving mode). From this, we conclude[18] that the walls under discussion are superconducting. Moreover, each fermion (antifermion) type gives one left-mover and one right-mover and the superconductivity of the walls is normal, i.e., there is no apparent violation of any conservation laws[18].

These walls could, in principle, survive till cosmic time of the order of 10^5 sec and provide perhaps interesting effects due to their superconductivity. Unfortunately, keeping these walls till 10^5 sec requires that C breaks at about 1 TeV and production of baryon asymmetry may then be difficult.

One can easily construct other models with superconducting walls[18] which have no problems with baryon asymmetry. For instance, the breaking

$$E_6 \xrightarrow[\langle\varphi_{351'}\rangle]{} \text{Spin}(10) \times Z_2 \qquad (28)$$

produces Z_2-strings. Subsequent breaking of the Z_2 in eq.(28) (say with a $\langle\varphi_{27}\rangle$) produces Z_2-walls bounded by the Z_2-strings. These walls can be shown[18] to be superconducting and can survive till 10^5-10^7 sec with no problems with the baryon asymmetry of the universe.

REFERENCES

1) M.B.Green and J.H.Schwarz, Phys.Lett.149B(1984)117.

2) D.J.Gross, J.Harvey, E.Martinec and R.Rohm, Phys. Rev.Lett.54(1985)502.

3) E.Witten, Nucl.Phys. B249(1985)557.

4) A.Vilenkin and A.Everett, Phys.Rev.Lett.48(1982)1867;
   G.Lazarides and Q.Shafi,Phys.Lett.115B(1982)21.

5) G.Lazarides and Q.Shafi, Phys.Lett. 151B(1985)123;
   C.Callan and J.Harvey, Princeton Univ.preprint(1984).

6) P.Sikivie, Phys. Rev.Lett. 48(1982)1156.

7) E.Witten, Phys.Lett.153B(1985)243.

8) R.Jackiw and P.Rossi, Nucl.Phys. B190[FS3] (1981) 681.

9) E.Weinberg, Phys. Rev. D24(1981)2669.

10) R.Peccei and H. Quinn, Phys.Rev.Lett.38(1977)1440.

11) P.Sikivie, Phys.Lett.137B(1984)353.

12) V.I.Ogievetsky and I.V. Polubarinov, Sov. J.Nucl. Phys.4(1967)156;
    M.Kolb and P.Ramond, Phys.Rev.D9(1974)2273.

13) E.Witten, Phys.Lett.149B(1984)351.

14) S.Randjbar-daemi, A.Salam, E.Sezgin and J.Strathdee, Phys. Lett.151B(1985) 351.

15) P.G.O.Freund and F.Mansouri, Z. Phys. C14(1982)279.

16) Ya.B.Zeldovich, I.Yu. Kobzarev and L.B. Okun, JETP(Sov.Phys.)40(1975)1.

17) T.W.B.Kibble, G.Lazarides and Q.Shafi, Phys.Rev.D26(1982)435.

18) G.Lazarides and Q.Shafi, Phys.Lett. 159B(1985)261.

QUARK AND LEPTON MASSES IN A SIX DIMENSIONAL SO(12) MODEL

C. WETTERICH

Deutsches Elektronen-Synchrotron DESY, Notkestr. 85, 2000 Hamburg 52, FRG

I calculate Yukawa couplings in a six dimensional Einstein-Yang-Mills theory based on SO(12). I argue that the fermion masses may follow the pattern $m_t \gg m_b, m_c, m_\tau \gg m_s, m_\mu \gg m_d, m_u, m_e$.

There have been many attempts to predict the spectrum of fermion masses and their mixing angles in a four dimensional framework, as for example by discrete symmetries, grand unification, family unification, technicolor or constituent models. So far, no conclusive model has emerged which predicts more than (at best!) a few relations between fermion masses like $m_b(M_x) = m_\tau(M_x)$. Higher dimensional models can naturally account for several fermion generations, the number of generations being given by the chirality index[1] related to topology and symmetry of internal space. Since the hierarchical structure of the fermion mass matrices seems to suggest a generation pattern, one may conjecture that this is again related to properties of internal space in the context of a higher dimensional theory.

On the other hand, the fermion mass matrices may be among the few testable predictions of higher dimensional models. In contrast to usual four dimensional models, higher dimensional theories often have only one fundamental spinorial quantity. This eliminates the possibility of many free Yukawa couplings and typically leads to models with only very few or no free parameters describing the coupling of fermions. In a theory of gravity and gauge interactions, for example, the fermion interactions are contained in the minimal covariant kinetic term

$$\mathcal{L}_\psi = i\bar{\psi}\hat{\gamma}^{\hat{\mu}} D_{\hat{\mu}} \psi \tag{1}$$

In addition, four dimensional gauge fields and scalar fields are often different components of the same higher dimensional field. Relations between the fermion masses and the mass of the W-boson should therefore not surprise us.

In this talk I report on a calculation of Yukawa couplings for quarks and leptons in a six dimensional SO(12) gauge theory coupled to gravity[2]. After spontaneous symmetry breaking of $SU(2)_L \times U(1)_Y$ this would result in predictions for ratios of fermion masses compared to the W-boson mass.

Let me start with the action

$$S = - \int d^6 z \, \hat{g}_6^{\frac{1}{2}} \, \{\delta \hat{R} + \frac{1}{8} G^{AB}_{\hat{\mu}\hat{\nu}} G^{AB\hat{\mu}\hat{\nu}} + \varepsilon - L_\psi\} \quad (2)$$

$$L_\psi = i\bar{\psi}_1 \gamma^{\hat{\mu}} D_{\hat{\mu}} \psi_1 + i\bar{\psi}_2 \gamma^{\hat{\mu}} D_{\hat{\mu}} \psi_2 \quad (3)$$

with $G^{AB}_{\hat{\mu}\hat{\nu}}$ the six dimensional SO(12) gauge field strength and ψ_1 and ψ_2 Majorana-Weyl spinors with opposite six dimensional helicity and belonging to the two inequivalent 32 dimensional spinor representations of SO(12). (For details and conventions see ref.2.) This model has no anomalies[3]. I investigate solutions of the field equations with four dimensional Poincaré symmetry and unbroken gauge symmetry $SU(3)_C \times SU(2)_L \times U(1)_Y$. For simplicity I assume that spacetime is a direct product of Minkowski space M^4 and a two dimensional internal space. Then the most general solution has internal space S^2 with gauge fields in an SO(3) symmetric monopole configuration on S^2:[4]

$$A_\varphi = -\frac{i}{2g} (\pm 1 - \cos\vartheta) \times \begin{pmatrix} om \\ -mo \\ & om \\ & -mo \\ & & op \\ & & -po \\ & & & op \\ & & & -po \\ & & & & op \\ & & & & -po \\ & & & & & on \\ & & & & & -no \end{pmatrix}$$

$$A_\vartheta = A_\mu = 0 \quad (4)$$

Here m,n and p are integers with n+p even. They label the different monopole configurations which are all topologically trivial according to $\Pi_1(Spin(12))=0$. This means that all solutions can be obtained from each other by continuous deformations. For generic m,n and p the four dimensional gauge symmetry is

$$K = SU(3)_C \times SU(2)_L \times U(1)_R \times U(1)_{B-L} \times U(1)_q \times SU(2)_G \quad (5)$$

Here $U(1)_q$ is the abelian group commuting with the SO(10) subgroup of SO(12) and $SU(2)_G$ corresponds to isometry transformations on S^2. For specific values of the monopole numbers the symmetry may be larger, like $SU(5) \times U(1) \times U(1)_q \times SU(2)_G$ for the case m = p or $SO(10) \times U(1)_q \times SU(2)_G$ for m = p = 0. Since the solutions with lower symmetry can be obtained continuously from the SO(10) symmetric solutions, one may interprete them as spontaneous symmetry breaking of SO(10).

I will concentrate on an SU(5) symmetric example with n=3, m=p=1. Harmonic expansion of the six dimensional spinors leads to three massless chiral fermion generations. Under the generation group $SU(2)_G \times U(1)_q$ the fermions of the type u, d, u^c and e^c in the 10 of SU(5) transform as a doublet with charge $q = \frac{1}{2}$ plus a singlet with $q = -\frac{1}{2}$. In contrast, the fields e, ν and d^c in the $\bar{5}$ of SU(5) transform as a triplet with $q = -\frac{1}{2}$. (In addition the model leads to massless right handed neutrinos at this stage.) One observes that quarks and leptons within the same generation transform differently with respect to the generation group $SU(2)_G \times U(1)_q$. This feature is generic for all $m, p \neq 0$. It has important consequences for predictions on mass ratios, since Yukawa couplings will depend on $SU(2)_G \times U(1)_q$ transformation properties.

For a calculation of Yukawa couplings one needs to identify candidates for the weak Higgs doublet responsible for breaking of $SU(2)_L \times U(1)_Y$. They must be contained in the SO(12) gauge fields, noting that the adjoint of SO(12) contains a complex SO(10) - 10 plet with $q = \pm 1$. One therefore has two independent $SU(2)_L$ doublets H_1 and H_2 within the six dimensional gauge fields. The internal components of six dimensional gauge fields transform as four dimensional scalars and their harmonic expansion leads to an infinite tower of scalar doublets with various $SU(2)_G$ quantum numbers. However, only the lowest modes can have Yukawa couplings to the chiral fermions whereas all higher modes decouple due to $SU(2)_G$ symmetry. At this stage, the low energy Higgs field must be a combination

$$\varphi = \gamma_i H_i \qquad (6)$$

with $\sum_i |\gamma_i|^2 = 1$.

In my example (n=3, p=m=1) the lowest modes transform as a $SU(2)_G$ singlet (H_1) and a triplet (H_2^-, H_2^o, H_2^+). Comparison with the quantum numbers of quarks and leptons immediately implies that H_1 can only couple to the bilinear of the $SU(2)_G$ singlet, charge two third quark and antiquark, which I will identify with the top quark:

$$L_{Yuk,1} \sim t_L t_L^c H_1 + \text{h.c.} \qquad (7)$$

A similar analysis for H_2 gives the general form

$$L_{Yuk,2} \sim (b_L b_L^c + \tau_L \tau_L^c) H_2^-$$
$$+ (c_L, u_L) \begin{pmatrix} H_2^- & H_2^o \\ H_2^o & H_2^+ \end{pmatrix}^* \begin{pmatrix} c_L^c \\ u_L^c \end{pmatrix} + \text{h.c.} \qquad (8)$$

where the different coupling constants still have to be calculated. No Yukawa couplings are allowed for s, μ, d and e as a consequence of the symmetries of the model.

Since the different doublets have all different $SU(2)_G$ quantum numbers, their mixing will be small if $SU(2)_G$ is broken at a scale M_G somewhat below the scale M_C characteristic for spontaneous compactification[5]. Let me assume that the leading contribution to the low energy doublet φ comes from H_1 and that the admixtures from H_2^-, H_2° and H_2^+ are suppressed by coefficients γ_i involving different powers of the ratio M_G/M_C so that

$$\langle H_1 \rangle \gg \langle H_2^- \rangle \gg \langle H_2^\circ \rangle, \langle H_2^+ \rangle \tag{9}$$

Once φ gets a vacuum expectation value, this would result in a mass hierarchy

$$m_t \gg m_b, m_\tau, m_c \gg m_s, m_\mu, m_u, m_d, m_e \tag{10}$$

resembling already a little bit the observed pattern.

One may go one step further and explicitly calculate the Yukawa couplings of H_1 and H_2. One finds that they are all proportional to the four dimensional gauge coupling with ratios being given by ratios of $SU(2)_G$ - Clebsch-Gordon-coefficients. The approximation $\langle H_2^\circ \rangle = \langle H_2^+ \rangle = 0$ results in the following predictions for fermion masses (M_W in the W-boson mass)

$$m_t = 2M_W \tag{11}$$

$$m_b = m_\tau \tag{12}$$

$$m_c = \sqrt{\frac{4}{3}} \, m_b \tag{13}$$

$$m_\mu = m_s = m_d = m_u = m_e = 0 \tag{14}$$

Especially eq.(13) can certainly not be considered a good approximation to a realistic spectrum.

In any case, the monopole solution (4) should not be taken as a realistic ground state (even in the limit of unbroken $SU(2)_L \times U(1)_Y$). Its symmetry is too large - $SU(5)$ and the generation group have to be spontaneously broken - and the spectrum of bosonic fields contains tachyons indicating an instability. To obtain a more realistic model with a possible approximate ground state where all gauge symmetries except $SU(3)_C \times SU(2)_L \times U(1)_Y$ are spontaneously broken, one may introduce[2] a six dimensional scalar field in the fifth rank totally antisymmetric tensor representation 792 of $SO(12)$. With respect to the $SO(10) \times U(1)_q$ subgroup this field decomposes

$$792 \to 126_\circ + \overline{126}_\circ + 210_{\pm 1} + 120_\circ \tag{15}$$

The $SU(3)_C \times SU(2)_L \times U(1)_Y$ singlets in 126 and 210 can be responsible for spontaneous symmetry breaking, including the generation group. In the approximation of an unbroken abelian symmetry $U(1)_{\tilde{q}}$ (which is a combination of $U(1)_q$, $U(1)_G$, $U(1)_R$ and $U(1)_{B-L}$ with charge $\tilde{q} = I_{3G} + \frac{1}{2}q + aI_{3R}$) one can again perform a quantum number analysis for the various Higgs doublets. One finds that the only two sorts of fields which can mix with H_1 are H_2^- and h, a combination of scalar doublets from the 792-scalar. The mass matrices for the quarks and charged leptons are

$$M_U = \begin{pmatrix} a_{33}H_1 & a_{32}h & 0 \\ a_{23}h & a_{22}H_2^{-*} & 0 \\ 0 & 0 & 0 \end{pmatrix} \qquad (16)$$

$$M_D = \begin{pmatrix} b_{33}H_2^- & 0 & b_{32}h^* \\ 0 & b_{22}h^* & 0 \\ 0 & 0 & 0 \end{pmatrix} \qquad (17)$$

$$M_E = \begin{pmatrix} c_{33}H_2^- & 0 & 0 \\ 0 & c_{22}h^* & 0 \\ c_{13}h^* & 0 & 0 \end{pmatrix} \qquad (18)$$

The non zero Yukawa couplings a_{ij}, b_{ij}, c_{ij} will be calculable for any given ground state with $SU(3)_C \times SU(2)_L \times U(1)_Y \times U(1)_{\tilde{q}}$ symmetry. This symmetry implies that the first generation remains massless

$$m_d = m_u = m_e = 0 \qquad (19)$$

For $H_1 \gg H_2^- \gg h$ one finds the pattern

$$m_t \gg m_b, m_c, m_\tau \gg m_\mu, m_s \gg m_u, m_d, m_e \qquad (20)$$

In addition, the admixture of h leads to non vanishing mixing angles and CP violating phase. This scenario predicts mixing angles of order $h/H_2^- \sim m_s/m_b \sim$ a few percent for the third generation. Finally, the fermions of the first generation acquire masses once $U(1)_{\tilde{q}}$ is spontaneously broken. These masses put a lower limit on the corresponding symmetry breaking scale. In a realistic model, corrections from $U(1)_{\tilde{q}}$ breaking would replace the zeros in (16)-(18) by entries of order 1-50 MeV. The qualitative structure of these mass matrices is therefore fixed and it would be interesting to investigate if this is consistent with the observed phenomenology for mixing angles and CP violation.

To conclude, let me resume the general scenario how I imagine that higher dimensional theories could explain the hierarchical structure of fermion masses. One observes four groups for the masses of fermions:

$$\begin{array}{l} m_t \\ m_b, m_c, m_\tau \\ m_s, m_\mu \\ m_d, m_u, m_e \end{array} \qquad (21)$$

The overall scale changes from one group to the next roughly by a factor 20. I emphasize that these groups do <u>not</u> correspond exactly to the generations. For example, the second group contains quarks of the second and the third generation. In four dimensional models where generations appear essentially as repetitions, this feature is difficult to understand. I propose an explanation where these groups are induced by quantum numbers of a generation group. Quarks and leptons within the same generation may transform differently with respect to this generation group, as I have shown above.

In usual four dimensional gauge theories Yukawa couplings are free parameters and one may take them as small as one needs. This gives only little predictivity, but these models can accomodate the observed spectrum. In higher dimensional models, four dimensional Yukawa couplings are not free. Typically, they turn out to be of the same order of magnitude as the gauge coupling g or they vanish due to symmetries. At first sight, this seems to run into conflict with a hierarchical mass pattern. However, higher dimensional models typically contain several or many fields with the quantum numbers of the Higgs doublet. As we have seen above, different doublets may couple to different fermions. The idea is that there are essentially four sorts of fields H_1, H_2, H_3, H_4 where H_1 couples only to top and H_2 to the second group, whereas H_3 does not couple to the first generation. If the low energy doublet is mainly H_1 with small admixtures of H_2, H_3 and H_4 obeying

$$\langle H_1 \rangle : \langle H_2 \rangle : \langle H_3 \rangle : \langle H_4 \rangle \sim 20 \qquad (22)$$

the qualitative structure of the mass hierarchies would be obtained. Already at this stage the scenario is very restrictive for higher dimensional model building. It is not easy to find models with appropriate quantum numbers for the doublets H_i so that the above pattern of couplings is realized. However, if electron and top quark are allowed to couple to the same scalar field, one typically expects a prediction $m_e \approx m_t$ since all Yukawa couplings are of the same order. Thus a serious problem arises for models not consistent with the above scenario.

It may seem that I only have replaced the problem of explaining small Yukawa couplings by a similar one for the small mixings between scalar fields. The second one, however, will be easier to solve, at least if the model exhibits somewhat different length scales. If the generation group is spontaneously broken at a scale M_G smaller than the compactification scale M_C ($M_G/M_C \sim 1/20$, for example), the mixing of fields with different generation quantum numbers would be suppressed by different powers of M_G/M_C. The hierarchical structure of fermion masses would then be explained by a fine structure of scales near the compactification scale.

ACKNOWLEDGEMENT

I would like to thank the organizers for their work making this pleasant conference possible and successful.

REFERENCES

1) C. Wetterich, Nucl. Phys. B223 (1983) 109;
   E. Witten, Proc. 1983 Shelter Island II Conf. (MIT press, 1984).

2) C. Wetterich, Nucl. Phys. B244 (1984) 359; Nucl. Phys. B260 (1985) 402;
   CERN preprint TH 4154 (1985), to appear in Nucl. Phys. B.

3) L. Alvarez-Gaumé and E. Witten, Nucl. Phys. B234 (1984) 269.

4) E. Cremmer and J. Scherk, Nucl. Phys. B108 (1976) 409; Nucl. Phys. B118 (1977) 61.
   Z. Horvath, L. Palla, E. Cremmer and J. Scherk, Nucl.Phys. B127 (1977).
   S. Randjbar-Daemi, A. Salam and J. Strathdee, Nucl. Phys. B214 (1983) 491; Phys. Lett. 124B (1983) 345.
   S. Randjbar-Daemi and C. Wetterich, Phys. Lett. 148B (1984) 48.

5) G. Lazarides, Q. Shafi and C. Wetterich, Nucl.Phys. B181 (1981) 287.

ATTRACTOR IN A HIGHER-DIMENSIONAL COSMOLOGY

Kei-ichi MAEDA

International School for Advanced Studies, Strada Costiera 11, 34014 Trieste, Italy

Instead of the "cosmological principle" in 4-dim theory, the idea of an "attractor" is introduced in a higher-dimensional cosmology in order to understand why our universe is the present one, i.e. (the 4-dim Friedmann universe; F^4)x(a constant internal space; K). Two examples are presented. One is the S^2-monopole compactification in 6-dim, N=2 supergravity coupled to Yang-Mills and matter fields, and the other is the Calabi-Yau compactification in 10-dim, N=1 supergravity coupled to the Yang-Mills field with the curvature squared term ($R^2_{\mu\nu\rho\sigma} - 4R^2_{\mu\nu} + R^2$). In both cases, F^4 x K is the attractor in the later stage of the universe and it is asymptotically unique apart from the time reversal ones.

1. INTRODUCTION

Higher-dimensional theories including a superstring theory are certainly very attractive candidates for a unified theory. Since the behaviour of the universe in the very early stage (the Planck era) must be quite different from that in the conventional 4-dim cosmology, the study of a higher-dimensional cosmology is also very interesting and important. Many works have been done to solve the cosmological problems (horizon-, flatness-, entropy-, singularity- problems etc.) instead of the GUT scenario. (See the overview by Kolb[1]). However, before going into such discussions, first of all, we have to explain why our present universe is (the 4-dim Friedmann universe; F^4)x(a constant internal space; K). Even if we assume some compactification which provides the topological separation of the internal space from the 4-dim space-time, the above question is not trivial. In the conventional 4-dim theory, an isotropy and a homogeneity of space-time, which may be deduced from the cosmological principle or from an inflationary scenario, guarantee our universe to be the Friedmann one. While, in a higher-dimensional cosmology, since the present universe is truly anisotropic in higher dimensions, i.e. the physical 3-space is very large and expanding but the internal space must be extremely small and nearly static, the above "cosmological principle" does not guarantee our universe to be the present one. We may need another "principle". The dynamics of the universe may give the answer. Here, the idea of an "attractor" is proposed as such a "principle"[2]. That is, if $F^4 \times K$ is the attractor in our dynamical system, we can easily understand why our universe is the present one. Two examples are presented here, which are the S^2-monopole compactification in 6-dim, N=2 supergravity coupled to Yang-Mills and matter fields (§2) and the Calabi-Yau compactification in 10-dim N=1 supergravity coupled to the Yang-Mills field with the curvature squared term ($R^2_{\mu\nu\rho\sigma} - 4R^2_{\mu\nu} + R^2$) (§3).

2. 6-DIM, N=2 SUPERGRAVITY

The first example[3] is the 6-dim, N=2 supergravity theory coupled to Yang-Mills and matter fields[4]. The compactification of 6-dimensions into (a 4-dim conformally flat space-time)×(a 2-dim sphere with a time-dependent radius b(t)) is obtained by the following vacuum expectation values (VEV's) of "dilaton" and Yang-Mills field $F_{\mu\nu}$ (so-called monopole compactification);

$$<\phi> = \phi(t) \tag{1}$$

$$<F^3_{\mu\nu}> = F_{\mu\nu} = \begin{cases} F_{MN} = -\frac{1}{2g_0 b^2} \epsilon_{MN} & (M,N=5,6) \\ 0 & \text{otherwise} \end{cases} \tag{2}$$

where g_0 is a gauge coupling constant and F_{MN} is given by a monopole solution;

$$A_M dx^M = \frac{1}{2g_0}(\cos\theta \pm 1) d\phi . \tag{3}$$

The metric of 6-dim space-time is

$$ds^2 = g_{\mu\nu} dx^\mu dx^\nu = -dt^2 + a^2(t) d\vec{x}^2 + b^2(t)[d\theta^2 + \sin^2\theta \, d\phi^2] . \tag{4}$$

We can easily check that the Yang-Mills equation,

$$\nabla_\nu (e^{\sqrt{2}\kappa\phi} F^{\mu\nu}) = 0 , \tag{5}$$

is automatically satisfied, where $\kappa^2 = 8\pi G_6$ and G_6 is a 6-dim gravitational constant.

By this compactification, the Lagrangian for the background fields

$(g_{\mu\nu}, \phi, A_\mu)$ is

$$L_{BG} = \frac{\sqrt{-g}}{2\kappa^2}[R - (\nabla\sigma)^2 - V_{BG}] \qquad (6)$$

$$V_{BG} = \frac{\kappa}{2}\frac{e^\sigma}{b^4} + \frac{2}{\kappa}e^{-\sigma}, \qquad (7)$$

where

$$\sigma \equiv \kappa\phi - 2\ln g_0 - \tfrac{1}{2}\ln 2. \qquad (8)$$

e^σ/b^4 is the VEV of the Yang-Mills field ($e^\sigma F^2_{\mu\nu}$) and $e^{-\sigma}$ is the cosmological "constant", which is deduced from the requirement of SUSY in a "gauging" process. This cosmological constant in 6-dimensions guarantees the Minkowski space-time in 4-dimensions without fine tuning.

In cosmology, besides the above Lagrangian, we must take into account a finite temperature effect, i.e. excitation of fundamental particles contained in our theory. If these particles are in a thermal equilibrium state, these are described by the equation of state of "radiation". In the later stage of the universe, the temperature is in the range of $1/b > T > 1/a$. Then, the equation of state is that of 4-dim radiation ($P = \rho/3$).

Adding this term, the basic equations are now:
(1) the Einstein equations
   (i) constraint equation

$$3p^2 + 6pq + q^2 = \kappa^2\rho + r^2/2 + \underline{\kappa^{-1}e^{-\sigma}(1-e^{-w})^2} \qquad (9)$$

   (ii) dynamical equations

$$\dot{p} + (3p + 2q)p = \kappa^2\rho/3 + \underline{(2\kappa)^{-1}e^{-\sigma}(1-e^{-2w})} \qquad (10)$$

$$\dot{q} + (3p + 2q)q = (2\kappa)^{-1}e^{-\sigma}\underline{(1-e^{-w})(1-3e^{-w})}, \qquad (11)$$

(2) the equation for the dilaton field

$$\dot{r} + (3p + 2q)r = \underline{\kappa^{-1}e^{-\sigma}(1-e^{-2w})}, \qquad (12)$$

where

$$p = \dot{a}/a, \quad q = \dot{b}/b, \quad r = \dot{\sigma} \quad \text{(An over dot denotes the time derivative)} \qquad (13)$$

and

$$w \equiv 2\ell nb - \sigma - \ell n(\kappa/2). \qquad (14)$$

If $w = 0$, the underlined terms in Eqs. (9)$\sim$(12) vanish, then the basic equations are the same as those of the trivial torus-compactification. One can solve these equations analytically. All solutions with $\rho(\neq 0)$, apart from the time reversal ones, approach $F^4 \times S^2$ ([the 4-dim Friedmann universe]×[a constant 2-sphere]) for $t \to \infty$.

In order to show that $F^4 \times S^2$ is the attractor, we must investigate the case of $w \neq 0$ as well. The equation for w is rewritten from Eqs.(10)$\sim$(12) as

$$\dot{E} = -\dot{\xi}\dot{w}^2/2 \qquad (15)$$

where the "energy" E and the "potential" V of the dynamical system for w are defined by

$$E = \tfrac{1}{2}\dot{w}^2 + V(w) \qquad (16)$$

and

$$V(w) = \tfrac{1}{2}\kappa \, (1-e^{-w})^2 \, . \qquad (17)$$

From (9),

$$\dot{\xi}^2 = 2\dot{w}^2 + 3\dot{\eta}^2 + 8[\kappa^{-1} e^{-\sigma}(1-e^{-w})^2 + \kappa^2 \rho] \qquad (18)$$

where

$$\dot{\xi} \equiv 6p + 4q - r \quad \text{and} \quad \dot{\eta} \equiv 2p - r \, . \qquad (19)$$

Since Eq. (18) is positively definite, $\dot{\xi}$ is always positive if it is initially positive, which means that the system is always dissipative, then, the universe approaches the minimum of the potential V at $w = 0$ losing the energy (see Fig.1). The universe approaches $w = 0$, i.e. $F^4 \times S^2$, asymptotically. A numerical example is shown in Figs. 2 and 3. One can see a nice attractor in Fig. 2. In Fig. 3, the time evolution of scale factors, a and b, is shown.

$F^4 \times S^2$ is the attractor for any initial conditions (apart from the time reversal ones, i.e. except for the case of $\dot{\xi} < 0$). Since $w = 0$ corresponds to zero-cosmological constant in 4-dimensions, in our theory the evolution of the universe explains why $\Lambda_4 \approx 0$ without fine tuning.

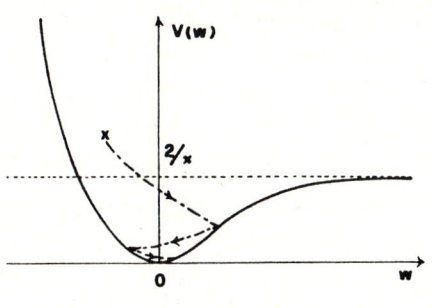

FIGURE 1
The potential $V(w)$. Losing the energy, the universe approaches $w=0$.

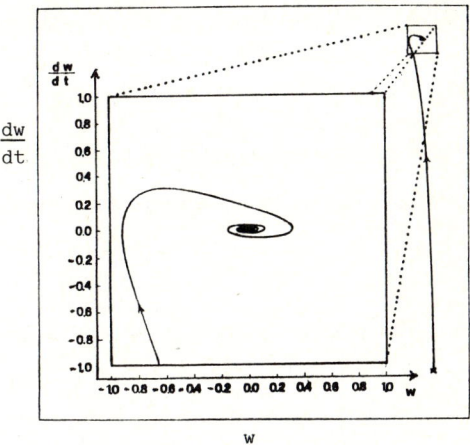

FIGURE 2
The phase diagram of dynamical system w. The origin ($w=\frac{dw}{dt}=0$) corresponds to $F^4 \times S^2$.

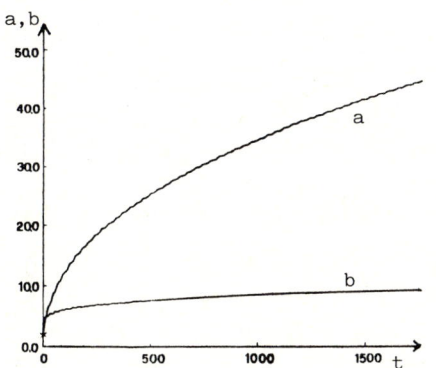

FIGURE 3
The behaviour of scale factors a and b. a approaches the scale factor of the Friedmann universe ($t^{\frac{1}{2}}$) and b becomes some constant.

3. 10-DIM, N=1 YANG-MILLS SUPERGRAVITY

The second example[5] is the 10-dim, N=1 Yang-Mills supergravity theory[6] which is deduced from the superstring theory in the field theory limit[7]. Adding the Riemann curvature squared term, Candelas, Horowitz, Strominger and Witten[8] have found Calabi-Yau compactification in this theory. As for the curvature squared terms, we shall adopt the special combination of curvature squared terms,

$$R^2_{\mu\nu\rho\sigma} - 4R^2_{\mu\nu} + R^2 \quad , \tag{20}$$

which is proposed by Zwiebach and Zumino[9].

The Lagrangian of the bosonic part of the theory is

$$L_{10} = \frac{\sqrt{-g}}{2\kappa^2}\left[R - \tfrac{1}{2}(\nabla\Phi)^2 - \frac{\alpha'}{2}e^{-\frac{\Phi}{2}}\operatorname{tr} F^2_{\mu\nu}\right], \tag{21}$$

where $\kappa^2 = 8\pi G_{10}$, G_{10} is a 10-dim gravitational constant, α' is the inverse string tension. We have assumed $H_{\mu\nu\rho} = 0$. The metric of 10-dimensions in our system is

$$ds^2 = -dt^2 + a^2(t)d\vec{x}^2 + b^2(t)\,ds^2_{CY} \,. \tag{22}$$

ds^2_{CY} is the metric of a Calabi-Yau manifold and it does not depend on the time t. We consider only a change of the scale of the Calabi-Yau space but not any deformation of it. Suppose $\tilde{F}_{\mu\nu} = (0, \tilde{F}_{MN})$ is a "Calabi-Yau" solution of the Yang-Mills equation for a static background, $d\tilde{s}^2 = -dt^2 + d\vec{x}^2 + ds^2_{CY}$. From the equation for ϕ, $\mathrm{tr}\,\tilde{F}^2_{MN} = \tilde{R}^2_{MNPQ}$, where $\tilde{R}_{MNPQ}$ is the Riemann tensor of ds^2_{CY}. Using the conformal transformation, one can easily show that

$$F_{\mu\nu} = \begin{cases} \tilde{F}_{MN} & \text{for } \mu,\nu = M,N \\ 0 & \text{otherwise} \end{cases} \tag{23}$$

is also a solution of Yang-Mills equations

$$g^{\mu\nu}\nabla_\nu (e^{-\frac{\phi}{2}} F_{\mu\rho}) = 0 \tag{24}$$

for the time-dependent background (22) and

$$\mathrm{tr}\,F^2_{\mu\nu} = b^{-4}\,\mathrm{tr}\,\tilde{F}^2_{MN} = b^{-4}\,\tilde{R}^2_{MNPQ} \,. \tag{25}$$

Through the VEV's of the Yang-Mills field, (23), 10-dim space-time is compactified into (a 4-dim conformally flat space-time)×(a 6-dim Calabi-Yau manifold with one time-dependent scale b(t)).

Using the properties of Calabi-Yau manifold ($\tilde{R}_{MN} = \tilde{R} = 0$), the curvature squared term (20) and $F^2_{\mu\nu}$ -term are described as

$$\frac{\alpha'}{2}\sqrt{-g}\,e^{-\frac{\phi}{2}} [R^2_{\mu\nu\rho\sigma} - 4R^2_{\mu\nu} + R^2 - \mathrm{tr}\,F^2_{\mu\nu}]$$

$$= \frac{\alpha'}{2}\sqrt{-g}\,e^{-\frac{\phi}{2}} b^{-4}(\tilde{R}^2_{MNPQ} - \mathrm{tr}\,\tilde{F}^2_{MN}) - \alpha'\sqrt{-g}\,e^{-\frac{\phi}{2}} I(p,q,r)$$

$$+ \text{(totally divergent term)} \,, \tag{26}$$

where

$$I(p,q,r) = 12q(2p^3 + 15p^2 q + 20pq^2 + 5q^3) - 2r(p^3 + 18p^2 q + 45pq^2 + 20q^3) \tag{27}$$

with

$$p = \dot{a}/a,\ q = \dot{b}/b \text{ and } r = \dot{\phi}. \tag{28}$$

From (25), the first term of the r.h.s. of Eq. (26) vanishes automatically, then the Lagrangian for background fields (a,b, and Φ) is

$$L = \frac{\sqrt{-g}}{2\kappa^2} [-6(p^2 + 6pq + 5q^2) + \tfrac{1}{2}r^2 - \alpha' e^{-\frac{\Phi}{2}} I(p,q,r)]. \tag{29}$$

From this, we obtain the basic equations as follows:

(1) the Einstein equations

    (i) constraint equation

$$3(p^2 + 6pq + 5q^2) = \kappa^2 \rho + \frac{r^2}{4} - \frac{3}{2} \alpha' e^{-\frac{\Phi}{2}} I \tag{30}$$

    (ii) dynamical equations

$$\dot{F} + 3(p+2q)F = \kappa^2 \rho/3 - \frac{\alpha'}{8} e^{-\frac{\Phi}{2}} I \tag{31}$$

$$\dot{G} + 3(p+2q)G = -\frac{\alpha'}{8} e^{-\frac{\Phi}{2}} I \tag{32}$$

with

$$F \equiv p + \frac{\alpha'}{48} e^{-\frac{\Phi}{2}} [-5\frac{\partial I}{\partial p} + 3\frac{\partial I}{\partial q}] \tag{33}$$

and

$$G \equiv q + \frac{\alpha'}{48} e^{-\frac{\Phi}{2}} [3\frac{\partial I}{\partial p} - \frac{\partial I}{\partial q}] \tag{34}$$

(2) the equation for the dilaton Φ

$$\dot{H} + 3(p+2q)H = \frac{\alpha'}{2} e^{-\frac{\Phi}{2}} I \tag{35}$$

with

$$H \equiv r - \alpha' e^{-\frac{\Phi}{2}} \frac{\partial I}{\partial r} \tag{36}$$

Here, we have again added "4-dim radiation" ρ, which may be justified in the later stage of the universe.

The terms which contain $I(p,q,r)$ are from the curvature squared term. (We call them I-terms). First, since α' is always in front of I-terms, the order of magnitude of those terms is much smaller than the other terms by $O([t_{PL}/t_{exp}]^2)$ in the later stage of the universe ($t_{exp} \gg t_{PL}$). Here, we have defined the expansion time scale t_{exp} by $t_{exp} \equiv \min[a/\dot{a}, b/\dot{b}, 1/\dot{\Phi}]$, and supposed $\alpha' e^{-\Phi/2} \sim t_{PL}^2$, where t_{PL} is the Planck time. Secondly, since $I(p,q,r)$ consists only of the first-order time derivatives of metric, any higher-order derivative terms such as $\ddot{a}$ do not appear in the basic equations (30)~(36). Therefore, the dynamical structure of our system (e.g. stability) does not change drastically by the addition of I-terms. Notice that in the case with higher-order derivative terms it must change drastically even if the order of magnitude of those terms is very small.

From the above two facts, I-terms may be treated as small perturbations, at least in the later stage. Therefore, first, we shall look at solutions of Eqs. (30)∼(36), neglecting I-terms. These equations are again the same as the trivial torus-compactification, and then, one can easily find general solutions. For the case with "4-dim radiation", all solutions (apart from the time reversal ones) approach $F^4 \times K^6$ (a constant Calabi-Yau space)[5] like the case of 6-dim theory (§2). Using these solutions, one can easily check that I-terms are decreasing with time asymptotically more rapidly than the other terms. For example $I(p,q,r) \alpha 1/t^4$ and $\rho \alpha 1/t^2$. Therefore, the above solutions are asymptotic solutions of our system.

One can also show that those solutions are stable against small perturbation. Then, if I-terms can be treated as small perturbations at the initial stage, which is likely in the field-theory limit of superstring theory, we would say that $F^4 \times K^6$ is the attractor and it is asymptotically unique in the later stage of the universe apart from the time reversal ones.

REFERENCES

1) E. Kolb, this volume.

2) K. Maeda, SISSA preprint, Trieste, 34/85/A.

3) K. Maeda and H. Nishino, Phys. Lett. 154B (1985), 358; SISSA preprint 13/85/A.

4) H. Nishino and E. Sezgin, Phys. Lett. 144B (1984), 187; A. Salam and E. Sezgin, Phys. Lett. 147B (1984), 47; I.G. Koh and H. Nishino, Phys. Lett. 153B (1985), 45; S. Randjbar-Daemi, A. Salam, E. Sezgin and J. Strathdee, Phys. Lett. 151B (1985), 351.

5) K. Maeda, SISSA preprint, Trieste, 51/85/A (1985).

6) A.H. Chamseddine, Nucl. Phys. B 185 (1981), 403; E. Bergshoeff, M. de Roo, B. de Wit and P. van Nieuwenhuizen, Nucl. Phys. B 195 (1982), 97; G.F. Chapline and N.S. Manton, Phys. Lett. 120B (1983), 105.

7) G.G. Callan, E.J. Martinec, M.J. Perry and D. Friedan, Princeton preprint (1985).

8) P. Candelas, G. Horowitz, A. Strominger and E. Witten, Santa Barbara preprint (1985).

9) B. Zwiebach, Phys. Lett. 156B (1985), 315; B. Zumino, LBL preprint, UCB-PTH-85/13, LBL-19302 (1985).

SEARCHING FOR CYGNETS

Edward W. KOLB

Fermi National Accelerator Laboratory, Batavia, IL, 60510 U.S.A.

The reported observation in underground detectors of high-energy muons from the direction of the compact binary X-ray source CYG X-3 (2030+4047) cannot be explained by conventional physics. In this paper some explanations for the effect based upon unconventional physics are reviewed.

Cygnus X-3 is believed to be a compact X-ray source in the Cygnus constellation. It is observed to be a very robust source of radiation from infrared to UHE (ultra high energy, $E > 1$ TeV). All radiation above infrared is modulated with a 4.8 hour period. This 4.8^h period is thought to be the orbital period of a binary system composed of a neutron star and a companion star of about 4 $M_\odot$.

The observed spectrum of radiation from CYG X-3 can be fit by a single power law over 13 decades in energy, from 10^3 eV to 10^{16} eV:[1)]

$$\frac{dN_\gamma}{dE} = 3\times10^{-10} \left(\frac{E}{1\text{TeV}}\right)^{-2.1} \text{cm}^{-2}\text{s}^{-1}\text{TeV}^{-1}. \tag{1}$$

This spectrum has roughly equal luminosity per decade of energy. Assuming a distance of 12 kpc, the total luminosity of CYG X-3 above 1 GeV is in excess of 10^{38} erg s^{-1}, making it the brightest γ-ray source in our galaxy.

It is possible to construct an astrophysical model for the CYG X-3 system by using the phase information of the radiation. A 'theorist's rendition' of the phase information is shown in Figure 1. The origin of the X and γ rays is thought to be the neutron star, and the minimum of the X-ray and γ-ray radiation occurs when the neutron star is eclipsed by the main sequence star between phase -0.25 and +0.25. The absence of a zero-flux minimum for the X-rays can be understood if there is a cocoon of optical depth order unity for X-rays surrounding the system. The cocoon can back scatter X-rays during the eclipse, giving a reduced, but non-zero, X-ray flux. The cocoon will be transparent to γ-rays, giving a zero minimum γ-ray flux during eclipse. As seen in Figure 1, the UHE radiation has a 4.8^h period, but a phase structure much different than the X-rays or γ-rays. Vestrand and Eichler[2] and also Stecker[3] and Stenger[4] have used the phase structure to model the UHE emission. They assume the pulsar is a source of an energetic beam of primary protons. The primary protons hit the star and produce secondary π^0's (among other things), which decay to the γ's detected as the UHE flux. This mechanism will produce detectable UHE γ's if the primary beam passes through enough material to produce π^0's but not enough material to completely absorb the γ's from π^0 decay. This condition will be met for only a small fraction of the orbital period, around $\psi = \pm 0.25$ when the neutron star is at grazing incidence. The primary proton beam is not detected because it is dispersed by galactic magnetic fields before reaching the solar system. Models of the acceleration mechanism for the primary beam are quite complicated. They are thought to involve large $\vec{v} \times \vec{B}$ electric fields in the vicinity of the pulsar, but a completely self-consistent picture for the acceleration mechanism is very difficult to construct.[5] Most of our results will be independent of the details of the acceleration mechanism of the primary beam.

In order to understand the signal for new physics it is first necessary to calculate the expected flux of neutrinos from CYG X-3. The calculation reported here was done by several groups with very similar results.[6,7] The basic assumption is that the primary beam makes charged mesons, in addition to the neutral mesons, in the collision with the star. The decay of the charged mesons will result in a neutrino flux from the system.

If the UHE γ-rays originate from a source spectrum with the power law form

$$\frac{dS_\gamma}{dE_\gamma} = AE^{-n} , \qquad (2)$$

the π^0 source spectrum should be of the form

$$\frac{dS_{\pi^0}}{dE_\pi} = A \, 2^{n-2} E^{-n} , \qquad (3)$$

where the factor 2^{n-1} is from 2 photons of energy $E_{\pi^0}/2$. There should be $\pi^\pm$'s produced also, and the source spectrum for the $\pi^\pm$'s should be

$$\frac{dS_{\pi^+ + \pi^-}}{dE_\pi} = 2^{n-1} AE^{-n} = 2^{n-1} \frac{dS_\gamma}{dE_\gamma} . \qquad (4)$$

The energy of the neutrinos produced by $\pi^\pm$ decay will depend upon whether the

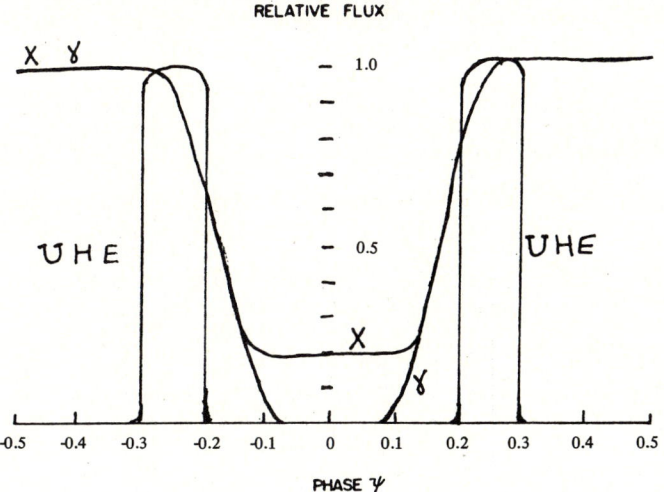

FIGURE 1 : PHOTON FLUX

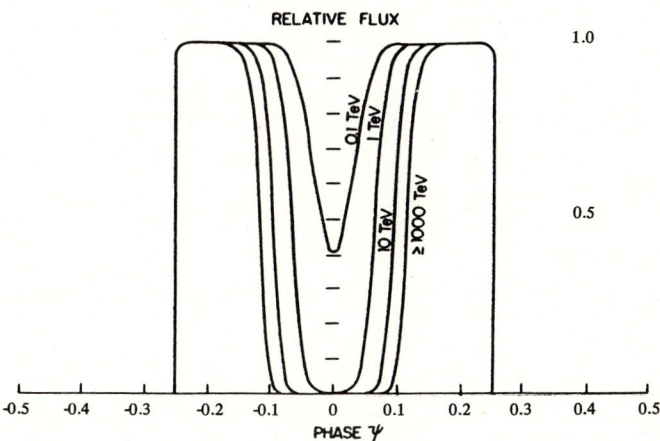

FIGURE 2 : NEUTRINO FLUX

$\pi^{\pm}$'s decay in flight or interact before decay. If the $\pi^{\pm}$'s decay in flight the neutrino source spectrum should be related to the photon source spectrum by

$$\frac{dS_\nu}{dE_\nu} = \frac{(1-m_\mu^2/m_\pi^2)^{n-1}}{2} \frac{dS_\gamma}{dE_\gamma} \qquad (5)$$

Propogation of the neutrino and photon source spectra through the star will result in the flux detected terrestrially. The absorption of neutrinos and photons depends upon the material seen by the particle traversing the star, which, in turn, depends upon the phase of the orbit. The γN cross section at high energies is nearly energy independent, so the phase structure of UHE photons should be energy independent. Unlike photons, the neutrino cross section is proportional to the energy (for $E_\nu <$ 100 TeV), and the phase diagram for the neutrino flux will depend upon E_ν. The neutrino light curve found by propagating neutrinos through a $4M_\odot$ main sequence star is shown in Figure 2. It should be remembered that the _relative_ flux is shown in Figure 2 - the _absolute_ flux relating different energies falls as $E^{-2.1}$. The details of the phase structure is most sensitive to the central density of the star. Observation of the neutrino light curve would offer a unique tool to probe the central density of the star.

In the above calculations we have assumed the mesons decay before interacting. Since the decay length of the mesons is proportional to E/m, sufficiently energetic mesons will decay before interacting, and neutrino production will be via a beam dump mode. The decay lengths of $\pi^\pm$ and $K^\pm$ in the star are

$$(\gamma c\tau)_{\pi^\pm} = 5\times10^6 \ (E/1 \ \text{TeV})\text{cm}$$
$$(\gamma c\tau)_{K^\pm} = 8\times10^5 (E/1 \ \text{TeV})\text{cm} \qquad (6)$$

The cross section for interaction of the mesons is about 3×10^{-26} cm^2 at $E >$ 1 TeV. The typical density in the envelope of the star is about 10^{-6}g cm^{-3}, so we parameterize the envelope density as $\rho = 10^{-6}\rho_{-6}$g cm^{-3}. There will be a threshold energy above which the mesons will decay before interacting, given by

$$E_T = \begin{cases} 300 \, (\rho_{-6})^{-1} \text{ TeV} & (K^{\pm}) \\ 30 \, (\rho_{-6})^{-1} \text{ TeV} & (\pi^{\pm}) \end{cases} \qquad (7)$$

The existence of this threshold should result in a feature in the neutrino spectrum at E_T, providing a unique tool to examine the density of the stellar envelope.

Using the observed UHE photon spectrum and phase width, the phase-averaged neutrino flux is expected to be

$$\frac{dN_\nu}{dE} = 2 \times 10^{-10} \left(\frac{E}{1\text{TeV}}\right)^{-2.1} \text{cm}^{-2}\text{s}^{-1} \qquad (E < E_T)$$

$$= 2 \times 10^{-12} \left(\frac{E}{1\text{TeV}}\right)^{-2.1} \text{cm}^{-2}\text{s}^{-1} \qquad (E > E_T). \qquad (8)$$

The normalization of the neutrino flux is uncertain by at least an order of magnitude due to uncertainties in the photon phase width, absorption of photons, etc. However the normalization is unlikely to be off by more than two orders of magnitude, and the slope of the power law spectrum should be close to -2.1.

The neutrino flux can be detected in underground detectors either by observing a $\nu_\mu \to \mu$ conversion in the detector (which we call a contained event) or by observing a muon from a $\nu_\mu \to \mu$ conversion in the surrounding rock passing through the detector (which we call an external event). The probability that the neutrino converts in a detector of linear dimension $\ell \approx 10$ m is given by (all distances in the paper are given in terms of water equivalent)

$$P_c = n\sigma\ell$$

$$= 4 \times 10^{-9} (E/1 \text{ TeV}) \qquad (E \lesssim 100 \text{ TeV})$$

$$= 7 \times 10^{-8} \ln(E/1 \text{ TeV}) \qquad (E \gtrsim 100 \text{ TeV}), \qquad (9)$$

where σ is the cross section for $\nu_\mu N \to \mu^- + X$ (for simplicity we have assumed equal cross section for ν and $\bar\nu$). The event rate for contained events is given by

$$\Gamma_c = A_D \int P_c \frac{dN_\nu}{dE} dE \tag{10}$$

$$= 6\times10^{-12} \left(\frac{E_{min}}{GeV}\right)^{-0.1} \text{sec}^{-1}$$

where E_{min} is the larger of the detector threshold and the low energy cutoff in the neutrino spectrum, and A_D is the detector area, assumed to be $4\times10^6 \text{cm}^2$.

The probability that a neutrino interacts in the earth and produces a muon which passes through the detector depends upon the range of the muon. With a muon energy loss rate of

$$\frac{-dE}{dX} = 1.9\times10^{-6} \text{ TeVcm}^{-1} + 4\times10^{-6} \text{ cm}^{-1} E , \tag{11}$$

the range of the muon is

$$R(E) = 3\times10^5 \ln(1 + 2E/1\text{TeV}) \text{cm}. \tag{12}$$

For external events, the range of the muon replaces the detector linear dimension in Eq. (9). If we assume the muon has half the incident neutrino energy, then the probability of an external event is

$$P_E = 10^{-6} (E/1\text{TeV}) \ln(1 + E/1\text{TeV}). \qquad (E < 100 \text{ TeV}) \tag{13}$$

Notice that for E greater than a few GeV, $P_E > P_c$. The rate for external events is

$$\Gamma_E = A_D \int P_E \frac{dN}{dE} dE$$

$$= 3\times10^{-8} \text{ sec}^{-1}. \tag{14}$$

The fact that the external events dominate the contained events is a result of the slope of the spectrum. The total detection rate is about one per year, but as discussed above, that estimate could be off by one or possibly two orders of magnitude. For the external events, the effective size of the target is limited either by the muon range, or by the distance to the surface of the earth. It is clear that as the zenith angle of CYG X-3 increases, the muon signal due to primary neutrinos should not decrease.

Recently two experimental groups, Soudan and NUSEX, have reported an excess of high energy muons from the direction of CYG X-3, with a distribution of arrival times modulated with a 4.8^h period. The number of muons seen by the experiments, 84 ± 20 events in 0.96 years in Soudan and 32 events in 2.4 years in NUSEX, is much larger than the above estimates for neutrino-induced events, if one takes into account the relatively small size of the detectors. In fact, the detected muon flux is comparable to the total "photon" flux. The most striking characteristic of the signal is that the muons are not seen at large zenith angles. The zenith angle dependence of the signal strongly suggests that the muons have an atmospheric origin (or perhaps an origin in the first few hundred meters of rock). The Soudan and NUSEX results seem to confirm previous results from the Kiel air shower experiment[10] of excess muons in the air showers from CYG X-3. The magnitude and zenith angle dependence of the muons rule out neutrinos as a source of the muons. If the primary particles in the air showers are photons, conventional calculations of muon production in the shower cannot account for the observed muon flux.[11] (Even if the air shower primaries are protons, the observed underground muon flux is too large to be accounted for by the observed air shower flux.[12])

The data suggest that the initiating particle must: 1) be neutral, in order to reach the solar system without being dispersed by galactic magnetic fields; 2) be light (less than a few GeV), in order to keep phase coherence with the photons over the 12 kpc distance to CYG X-3; 3) be long-lived, with a lifetime greater than weeks or months depending on γ; 4) shower in the atmosphere like a hadron, i.e., produce muons efficiently; and 5) have a flux comparable to the air shower flux of $3\times10^{-10}(E/1TeV)^{-2.1}$ cm^{-2} sec^{-1} TeV^{-1}. It is clear that the above profile for a particle cannot be fit by any known particle. The unknown particle postulated to fit the above profile has been given the name cygnet.[16] In the rest of this paper I will mention some recent proposals for cygnet candidates.

QUARK NUGGETS[12,13] It is conceivable that there is a separate, stable phase of matter at greater than nuclear matter density. Quark nuggets would have a very small Z/A ratio, and can exist as a stable hadronic system with large A. It is not inconceivable to imagine the neutron star is, in fact, a quark star, and a source of quark nuggets. A problem with this scenario is that the conventional quark nuggets are only stable for large A, and are probably too massive to account for the phase coherence. Quark matter does lead to an enhancement over normal nuclear matter in muon production when the primary showers in the atmosphere. But the enhancement is only about a factor of two.[12]

R-ODD PARTICLES FROM SUPERSYMMETRY[14,15] In supersymmetric theories with an unbroken R-parity, the lightest R-odd particle is stable. The photino is a good candidate for the lightest R-odd particle. It has been proposed that gluinos are produced in pp collisions along with the neutral and charged mesons responsible for the γ and ν flux. The gluinos will decay to photinos before interacting. If threshold for gluino production is low enough, which requires gluino masses less than a few GeV, the photino flux from CYG X-3 could be comparable to the photon flux at high energies. Although the photino would be a relatively light, neutral, stable particle, there are problems with this scenario. First, the photino most likely will not interact in the atmosphere, and the zenith angle dependence for muons produced by photino primaries should probably resemble that for neutrinos. Second, the mass needed to push threshold for gluino production low enough for an appreciable photino flux is very close to being ruled out by experimental data, if it has not already been ruled out. Although photinos are unlikely to be cygnets, they may prove to be detectable in future underground experiments.

H-PARTICLES[16] The H-particle proposed by Jaffe[17] is a metastable neutral strange dibaryon. The H would be a tightly bound six quark state (uuddss). The color-spin wavefunction of the H is the most symmetric, which should maximize the QCD hyperfine interaction leading to a more attractive potential than in other dibaryon systems. If the mass of the H is below pΛ threshold, the H can only decay via double beta decay, and can have a lifetime long enough to reach the solar system from Cygnus. The H is almost unique in matching the first four profiles for a cygnet. Whether the H flux can be comparable to the total air shower flux depends upon the mechanism for cygnet production. If the incident beam from the neutron star is a proton beam, H production will be suppressed, as the pp → HX cross section is smaller than the pp → πX cross section because it is necessary to create two units of strangeness and an A = 2 system. Under these conditions, it is difficult to imagine an H flux comparable to the γ flux. If, however, the primary beam has a large strangeness fraction and consists of particles with A > 1, then there will not be any large supression factors in H/γ production, and the flux of cygnets could be a large fraction, perhaps O(1), of the total air shower flux.

There are several potential problems with the H explanation for cygnets. The mass of the H may not be below pΛ threshold (or, perhaps, not even below ΛΛ threshold). The mass of the H is a question that can be settled by experiment. Even if the H flux is comparable to the total air shower flux, the secondary muon flux would be smaller than that reported by a factor of 2-10. This is a problem for any explanation of the observation, not just for the H scenario. One possible reason for the discrepancy could be an intrinsic variability in the source. The astrophysical environment of CYG X-3 is much more complicated than the simple picture presented here. If anything, it is surprising that the system is as stable as observed. The flux may change between the measurements of the air shower flux and the detection of the underground neutrinos. Finally, there is no explanation for the angular spread seen in the data. The NUSEX signal is seen in a 10° × 10° window in celestial coordinates, much larger than the 0.5° expected angular resolution. Again, this is a problem for any explanation of the signal.

This work was supported in part by NASA and DOE. I would like to acknowledge my collaborators G. Baym, R. Jaffe, L. McLerran, M. Turner, and T. Walker.

REFERENCES

1) A review of the CYG X-3 observations can be found in N. Porter, Nature 305, 179 (1983).
2) W. T. Vestrand and D. Eichler, Ap. J. 261, 251 (1982).
3) F. W. Stecker, Ap. J. 228, 919 (1979).
4) V. J. Stenger, Ap. J. 284 (1984).
5) See, e.g. M. A. Ruderman and P. G. Sutherland, Ap. J. 196, 51 (1975).
6) E. W. Kolb, M. S. Turner, and T. P. Walker, Phys. Rev. D15, to appear (1985).
7) M. V. Barnhill, T. K. Gaisser, T. Stanev, and F. Halzen, preprint MAD/PH/243; V. S. Berezinsky, C. Castagnole, and P. Galeotti, preprint (1985).
8) M. L. Marshak, etal., Phys. Rev. Lett. 54, 2079 (1985).
9) G. Battistoni, etal., Phys. Lett. 155B, 465 (1985).
10) M. Samorski and W. Stamm, Ap. J. 268, L17 (1983).
11) T. Stanev and Ch. Vankov, Phys. Lett., to appear (1985).
12) M. V. Barnhill, T. K. Gaisser, T. Stanev, and F. Halzen, preprint MAD/PH/259 (1985).
13) G. L. Shaw, G. Benford, and D. J. Silverman, preprint 85-14 (1985).
14) V. J. Stenger, preprint HDC-3-85 (1985).
15) R. W. Robinett, Phys. Rev. Lett. 55, 469 (1985).
16) G. Baym, E. W. Kolb, L. McLerran, T. P. Walker, and R. L. Jaffe, preprint 85/98-A (1985).
17) R. Jaffe, Phys. Rev. Lett. 38, 195 (1977).

FRACTIONAL STATISTICS, EXCEPTIONAL PREONS, SCALAR DARK MATTER, LEPTON NUMBER VIOLATION, NEUTRINO MASSES, and HIDDEN GAUGE STRUCTURE $

A. ZEE

Department of Physics FM-15, University of Washington, Seattle, Washington 98195, USA

I would like to talk about what I did over the last year or so. Since I probably lack the strength of character (which should not be confused with strength of conviction) to work on superstrings, I worked on a variety of topics instead in what may be called the dim-sum approach to physics research.
NOTE: This write-up represents an abbreviated version of my talk and is intended to provide a sort of headline service. The reader is referred to the papers cited for details. Needless to say, I do not provide a complete list of references here.

I. FRACTIONAL STATISTICS

My interest in fractional statistics continued into this past year. Since presumably not everyone is conversant with this fascinating subject I thought I would give an extremely brief review of the basics before going on to my more recent work.

The basic physics underlying the phenomenon of fractional statistics [1] in (2+1)-dimensional spacetime is exceedingly simple: it is none other than the Dirac-Bohm Aharonov effect. Consider a theory with a conserved current J_μ coupled to an abelian gauge potential A_μ. Instead of the usual Maxwell action for A_μ, let us add the so-called Chern-Simon action $\int d^3x g^{-1} \epsilon^{\mu\nu\lambda} F_{\mu\nu} A_\lambda$ to the action. We then obtain the equation of motion $\epsilon^{\mu\nu\lambda} F_{\nu\rho} = g J^\mu$.

To make life simple, let us specialize to a point particle sitting at rest at the origin so that $J^i = 0$ and $J^0 = \delta^{(2)}(\vec{x})$. The equation of motion then reads $\epsilon^{ij} F_{ij} = g \delta^{(2)}(\vec{x})$

Now here is an equation which practically anyone can solve: it just says that away from the origin, the gauge field vanishes. Thus, the potential is (locally) a pure gauge: $A_i = \partial_i(\text{something})$. But on the other hand, if we integrate along a closed curve encircling the origin, we have by Stokes' Theorem $\oint A_i dx^i = \int \epsilon^{ij} F_{ij} d^2x = g \int \delta^{(2)}(\vec{x}) d^2x = g \neq 0$. Thus, "something" must be topologically non-trivial and is in fact just the azimuthal angle. We conclude that $A_i = \frac{1}{2\pi} g \delta_i \phi$.

So far, there is nothing quantum mechanical about the discussion. We now introduce quantum physics. A basic quantum principle states that in the presence of a gauge potential, if we move a particle around generating the current J_μ, the particle's wave function acquires

---

$ Partially supported by the Department of Energy No. DE-AC06-81ER40048

the phase $e^{i\int d^3x J_\mu A^\mu}$. Thus, in a quantum world, if we move a second particle around the particle sitting at the origin the wave function of the system changes by a phase factor proportional to g and to $\Delta\phi$, the azimuthal angle through which we move the second particle.

To determine the statistics obeyed by these particles, we interchange two particles. The interchange may be effected by rotating a system of two particles around one of the particles through π and by translating.

It is clear from the preceding discussion that in gerneral, when we rotate a two-particle system through $\Delta\phi$, the wave function acquires a phase $e^{i\Theta\Delta\phi/\pi}$. Here Θ is a parameter of convenience, defined proprtionally in terms of g.

Nothing in the discussion fixes g and hence Θ is a totally arbitrary parameter. If Θ happens to be 0 (or an integer multiple of 2π), the particles are bosons. If Θ happens to be π (or an odd integral multiple of π), The particle are fermions. But in general, the particles obey a new kind of statistics interpolating between Bose-Einstein and Fermi-Dirac statistics.

Whether or not fractional statistics is realized in actual physical system is a question of great interest, but there is no question that fractional statistics occurs theoretically. It demonstrably exists as a field theory phenomenon. (A model can be constructed[2] with solitons obeying fractional statistics and possessing fractional spin.) The spin-statistics theorem holds.[2,3]

The statistical mechanics of a system of particles obeying fractional statistics poses a fascinating problem in mathematical physics. Thus far, we have been able only to work in the low density regime.[4] The interesting feature is that the second virial coefficient as a function of Θ has cusps when Θ is equal to an odd integral multiple of π.

It is a challenge to go beyond the low density regime. We would like to answer such questions as what happens to Bose condensation as Θ varies.

The Chern-Simon term responsible for fractional statistics may be absorbed into the quantum wave function or functional of the system. Fractional spin and statistics are then manifest in the wave functional.[5]

The construction outlined above for a point particle can be generalized[6] to extended objects. The key is to use a "potential" with m-indices $A_{\mu_1...\mu_m}$ in $(2m+1)$-dimension. The construction generates a phase interaction between extended bojects. The effect of the interaction cannot be interpreted as statistics however, since the phase obtained as we move one extended object around another depends on their relative orientations.

I like to mention a work which is not directly related to fractional statistics but in some rather vague way evolved out of the preceding considerations. The effective 10-dimensional field theory obtained from the superstring contains a gauge potential with two indices $A_{\mu\nu}$ and a corresponding field strength $H_{\mu\nu\lambda}$. We have studied the consequences of a generalized Freund-Rubin Ansatz for H in compactifying this field theory.[7]

It was discovered[8] that in field theory if the parameter Θ is set equal to zero it remains zero to one-loop order. Bernstein and my student Lee found subsequently that the parameter Θ keeps its tree value to one-loop order. Recently, a non-renormalization theorem has been proved to all orders.[10]

The study of fractional statistics on topologically interesting spaces may be relevant for actual physical systems.[11,12] Consider N particles obeying fractional statistics living on the sphere S^2. Imagine moving a particle along a closed path C which has the following property: of the two pieces of the spheres which C divides the sphere into, one does not contain any particles. In general, we will pick up a different phase according to which of the two pieces we think of as being enclosed by C. Consistency implies that Θ is quantized[11,13] in terms of the number of particles present: $\Theta = 2\pi k/(N-1)$, k an integer.

I may also mention here a speculative suggestion[14] that in (3+1)-dimensional spacetime point fermions may be constructed out of point bosons by a construction analogous to that producing fractional statistics in (2+1)-dimension. The difficulty here is that the particles here will be carrying two Dirac tails extending to infinity and it is not clear that the theory can be made to be Lorentz invariant.

II. EXCEPTIONAL PREONS

In recent years, a number of people have played at constructing models of preons or rizoms, following what we may term the "least imaginative approach" of trying to re-live QCD, à la 't Hooft. The world at the rizom level is to be described by a hypercolor gauge group and a hyperflavor $G_{HC} \times G_{HF}$. One supposes G_{HC} to break G_{HF} into G'_{HF}. We are unwilling to give up grand unification and so we insist that G'_{HF} contains the grand unifying group G_{GUT}.

In order to have three fermionic preons bind into a quark or a lepton, we may assign the preon to the fundamental representation of a hypercolor group with a Z_3 center. Over the years, we, along with just about everyone else in this game, have played on and off with the "natural" choice $G_{HC} = E(2) = SU(3)$, familiar from QCD.

But notice the following curious group theoretic fact. The center of the groups $E(1)$, $E(2)$, $E(3)$, $E(4)$, $E(5)$, $E(6)$, $E(7)$, and $E(8)$ is respectively Z_2, Z_3, Z_4, Z_5, Z_4, Z_3, Z_2, and Z_1. Incidentally, the lower E groups $E(1)$, $E(2)$, $E(3)$, $E(4)$, and $E(5)$ have aliases; they are also known as $SU(2)$, $SU(3)$, $SU(4)$, $SU(5)$, and $SO(10)$ respectively. Thus, we recognize the E chain as the chain of interest for particle theory. (The E chain is defined by starting with the Dynkin diagram of $E(8)$ and by cutting off one root at a time on the long branch.) The centers wax and wane. (The fact that there is no such thing as Z_0 provides another indication that the chain terminates with $E(8)$.)

Perhaps we should try the other group with a Z_3 center, namely $E(6)$. This choice for G_{HC} has two other advantages: $E(6)$ is automatically free and its fundamental representation is complex (so that the preons cannot be given a mass.)

We thus propose a theory[15] with $G_{HC} \times G_{HF} = E(6) \times E(6)$ with preons in (27,27). We gauge $SO(10) \subset G_{HF}$ as the grand unifying group. We now notice the following curious numerological fact: $27 = 3^3$. When the anomaly matching condition, which has the form $3^3 = (\ldots)$, is solved, we find a solution with 3 families. The solution also has the appealing feature that no exotic complex representation of $SO(10)$ appears.

It is probably in bad taste to fling out the wild speculation that the $E(6) \times E(6)$ structure has something to do with superstrings.

III. SCALAR DARK MATTER

The $SU(3) \times SU(2) \times U(1)$ theory has now been enshrined as the standard theory but we must keep in mind that very little is actually known about the scalar sector of the theory. Over the years, we have played on and off at modifying the standard scalar sector. The possibilities are practically endless. Here we mention one.

Consider a scalar field X which transforms as a singlet under $SU(3) \times SU(2) \times U(1)$. (At the crudest level, we note that Nature includes fields transforming as singlets under the familiar gauge groups $U(1)_{em}$ and $SU(3)_c$. So why shouldn't there be singlet fields under the standard gauge group?)

The interesting point is that the particle X is naturally very feebly interacting since it can interact with ordinary matter only via the Higgs boson. We find that as a result the X can account for the dark matter of the universe.[16] (The stability of the X can be guaranteed by a discrete symmetry.) The point of this exercise is not so much to propose yet another candidate for the cosmic dark matter but to stress that the dark matter may well consist of a rather "mundane" particle such as the X rather than an exotic particle based on yet-unproven physical principles.

Assuming that the X saturates the mass density of the universe we find that with reasonable assumptions the galactic mass scale can emerge.

The annihilation of the X in pairs into $b + \bar{b}$ may account for the anti-protons observed in cosmic rays.

IV. SCALAR FIELDS AND VIOLATION OF INDIVIDUAL LEPTON NUMBERS

Following the philosophy expressed in the preceding section, we proposed[17] several years ago a charged scalar field which couples to two lepton doublets in the standard theory. Indeed, if one were to write down the list of the fermion representations in the standard theory and the representations which can be obtained by multiplying two fermion representations together, one sees that a rather large number of Yukawa terms are possible. (In fact, many of these Yukawa terms are required in a grand unified theory such as the SO(10) theory).

Out of this wealth of possible Yukawa terms only two are chosen in the standard theory. It is easy to speculate on the existence of the other Yukawa terms and to study their physical consequences.

In particular, we focussed on an SU(2) singlet charged scalar field h which couples to two lepton doublets. Fermi statistics implies that the coupling necessarily violate electron, muon, and tauon numbers.[18]

We have the striking consequence that the processes $\nu_\mu e^- \to \nu'\mu^-$ can now occur resonating in the s-channel on the h particle. This may be of potential importance for future experimental detection of ultra-high energy neutrinos from astrophysical sources as evidenced by the observations of energetic muons going through underground detectors. The resonance occurs for incident neutrino with energy $\sim 2 \cdot 3 \times 10^2 Tev(m_h/15Gev)^2)$.

The h is expected to have a very small width and so the resoncance looks like a sharp spike. (The spike is broadened out by a factor ~ 1.05 due to the motion of the electrons work.)

Naturally, we have to fold the energy spectrum of the incident neutrino into the resonance cross-section. If the spectrum falls rapidly, production of muons via this process may be overwhelmed by production via the more mundane process $\nu_\mu N \to \mu^- X$.

In principle, one can measure the energy of the muon to see the effect of resonance production.

There are a number of other phenomenological consequences which can be studied.

V. NEUTRINO MAJORANA MASSES

The introduction of the h field violates the individual lepton numbers but preserves the total lepton number. However, we can easily break total lepton number by coupling the h to two Higgs doublets. (This can occur only if the theory contains more than one Higgs doublet.) As a result, the neutrinos acquire (technically calculabble) Majorana masses.[17,19]

The mechanism of generating the neutrino Majorana masses involves the theory manufacturing a "composite" triplet Higgs field out of the Higgs doublets. An alternative model may be constructed by introducing an explicit triplet Higgs field "by hand". One then has to adjust the vacuum expectation value of this triplet field to be very small.

One interesting feature of our model is that in some versions a linear combination of the individual lepton numbers such as $L_e + L_\mu - L_\tau$ remains conserved.

VI. ABELIAN GAUGE FIELD HIDDEN INSIDE NON-ABELIAN GAUGE THEORY

In studying the classical mechanics of a point particle whose position is denoted by q, we can tell that an electromagnetic field is present if the Lagrangian contains a term linear in $\dot{q}$ and we identify the electromagnetic potential as the coefficient of $\dot{q}$, viz. $\dot{q}_i A_i(q)$.

Now, suppose we study the non-abelian gauge Lagrangian in the Schrdinger formalism in the gauge $A^\circ = 0$. Suppose the non-abelian gauge Lagrangian contains a topological term (i.e., the $\theta F \tilde{F}$ term for the (3+1)-dimensional theory, the Chern-Simon term for the (2+1)-dimensional theory, and so forth.) In the gauge chosen, the topological term has the form $\dot{A}_i \mathcal{A}_i(A)$. Thus, we are invited to think of the theory formally as describing the mechanics of a "point particle" whose position is denoted by $A_i^a(\vec{x})$ and which interacts with an abelian gauge potential A_i.

The interesting question is then what kind of electromagnetic field corresponds to A_i? To answer this question, we must project the theory down to gauge orbit space, in analogy with the projection to the radial wave function in quantum mechanics. Amusingly, we find that in the (3+1)-case, A_i corresponds to a vortex or flux tube, and in the (2+1)-case, to a Dirac magnetic monopole.[20]

The outlook and technique of this work evolved out of our previous work on cocycles and monopoles, on the topological structure of anomalies, and on the Chern-Simon term and fractional statistics. These topics are all related in some way.

REFERENCES

1) F. Wilczek, Phys. Rev. Lett. 49 (1982) 957.

2) F. Wilczek and A. Zee, Phys. Rev. Lett. 51 (1983) 2250.

3) D. Ravenel and A. Zee, Comm. Math. Phys. 98 (1985) 239.

4) D. Arovas, R. Schrieffer, F. Wilczek, and A. Zee, Nucl. Phys. B251 (1985) 117.

5) Y. S. Wu and A. Zee, Phys. Letters 147B (1984) 325.

6) R. Nepomechie and A. Zee, UW preprint, to appear in Festschrift for E. Fradkin.

7) R. Nepomechie, Y. S. Wu, and A. Zee, Phys. Lett. B, in press.

8) T. Kao and M. Suzuki, Berkeley preprint.

9) M. Bernstein and T. J. Lee, UW preprint.

10) S. Colemand and B. Hill, Harvard preprint, T. J. Lee, unpublished UW preprint.

11) D. Thouless and Y. S. Wu, Phys. Rev. B 31 (1985) 1191.

12) R. Chiao et al., Phys. Rev. Letters (1985)

13) Y. S. Wu and A. Zee, unpublished.

14) S. Coleman and A. Zee, unpublished.

15) V Silveira and A. Zee, Phys. Lett. 157B (1985) 191.

16) V Silveira and A. Zee, Phys. Lett. B to appear.

17) A. Zee, Phys. Letters 93B (1980) 389.

18) A. Zee, Phys. Letters B, in press.

19) A. Zee, Nucl. Phys. in press.

20) Y. S. Wu and A. Zee, Nucl. Phys. in press.

NON-ABELIAN INTERACTIONS IN RUBAKOV-CALLAN EFFECT

K. S. NARAIN

Rutherford Appleton Laboratory, Chilton, Didcot, Oxon OX11 0QX, UK

A systematic method to include non-abelian colour forces in Rubakov-Callan effect, in the limit of massless fermions, is presented, which is applicable to even quite complicated grand unified theories admitting monopole solutions. The method relies on a new bosonisation procedure for 2-dimensional non-abelian gauge theories and makes use of the flavour and Pauli-Gursey current algebra. The representation space of this algebra is shown to consist of a number of free massless scalar fields together with their solitons.

1. INTRODUCTION

Grand unified theories usually contain gauge fields that can mediate baryon number violating processes such as proton decay. Since these gauge fields are very heavy (of the order of GUT scale) they can arise rarely by quantum fluctuations, resulting in a very long life time for protons. However, in the early universe, they must have been abundant, and may have formed topologically stable magnetic monopoles which would survive as the universe cooled down. When such a monopole would go through the matter, they are expected to catalyse nucleon decay. Rubakov[1] and Callan[2] independently showed that the resulting decay is of the order of strong interaction rates.

The classical field of a GUT monopole belongs to a subgroup $SU(2)_M$ of the GUT group G. Outside the core non-abelian terms decay exponentially, and only a subgroup $U(1)_M$ of $SU(2)_M$ remains. The gauge fields of $U(1)_M$ have long range*, therefore they belong to the unbroken subgroup $(U(1)_e \times SU(3)_C)/Z_3$ of G. The image $U(1)_{3C}$ of $U(1)_M$ in $SU(3)_C$ defines the 3-direction in colour space. Thus the long range forces that are to be taken into account are

---

*By long range we shall always mean long compared to the monopole core size (and the weak interaction size) but smaller than 1 fm.

$$H = U(1)_M \times U(1)_Y \times SU(2)_C$$

where $U(1)_Y$ is the combination of $U(1)_e$ and $U(1)_{3C}$ which is orthogonal of $U(1)_M$. The fact that near the monopole $SU(3)_C$ colour is broken down to $U(1)_{3C} \times SU(2)_C$ is due to the colour orientation of the monopole, and precisely how the full $SU(3)_C$ is recovered at large distances is still not clear.

To see which catalysis reactions are possible, we have to group the fermions into representations of $SU(2)_M$ as the transitions will happen within these representations. Thus, for example, for standard $SU(5)$ theory we have the following $SU(2)_M$ doublets

$$\begin{bmatrix} e^+ \\ d_3 \end{bmatrix}_L \quad \begin{bmatrix} d_3^C \\ e^- \end{bmatrix}_L \quad \begin{bmatrix} u_1 \\ -u_2^C \end{bmatrix}_L \quad \begin{bmatrix} u_2 \\ u_1^C \end{bmatrix}_L$$

where the first two doublets are $SU(2)_C$ singlets and the last two form an $SU(2)_C$ doublet.

The original treatments[1,2] of this problem included only the abelian part of H ie. $U(1)_M$ and $U(1)_Y$. It was shown that there were basically two mechanisms for the catalysis reactions:

a) direct interactions with the heavy gauge bosons condensed inside the core which may cause transition between the components of each doublet. A single transition would leave electric charge at the core but another opposite transition in another doublet would lead to processes such as $u_{2L} d_{3R} \to u_{1L}^C e_R^+$

b) due to the anomaly which is proportional to $\int E \cdot B$. In the presence of a magnetic field, electric field of incoming fermions will disturb the fermi sea, so that incoming left handed fermions disappear into the sea while right handed ones come out, leading to processes such as $u_{1L} u_{2L} \to d_{3R}^C e_R^+$

These studies were made under drastic simplifications. Only s-wave fermions and the gauge fields were analysed which reduced the problem to a 2-dimensional gauge theory on a half plane. This is justified because only s-wave fermions can have appreciable overlap with the vicinity of the core where these effects are localised. In ref(3), non-abelian part of H, namely $SU(2)_C$, was also included. By summing up planar diagrams, it was shown that in the limit of massless fermions it led to new selection rules. Only $SU(2)_C$ singlet states could propagate near the core. The effect of this is to suppress processes of the type (a) relatively, whereas of type (b) proceed unhampered[4]. In ref (5) the same results were obtained by using current algebra techniques. In ref (6), it was shown that zero mass sector of the theory (in the s-wave system with massless fermions) is exactly solvable and all the relevant correlation functions could be exactly computed by means of solitonic representation of the current algebras. Here, I shall present briefly a generalisation of ref (6), that could be applied to arbitrary GUT group and representation contents. The details will appear elsewhere[7].

2. CURRENT ALGEBRA OF 2-DIMENSIONAL GAUGE THEORIES

As mentioned in the introduction, in the s-wave system the problem reduces to that of a 2-dimensional gauge theory, with $SU(2)_C$ non-abelian group. For different GUT groups one would get different number of $SU(2)_C$ doublets (For $SU(5)$ we have only one $SU(2)_C$ doublet whereas for $SO(10)$ monopole with charge 2 there are two $SU(2)_C$ doublets[8]). Therefore we consider a 2-dimensional gauge theory with non-abelian colour group $SU(N_C)$ and N_f massless flvours in the fundamental representation.

$$\ell_{fermion} = \sum_{i=1}^{N_f} \bar{\psi}_i \not{D} \psi_i$$

In the light cone coordinates $x^{\pm} = x^o \pm x^1$

$$\ell_{fermion} = \sum_i (\psi^*_{-i} D_+ \psi_{-i} + \psi^*_{+i} D_- \psi_{+i}) \quad (2.1)$$

where $\psi_{\pm} = \frac{1}{2}(1 \pm \gamma_5)\psi$

The global flavour symmetry is clearly $U(N_f)_L \times U(N_f)_R$ with conserved currents

$$J^a_{\pm} = \psi^{*\alpha}_{\pm i} T^a_{ij} \psi^{\alpha}_{\pm j} \quad (2.2)$$

where T^a are the generators of $U(N_f)$ and α is the colour index. They satisfy the conservation law

$$\partial_{\mp} J^a_{\pm} = 0 \quad (2.3)$$

For the special case $N_C = 2$, the global symmetry group is in fact larger. In addition to the colour singlet bilinears J^a of the mesonic type introduced above, there are also conserved baryon type bilinear currents

$$J_{\pm ij} = \psi^{\alpha}_{\pm i} \varepsilon_{\alpha\beta} \psi^{\beta}_{\pm j}$$

$$\bar{J}_{\pm ij} = \psi^{*\alpha}_{\pm i} \varepsilon_{\alpha\beta} \psi^{*\beta}_{\pm j} \quad (2.4)$$

The corresponding charges generate Pauli-Gursey type of transformations

$$\psi_i^\alpha \to a\psi^\alpha + b\psi^{*\beta}\varepsilon_{\alpha\beta}$$

The total number of conserved currents in left and right sector each is $N_f(2N_f + 1)$ and they form the Lie algebra of $Sp(N_f)_L \times Sp(N_f)_R$

At the quantum level, the equal time commutation relations between these currents is modified by a Schwinger term (central term).

$$\left[J_\pm^a(x)\ J_\pm^b(\eta)\right]_{x^0=y^0} = f^{abc}J_\pm^c(x)\,\delta(x-y) \pm \frac{ik}{4\pi}\delta^{ab}\delta'(x-y) \qquad (2.5)$$

where f^{abc} are the structure constants of the Lie algebra in question and the central charge k can be calculated from the 1-loop diagram connecting $J^a(x)$ and $J^b(y)$. It turns out that the flavour part $SU(N_f)$ (with standard structure constants f^{abc}) has a central charge $k = N_c$. This is not hard to see, as there would be N_c fermions running around the loop giving rise to a factor of N_c. In fact in the mathematical literature this algebra is known as Kac-Moody algebra and it describes a projective unitary representation of the symmetry group. It is also known that such representation exist if and only if k is an integer.

When the symmetry group is a product of simply laced groups (ie the groups whose dynkin diagrams have only single links - SU(N), SO(2N), E_6, E_7, E_8) an explicit representation of this algebra for the minimal value of k (in our normalisation k = 1) can be given in terms of solitons in a free bosonic theory[9]. In the following I shall generalise it for the Sp(N) groups that are relevant for SU(2) gauge theories (and hence for Rubakov-Callan effect). As a by product we will also get soliton representations of baryons for $SU(N_c)$ ($N_c > 2$) gauge theories (Note that for $N_c > 2$, baryons are not bilinears in fermions). This means that the entire massless sector of the gauge theory, which consists of mesons and baryons, can be reduced to a free bosonic theory.

3. SOLITON REPRESENTATIONS

Here I shall briefly sketch the construction, first for the known case of simply laced groups and then the generalisations.

a) Simply laced group G with minimal central charge

Define a free bosonic field theory of the maximal abelian subgroup H of G. H is given by a r-torus where r is the rank of G. We now impose periodic boundary conditions.

$$g(+\infty) = g(-\infty), \qquad g \in H \qquad (3.1)$$

Writing $g = e^{i\phi^a T^a \sqrt{2\pi}}$ with $\text{tr}(T^a T^b) = 2\delta^{ab}$

The above boundary condition implies

$$[\phi^a(+\infty) - \phi^a(-\infty)] = \sqrt{2\pi}\, \tfrac{1}{2} \text{tr}(T^a T_\alpha) \qquad (3.2)$$

where $T_\alpha \in$ Lie algebra of H with integer entries. If $\vec{h}_\alpha$ are the root vectors of G normalised so that $\vec{h}_\alpha^2 = 2$, then it turns out that the operators

$$J_\alpha(x) = N_\alpha : \exp i\sqrt{\pi}\vec{h}_\alpha \cdot \int_{-\infty}^{x} \partial_+ \vec{\phi} : \qquad (3.3)$$

creates a soliton with b.c. (3.2). Here : : means normal ordering and N_α are some constant quantities that can be represented in terms of certain Clifford algebras but are not relevant for the present discussion. The operators J_α's together with the diagonal generators $\partial_+ \phi_a$; $a = 1,\ldots,r$ define the currents of G as can be checked by evaluating the various commutators. It is crucial that $h_\alpha^2 = 2$ for all α, in order for this construction to work, and hence the restriction to simply laced groups.

b) $Sp(N_f)$ with minimal central charge.

This is the group which appears in $SU(2)_C$ gauge theory. Let us first look at the flavour part $U(N_f)$ of $Sp(N_f)$. $U(N_f)$ further decomposes into

$$U(N_f) = (U(1) \times SU(N_f))/Z_{N_f} \qquad (3.4)$$

where $U(1)$ is the baryon number operator and $SU(N_f)$ represents the flavour rotation. As mentioned following eq (2.5), the central charge for this $SU(N_f)$ subgroup is non-minimal ($k = 2$ in our normalisation). We now take two copies of $SU(N_f)$ each with $k = 1$ and obtain two copies of solitonic representations of the currents $J_\alpha^{(1)}$ and $J_\alpha^{(2)}$. The currents of $SU(N_f)$ ($k = 2$) are then given by

$$J_\alpha = J_\alpha^{(1)} + J_\alpha^{(2)} \qquad (3.5)$$

The next step in the construction is to relax the boundary condition (3.1). The new b.c. are $g(+\infty) = g(-\infty)g_0$, where $g_0 \in$ center of the group namely Z_N. The boundary conditions on ϕ's now become

$$(\phi^a(+\infty) - \phi^a(-\infty) = \sqrt{2\pi}\, \tfrac{1}{2}\, \text{tr}(T^a T_i) \qquad (3.6)$$

where

$$T_i = \pm \frac{1}{N_f} \begin{bmatrix} 1 & & & \\ & 1 & & \\ & & \cdot(1-N_f) & \\ & & & 1 \\ & & & & 1 \end{bmatrix} \ldots\ldots\ldots i^{th} \text{ row} \qquad (3.7)$$

The corresponding soliton operator $q_i(x)$ and $\bar{q}_i(x)$ (with + and - signs respectively in eq (3.7)) carry

$$\text{Spin} = \frac{1}{2}\left(1 - \frac{1}{N_f}\right) \qquad (3.8)$$

and are in the N_f and $\bar{N}_f$ representations of $SU(N_f)$. Since we have two copies of $SU(N_f)$, we have two copies of these fractional soliton operators $q^{(1)}$, $q^{(2)}$ and $\bar{q}^{(1)}$, $\bar{q}^{(2)}$. The operators $q^{(1)}_{(i}q^{(2)}_{j)}$ have the same $SU(N_f)$ flavour transformation properties as J_{ij} in eq (2.4) but they differ from J_{ij} in two respects. Firstly they do not carry baryon number and secondly they carry a fractional spin $= (1 - \frac{1}{N_f})$. Now, the baryon number current is the diagonal generator of $U(1)$ subgroup of $U(N_f)$ flavour group. Let $B(x)$ and $\bar{B}(x)$ be the soliton and anti soliton operators in this $U(1)$ subgroup that generate baryon numbers + 1 and - 1 respectively. Then it turns out that $B(x)$ carries spin $\frac{1}{N_f}$. Thus, we can define

$$J_{ij}(x) = B(x) q^{(1)}_{(i}(x) q^{(2)}_{j)}(x)$$

$$\bar{J}_{ij}(x) = \bar{B}(x) \bar{q}^{(1)}_{(i}(x) \bar{q}^{(2)}_{j)}(x) \qquad (3.9)$$

These operators carry spin 1 and have the correct transformation properties under $U(N_f)$ flavour group. A detailed calculation then shows that J_α, J_{ij} and $\bar{J}_{ij}$ generate the $Sp(N_f)$ Kac-Moody algebra.

c) Baryons in $N_c > 2$ theories.

From the construction given above for $Sp(N_f)$, generalisation to this case is straightforward. Now the $SU(N_f)$ flavour algebra carries a central charge N_c. So we take N_c copies of $SU(N_f)$ (k = 1), each represented by soliton operators J^A_α (A = 1, ... N_c). The currents

$$J_\alpha = \sum_{A=1}^{N_c} J^A_\alpha \qquad (3.10)$$

satisfy $SU(N_f)$ algebra with $k = N_c$. As above we define fractional soliton operators q^A_i, $\bar{q}^A_i$ each transforming as N_f and $\bar{N}_f$ representations of $SU(N_f)$ and carrying spin $\frac{1}{2}(1 - \frac{1}{N_f})$. Let $B(x)$ and $\bar{B}(x)$ be the soliton operators generating baryon numbers + 1 and - 1 respectively, then one can show that they carry a spin $\frac{1}{2} N_c/N_f$. (The factor N_c appears by comparing the Schwinger terms for the baryon number currents in the fermonic and the bosonic theory)

The baryon operators are then given by

$$B_{i_1, \ldots, i_{N_c}} = B(x) \, q^1_{(i_1} q^2_{i_2} \cdots q^{N_c}_{i_{N_c})}$$ (3.11)

They carry

$$\text{Spin} = \frac{1}{2} \frac{N_c}{N_f} + N_c \frac{1}{2} \left(1 - \frac{1}{N_f}\right) = \frac{N_c}{2}$$ (3.12)

as expected, and have the correct transformation properties under $U(N_f)$ flavour group. The details of these calculations will appear in ref (7).

4. CONCLUSION

We have shown that the massless sector of 2-dimensional non-abelian gauge theories with massless fermions in the fundamental representation can be exactly solved and all the correlation functions can be calculated in terms of a free bosonic theory. The entire physical Hilbert space is a product $H_o \otimes H_m$ where H_o consists of the massless sector involving mesons and baryons. $U(N_f)$ currents ($Sp(N_f)$ currents for $N_c = 2$) couple only to H_o, therefore H_m does not carry any flavour or baryon charges. The dynamics on H_o is completely independent from that in H_m and can be determined by using the bosonisation given above, which is sufficient for the purposes of low energy phenomenology in the presence of GUT monopoles. Solitonic representations of Kac-Moody algebras have played an important role in the construction of Heterotic strings[10]. It will be interesting to see if the generalisation presented here can have any application in String Theories.

REFERENCES

1. V.A. Rubakov, ZRETF, Fisma 33, (1981) 658; Nucl. Phys. B20, (1982) 311.

2. C.G. Callan, Phys. Rev. D25, (1982) 2142; D26, (1982) 2058.

3. N.S. Craigie, W. Nahm and V.A. Rubakov, Nucl. Phys. B241, (1984) 274.

4. N.S. Craigie, W. Nahm and K.S. Narain, Phys. Lett. 148B, (1984) 81.

5. N.S. Craigie and W. Nahm, Phys. Lett. 142, (1984) 64; 147B, (1984) 127.

6. N.S. Craigie, W. Nahm and K.S. Narain, Phys. Lett. 152B, (1985) 203.

7. N.S. Craigie, W. Nahm and K.S. Narain, (in preparation).

8. S. Dawson and A.N. Schellekens, Phys. Rev. D27, (1983) 2119.

9. I.B. Frenkel and V.G. Kac, Invent. Math. 62, (1980) 23.
   G. Segal, Comm. Math. Phys. 80, (1981) 301.
   P. Goddard and D. Olive, "Algebras, Lattices and Strings", DAMTP 83/22.

10. D. Gross, J. Harvey, E. Martinec and R. Rohm; Nucl. Phys. B256 (1985) 253.

JETS, $W^{\pm}$ AND Z^0 PRODUCTION IN UA2

The UA2 Collaboration

Giorgio GOGGI
University and INFN, Pavia

We present new results on the production of hadron jets and of Intermediate Vector Bosons at the CERN $\bar{p}p$ collider at $\sqrt{s}$ = 630 GeV. Comparisons are made with data previously collected at $\sqrt{s}$ = 546 GeV, and with theoretical predictions from QCD and the standard model of the electroweak interaction.

1. INTRODUCTION

The UA2 collaboration has already reported experimental results on the production of jets [1] and of the Intermediate Vector Bosons $W^{\pm}$ and Z^0 [2]. The data were collected at the CERN SPS $\bar{p}p$ Collider in the period 1981-1983 at $\sqrt{s}$ = 546 GeV, for a total integrated luminosity $\mathscr{L}$ = 142 nb^{-1}. In the subsequent run during Autumn 1984 the Collider operated at an increased energy, $\sqrt{s}$ = 630 GeV, and the total luminosity accumulated by the UA2 experiment was $\mathscr{L}$ = 310 nb^{-1}. The analyses presented here refer to both data samples, which are combined only for the determination of the intrinsic W and Z^0 parameters. The UA2 experimental apparatus [3] is essentially a higly segmented, tower structured calorimeter with complete cylindrical symmetry in azimuth and full coverage in the polar range ($20° < \theta < 160°$). In addition it provides for charged particle momentum measurement in the regions ($20° < \theta < 40°$; $140° < \theta < 160°$), where only electromagnetic calorimetry is present. Fig. 1 shows a schematic view of the longitudinal cross section of the detector in a plane containing the beam axis. In this apparatus electrons are measured with an energy resolution $\Delta E \approx 0.15 \sqrt{E}$ (E in GeV), with a systematic uncertainty of $\approx \pm 1.5\%$; the energy resolution for jets is $\Delta E^j/E^j \approx 0.32 (E^j)^{-.25}$ (E^j in GeV).

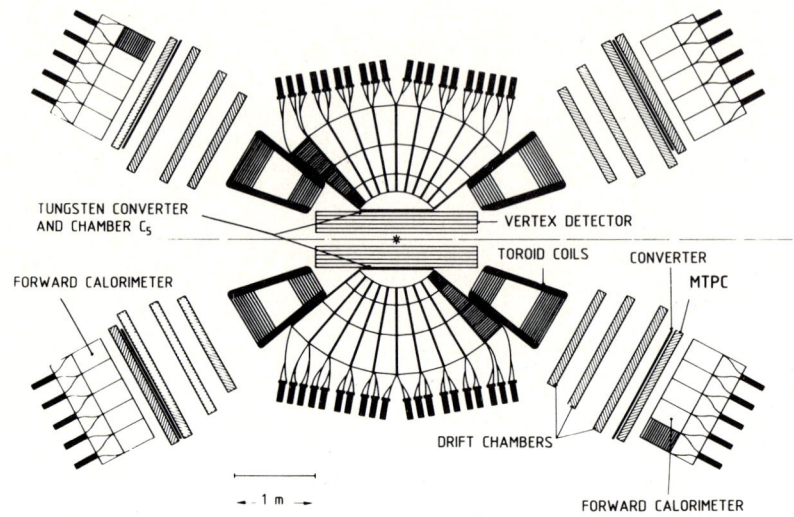

FIGURE 1
Longitudinal cross section of the UA2 detector in a plane containing the beam axis.

2. INCLUSIVE JET PRODUCTION

After rejecting a small background contamination, a clustering algorithm applied to the transverse-energy pattern of the calorimeter cells yields the transverse energy of hadronic jets. Jet directions are determined from the position of the weighted cluster centroids with respect to the center of the interaction region. The inclusive differential cross sections for jet production at $\eta = 0$ are shown in Fig. 2a as a function of jet transverse momentum p_T for both collider energies. An overall systematic uncertainty of ± 45% applies to the data.

The results are well described by a QCD calculation, shown by the curves in Fig. 2, which assumes $Q^2 = p_T^2$, $\Lambda = 200$ MeV and the structure functions by Eichten et al [4]. The ratio of the two cross sections, displayed in Fig. 2b, is a rising function of transverse momentum and is also well described by the calculation. Similar results are obtained for the invariant-mass distribution of inclusive jet-pair production, with both jets within $|\eta| < 0.85$.

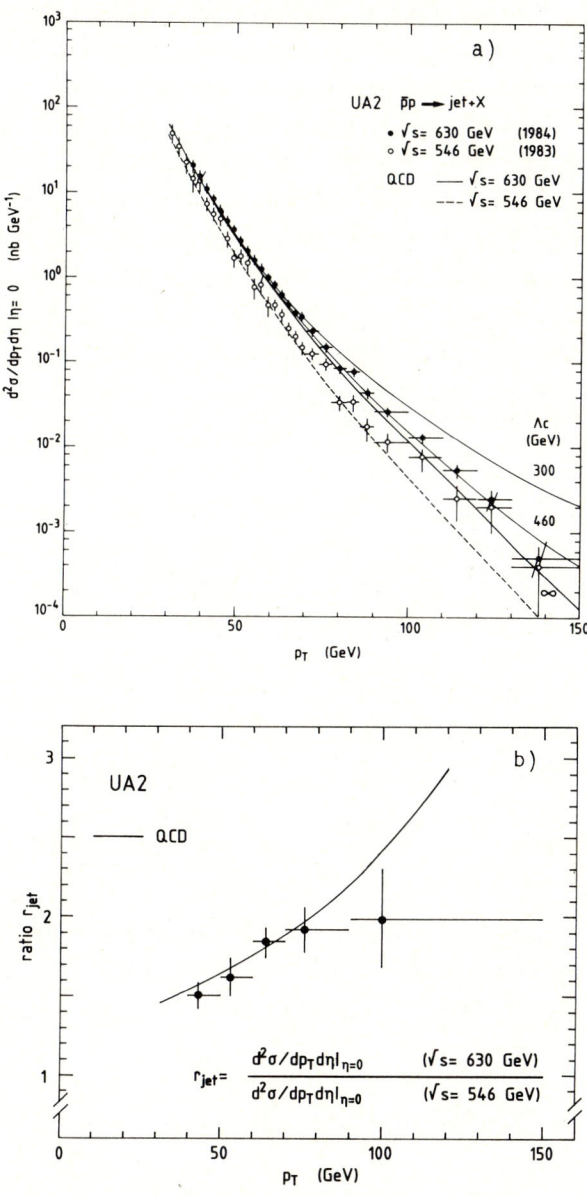

FIGURE 2

a) Inclusive jet production cross sections at $\sqrt{s}$ = 546 GeV and 630 GeV. The curves are a QCD calculation (see text).
b) Comparison of the ratio of inclusive jet cross sections with a QCD calculation (see text).

Possible deviations of these data from the QCD predictions at high values of p_T can be parametrized in terms of a new energy scale Λ_c, representing through a contact term the effective energy scale of a new interaction at the preon level [5]. The data require values of $\Lambda_c > 370$ GeV at the 95% confidence level (Fig. 2a).

In the 1984 data at $\sqrt{s} = 630$ GeV, searches were also performed for monojets and monophoton events, as well as for multijet events with large missing transverse momentum $\not{p}_T$. No significant signal could be isolated from background for either event type. Upper limits on the cross sections for the production of such types of events are as follows :

a) monojets :

$\sigma(\not{p}_T > 65$ GeV$) < 23$ pb (90% c.l.) or :

$\sigma(\not{p}_T > 50$ GeV$) < 73$ pb (90% c.l.),

b) monophotons :

$\sigma(p_T^\gamma > 45$ GeV, $\not{p}_T > 30$ GeV$) < 14$ pb (90% c.l.),

c) multijets + large p_T :

$\sigma(\not{p}_T > 35$ GeV$) < 27$ pb (90% c.l.).

The experimental method was checked by measuring the process ($\bar{p}p \to W+X$, $W \to e\nu$) through a study of events with an electromagnetic cluster and large $\not{p}_T$, and comparing them to the ones found in the electron analysis. Agreement within errors has been found for the two event samples as well as for the two measurements of the cross section.

3. $W^\pm$ PRODUCTION AND DECAY

Fig. 3a shows the distribution of the transverse momentum p_T^e for all electrons above 11 GeV/c identified in the standard UA2 analysis [2]. A significant population with $p_T^e \geq 25$ GeV/c is evident, as expected from $W^\pm$ decay. Other contributions include Z^0 decays, electrons from semileptonic decay of heavy quarks, electrons from the τ decay mode of $W^\pm$, and a contribution of jets misidentified as electrons.

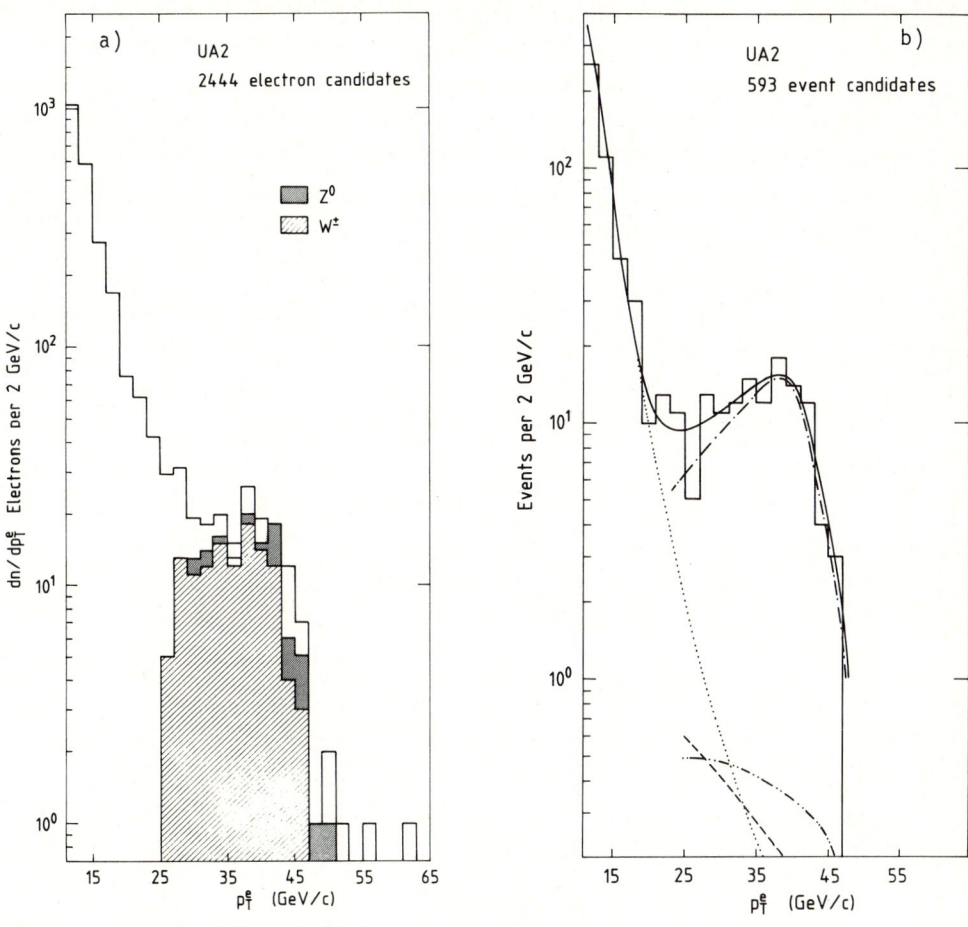

FIGURE 3
a) The electron p_T spectrum of all UA2 data.
b) The electron p_T spectrum of selected W events.

To extract from the data the $W^{\pm}$ signal we apply a topological cut using the quantity

$$\rho_{opp} = - \bar{p}_T^{\,e} \cdot \Sigma \bar{p}_T^{\,jet} / |\bar{p}_T^{\,e}|^2$$

where the sum extends to all energy clusters having an azimuthal separation $\Delta \phi > 120°$ to the electron candidate and $p_T > 3$ GeV/c. The sample is then split into two categories :

a) Events with $p_{opp} < 0.2$: this sample contains events which are unbalanced in p_T, and therefore include most $W \to e\nu$ candidates.

b) Events with $p_{opp} > 0.2$: this sample contains mostly balanced events, among which $Z^0 \to e^+e^-$. It is dominated by two-jet background whit one jet misidentified as an electron and it is used to estimate the background to the W sample.

The p_T^e distribution of the 592 events satisfying $p_{opp} < 0.2$ is shown in Fig. 3b. The background from QCD processes is superimposed (.....), together with expected contributions from decays $W \to \tau\nu$ (......) and $Z^0 \to e^+e^-$ with one electron escaping the detector acceptance (-----). The expected contribution of $W \to e\nu$ decays (-.-.-) is shown, together with the expected total from all processes. For $p_T^e > 25$ GeV/c the histogram contains 119 events, with a background of 5.8 ± 1.7 events.

A value of the W mass can be obtained from the electron candidates with $p_{opp} < 0.2$ using two different methods :

1. The p_T^e distribution of Fig. 3b is compared with that expected from $W \to e\nu$ decay. A Monte Carlo program is used to generate the electron differential distribution p_T^e for different values of the W mass, taking into account the detector response and using structure functions from Gluck et al. [6], the W transverse momentum distribution from Altarelli et al. [7] and a fixed W width, $\Gamma_W = 2.7$ GeV/c². A maximum likelihood fit to the data gives a value $M_W = 80.6 \pm 1.1$ GeV/c², where the quoted error is only statistical; an additional systematic error of ≈ 1 GeV/c² results from the uncertainty on the assumptions used to generate the Monte Carlo distributions.

2. The transverse mass distribution of the event sample (Fig. 4) is compared with distributions generated by Monte Carlo for various values of M_W. We find that the transverse mass distribution depends only weakly on p_T^W and suffers no distortions from the p_{opp} cut. The best fit value of M_W obtained by this method is $M_W = 81.1 \pm 1.0$ GeV/c², with an additional systematic error of ± .5 GeV/c².

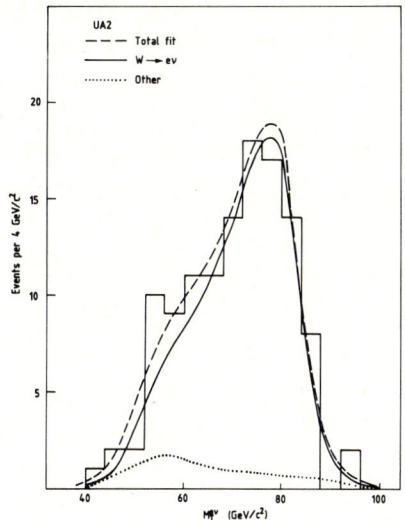

FIGURE 4

Transverse mass distribution for events with $p_T^e > 17$ GeV/c, $p_T^e > 25$ GeV/c.

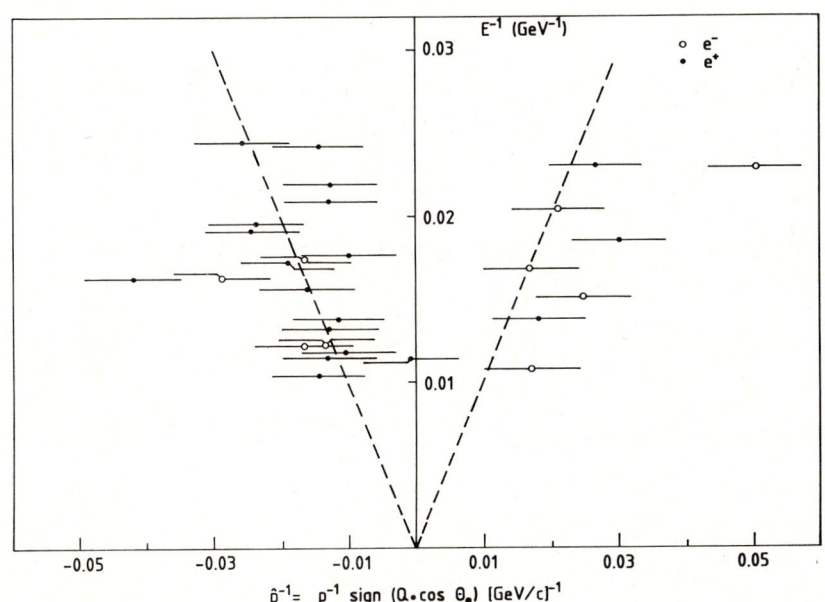

FIGURE 5

The W decay asymmetry for the electrons with $p_T^e > 20$ GeV/c detected in the forward regions of the UA2 apparatus. θ_e is the electron angle with respect to the proton direction.

Although the two methods give consistent results, we prefer to quote the mass value obtained by the latter method because of the smaller systematic error, which we add in quadrature to the statistical error. To summarize we quote

$$M_W = 81.2 \pm 1.1 \text{ (stat.)} \pm 1.3 \text{ (syst.) GeV/c}^2$$

where the systematic error is due to a ± 1.6% uncertainty on the calorimeter calibrations. As for the width of the W, we quote an upper limit $\Gamma_W < 7$ GeV/c² at the 90% confidence level.

As a consequence of the V-A coupling, the W is produced with almost full polarization along the direction of the incident p̄ beam, and a distinctive charge asymmetry can be observed in the decay W → eν.

In the UA2 detector a determination of the charge sign is only possible in the forward magnetic spectrometers. We consider all the electron candidates with $p_T^e > 20$ GeV/c and $\rho_{opp} < .2$ detected in the forward regions. Such a sample contains 28 events with an estimated background of 2 events. A comparison between the electron momentum p and the energy E, as measured in the calorimeter, is made in Fig. 5, which shows the position of these events in the plane (p^{-1}, E^{-1}), where p is the momentum with the sign of (q $\cos\theta_e$) and θ_e is the laboratory angle of the electron respect to the proton direction. The horizontal error bars represent the uncertainty on the measurement of p^{-1}, which is 0.007 (GeV/c)$^{-1}$. The measured asymmetry A = 0.43 ± .17 is in good agreement with the expected asymmetry for V-A coupling, A = 0.53 ± 0.06, as obtained by a Monte Carlo calculation which takes into account the background contamination and also an expected number of 0.6 events from Z° → e⁺e⁻ decay, with one of the two electrons undetected.

The main effects of non-leading corrections to W production are to increase the production cross section by a factor ~ 30% and to give the W a sizeable average transverse momentum p_T^W [7,8]. For relatively large values of p_T^W, the W bosons are expected to recoil against hadronic jets. To study these effects we cannot use the sample which satisfies the requirement $\rho_{opp} < .2$, because this cut rejects electrons in the presence of jets at opposite azimuthal angles. We use instead a sample of events containing an electron candidate with

$p_T^e > 15$ GeV/c and having at the same time a missing transverse momentum in excess of 25 GeV/c. This sample contains 133 events with a background estimate of 10.8 events. The distribution of W transverse momentum is shown in Fig. 6 together with a QCD predicion by Altarelli et al., [7] in good agreement with the data. The average value of p_T^W is $8.3^{+2.8}_{-3.5}$ GeV/c.

The cross section for $W^\pm$ production followed by the decay $W \to e\nu$ is measured to be :

$$\sigma_W^e = .50 \pm .09 \text{ (stat.)} \pm .05 \text{ (syst.) nb} \qquad \sqrt{s}=546 \text{ GeV}$$
$$\sigma_W^e = .53 \pm .06 \text{ (stat.)} \pm .05 \text{ (syst.) nb} \qquad \sqrt{s}=630 \text{ GeV}$$

The corresponding theoretical predictions [7] are $\sigma_W^e = 360^{+110}_{-50}$ and $\sigma_W^e = 460^{+140}_{-80}$ pb, where the errors reflect theoretical uncertainties. The increase of the W production cross section between the two $\sqrt{s}$ values is measured to be

$$r = \sigma_W^e(\sqrt{s}=630) / \sigma_W^e(\sqrt{s}=546) = 1.06 \pm 0.23$$

in good agreement with the theoretical prediction $r = 1.28$.

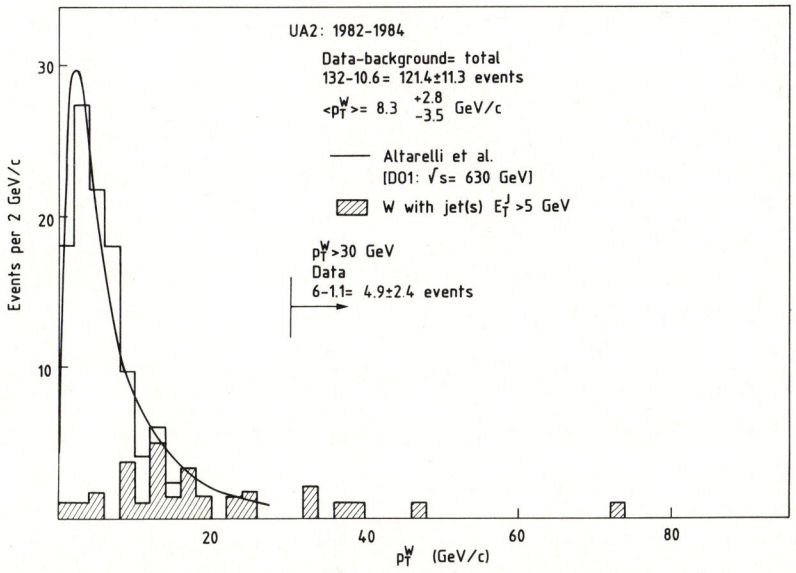

FIGURE 6

The W transverse momentum distribution.

4. Z⁰ PRODUCTION AND DECAY

A first selection of events on the 1984 data sample, requiring only two electromagnetic clusters in the calorimeter with invariant mass M_{ee} above 20 GeV/c², leaves a total of 1154 events. Their mass distribution is shown in Fig. 7a. A clear accumulation is already visible in the Z⁰ region. The additional requirement that at least one cluster satisfies all electron identification criteria [2] selects 54 events, whose mass distribution is shown in Fig. 7b. There is a clear peak in this distribution, consisting of 8 events with a mass value in excess of 75 GeV/c² with a background contamination of 0.21 ± 0.02 events. Fig. 8 shows the mass values and the measurement errors for the total Z⁰ sample of UA2; we note that the mass values of the 1982-1983 events were slightly readjusted following a recalibration of the calorimeter response. A fit to the Z⁰ mass yields :

$$M_Z = 92.5 \pm 1.3 \text{ (stat.)} \pm 1.5 \text{ (syst.) GeV/c}^2$$

where the systematic error reflects the ± 1.6% uncertainty on the absolute energy scale of the calorimeter.

The cross section for Z⁰ production followed by Z⁰ → e⁺e⁻ is measured to be :

$$\sigma_Z^e = 110 \pm 39 \text{ (stat.)} \pm 9 \text{ (syst.) pb} \quad \sqrt{s}=540 \text{ GeV}$$
$$\sigma_Z^e = 52 \pm 19 \text{ (stat.)} \pm 4 \text{ (syst.) pb} \quad \sqrt{s}=630 \text{ GeV}$$

Both values are consistent with theoretical predictions [7], which give $\sigma_Z^e = 42^{+13}_{-6}$ pb and 51^{+16}_{-10} pb at $\sqrt{s}$ = 540 and 630 Gev, respectively.

Fig. 9 shows the distribution of the Z⁰ transverse momentum, p_T^Z, for all 16 events, together with the result of a QCD calculation [7]. The average value of p_T^Z is 5.4 ± 1.0 GeV/c.

The Z⁰ width, Γ_Z, yields direct information on new decay channels, such as the number of new light neutrinos Δn_ν. Γ_Z can be either directly measured (method 1), or can be evaluated from the cross section ratio (method 2) [9] :

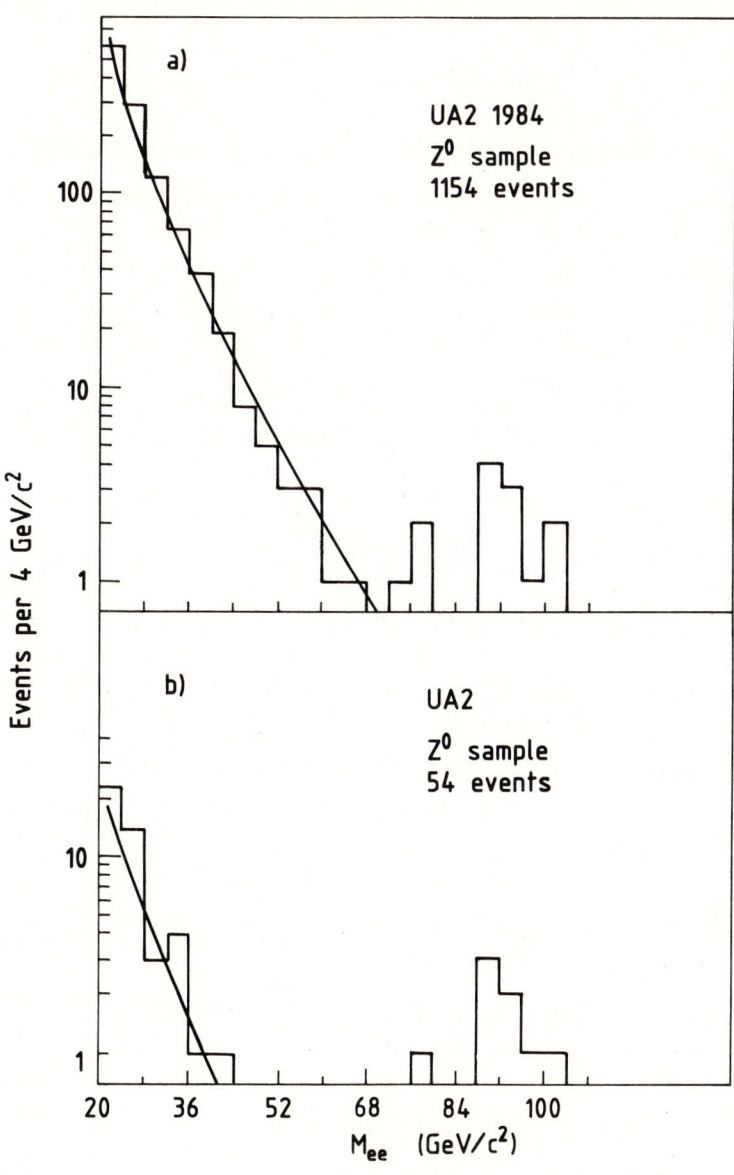

FIGURE 7

Electron pair mass spectrum from the 1984 data, a) after calorimeter cuts on both electron candidates, b) after requiring that at least one of the two candidates be a certified electron.

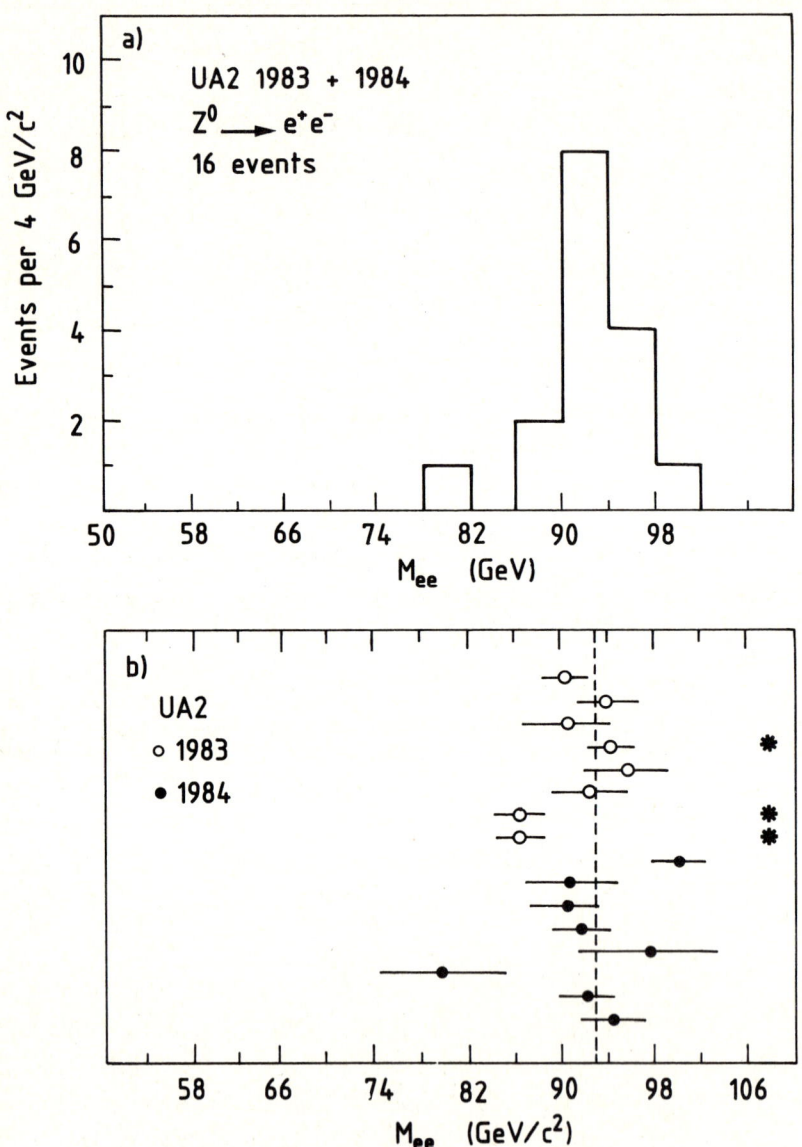

FIGURE 8
a) The Z° mass peak of the entire UA2 sample.
b) Mass values and errors of the individual Z° candidates.

$$R = \sigma_Z \, \Gamma(Z \to e e^-) \Gamma_W / \sigma_W \, \Gamma_Z \, \Gamma(W \to e\nu)$$

in which R is determined experimentally and the other factors are evaluated using QCD and the GWS model. The main advantage of this second method is that most detector-dependent systematic uncertainties cancel. The quoted UA2 results are :

$$\Gamma_Z \text{ (method 1)} < 3.3 \quad \text{GeV/c}^2 \text{ (90\% CL)}$$
$$\Gamma_Z \text{ (method 2)} = 2.19^{+0.7}_{-0.5} \text{ GeV/c}^2$$
$$\Delta n_\nu \text{ (method 1)} \le 2.6 \pm 1.7 \quad \text{(90\% CL)}$$

5. TESTS OF THE GWS MODEL

Using UA2 data alone, and assuming $\rho=1$, $\sin^2\theta_W$ can be defined independently of the energy scale

$$\sin^2\theta_W = 1 - M^2_W / M^2_Z = .229 \pm .005 \pm .008$$

Alternatively, $\sin^2\theta_W$ can be evaluated from the GWS model, after applying radiative corrections :

$$\sin^2\theta_W = (\pi\alpha/\sqrt{2}\, G_F)^2 \, M_W^{-2} \, (1-\Delta r)^{-1}$$

With the radiative correction estimate by Marciano and Sirlin [10], $\Delta r = 0.0696 \pm 0.0020$, we can evaluate $\sin^2\theta_W$ and ρ :

$$\sin^2\theta_W = 0.226 \pm 0.005 \pm 0.008$$
$$\rho = (M_W/M_Z \cos\theta_W)^2 = .996 \pm .033 \pm .009$$

Using $\rho=1$, Δr can be evaluated for UA2 data alone, or by combining the world average of $\sin^2\theta_W(\text{WA}) = 0.220 \pm 0.007$ and the measured values of M_W and M_Z :

$$\Delta r \text{ (UA2)} = .08 \pm .10 \pm .03$$
$$\Delta r \text{ (WA)} = .050 \pm .034 \pm .030$$

Fig. 10 compares the status of UA2 measurements with $\sin^2\theta_W$ (WA) and with expectations of the GWS model.

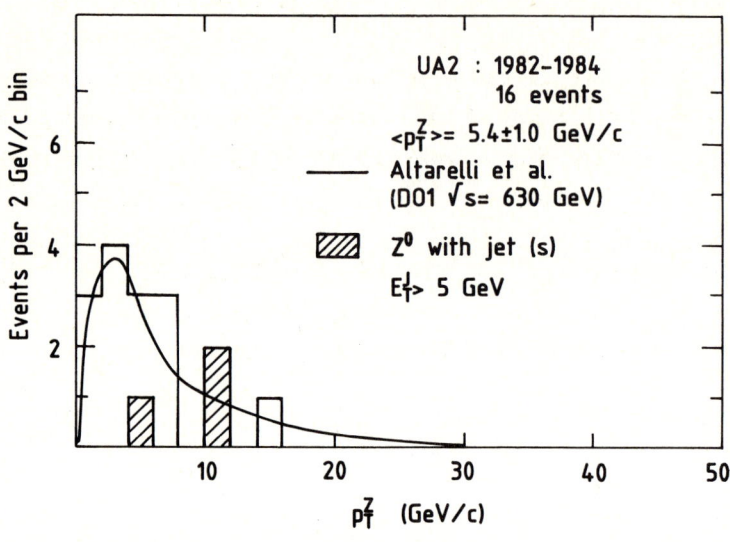

FIGURE 9

The Z^0 transverse momentum distribution.

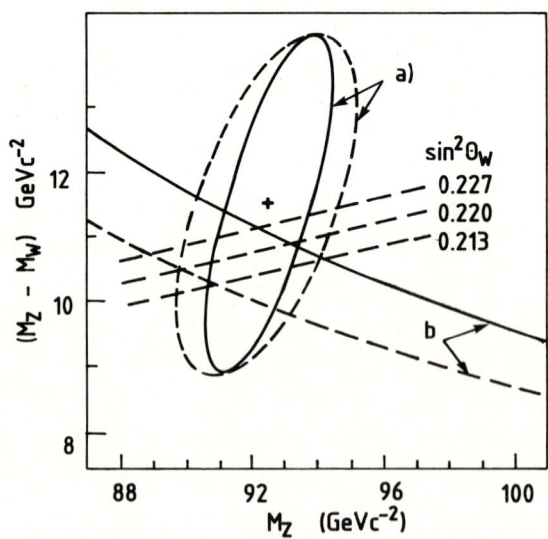

FIGURE 10

Distribution of M_Z versus (M_Z-M_W). a) statistical (-) and systematic (----) uncertainties of the UA2 measurements. b) expectations of the GSW model with (-) and without (---) radiative corrections. Also shown is the expectation based on the world average of $\sin^2\theta_W$ measurements.

6. CONCLUSIONS

Data collected at $\sqrt{s}$ = 630 GeV confirm previously published UA2 results at $\sqrt{s}$ = 546 GeV. The features of inclusive jet production are well described by QCD calculations. Concerning the Intermediate Vector Bosons, the production cross sections, the values of M_W, M_Z and Γ_Z, and the forward-backward asymmetry in W decays, all agree with the predictions of the minimal electroweak model and QCD. In conclusion all measurements on jet and electron production are in agreement with the standard $SU(3) \otimes SU(2) \otimes U(1)$ model of the strong and electroweak interactions.

REFERENCES

1. UA2, P. Bagnaia et al., Phys. Lett. 138B (1984) 430.
   UA2, P. Bagnaia et al., Phys. Lett. 144B (1984) 283.
   UA2, P. Bagnaia et al., Phys. Lett. 144B (1984) 291.
   UA2, P. Bagnaia et al., Z. Phys. C20 (1983) 117.

2. UA2, P. Bagnaia et al., Phys. Lett. 139B (1984) 105.
   UA2, M. Banner et al., Phys. Lett. 122B (1983) 476.
   UA2, P. Bagnaia et al., Phys. Lett. 129B (1983) 130.
   UA2, P. Bagnaia et al., Z. Pys. C24 (1984) 1.

3. B. Mansoulié : The UA2 apparatus at the CERN $\bar{p}p$ Collider, Proc. of the 3rd Moriond Workshop on $\bar{p}p$ physics (1983), p. 609 (éditions Frontières).
   M. Dialinas et al., The vertex detector of the UA2 experiment, LAL-RT/83/14 (1983).
   A. Beer et al., Nucl. Instr. Meth. 224 (1984) 360.
   C. Conta et al., Nucl Instr. Meth. 224 (1984) 65.
   K. Borer et al., Nucl. Instr. Meth. 227 (1984) 29.

4. E. Eichten et al., Rev. Mod. Phys. 56 (1984) 579.

5. E. Eichten et al., Phys. Rev. Lett. 50 (1983) 811.

6. M. Glück et al., Z. Phys. C13 (1982) 119.

7. G. Altarelli, R.K. Ellis, M. Greco and G. Martinelli, Nucl. Phys. B246 (1984) 12.

8. G. Altarelli, R.K. Ellis and, G. Martinelli, Z. Phys. C27 (1984) 329.
   F. Halzen and W. Scott, Phys. Lett. 78B (1978) 318.
   P. Aurenche and J. Lindfors, Nucl. Phys. B185 (1981) 274.
   P. Minkowski, Phys. Lett. 139B (1984) 431.

9. K. Hikasa, Phys. Rev. D29 (1984) 1939.

10. W.J. Marciano and A. Sirlin, Phys. Rev. D29 (1984) 945.

MAGNETIC MONOPOLES FROM STANFORD TO MACRO

W.P. TROWER

Physics Department, Virginia Polytechnic Institute and State University, Blacksburg, Virginia 24061, U.S.A.*

The existence of magnetic monopoles would have important consequences for our ideas in particle physics and cosmology. The Stanford event, the one serious monopole candidate, has lost its enthusiasts with time as no similar event has been seen. Here the current experimental situation is reviewed and several promising initiatives are discussed. One of the most ambitious of these, the Monopole, Astrophysics, and Cosmic Ray Observatory (MACRO), is described.

1. INTRODUCTION

Since a rationale for magnetic monopoles was given fifty years ago [1] much theoretical speculation has surrounded their nature while only a modest, and until recently, totally unfruitful experimental effort has been mounted [2]. The now fashionable Grand Unification Theories (GUTs) require that monopoles exist and most suggest that they have enormous mass, exceeding that of the proton by at least a million. The relevant consequences for experimentalists are that these monopoles can only have been created early in the Big Bang and so will be slow ($\beta \sim .001$); will ionize feebly and so will have great penetration in matter where they will seldom be trapped; and will always produce a persistent current change if they pass through a conducting ring.

Cabrera detected a solitary current change in a ring detector consistent with a passing magnetic monopole of a single Dirac charge [3]. No similar event has subsequently been seen in either this, since retired, device [4] or in the now operating, larger coincidence detector [5]. The previously

---

* This work supported in part by grants from the National Science Foundation.

detected smaller persistient current changes, which could be interpreted as near-miss events, now occur in only one of the three detector rings, and then on the average of once every other week. Although the original signal stands unimpuned, it becomes decreasingly credible as no other experiment has produced evidence for monopoles. If monopoles exist they are rarer than we thought in Spring 1982.

In the meantime, searches for magnetic monopoles have increased in number, sophistication, complexity, and ingenuity. In what follows I discuss the status and prospects for the discovery of GUT magnetic monopoles by their three principal detectable processes: induction, catalysis, and ionization.

2. INDUCTION SEARCHES

The idea of looking directly for magnetic charge using electromagnetic induction has a monopole threading a closed conducting ring [6]. Tiny induced current change in a room temperature ring will quickly die away but will persist in a superconducting ring. Superconducting QUantum Interferometer Devices (SQUIDs) and ambient magnetic field reducing techniques now allow dynamic detectors to be built. Cabrera's was the first of such detectors [7].

The candidate Stanford event [3] was recorded in the single-loop four-turn five-centimeter diameter superconducting niobium wire ring coupled to a SQUID mounted and operated in an ultra-low ambient magnetic field. The event was consistent with the passage of a single Dirac magnetic charge to within five percent, while the baseline noise was typically one percent of such a signal.

Induction detectors using gradiometers require less sophisticated shielding than the Stanford magnetometers. Further their multiple loops provide redundancy, rough direction information, and increased detection area. These second generation induction detectors, typically with areas of a tenth of a square meter and coincidence capabilities, have together established a flux limit of 10^{-12}/cm²/sr/s, over a thousand times the exposure of the first Stanford detector.

Three larger (more than a square meter), multiloop (4-10) detectors with high redundancy are now coming on line at Stanford, IBM, and Chicago. If the techniques on which they are based are proven in their current data runs, it should be possible to extrapolate them into a next generation. Here hundreds of coils providing multiple coincidences and having an area of about a hundred square meters could, in principal, be operational by the end of the decade.

Magnetic induction detectors rely only on the monopole's magnetic charge and so are independent of all other properties, such as mass, speed, direction, electric charge, ionization, and so on [8]. Thus induction detectors would be the clear choice for monopole searches if it weren't for their limited areas and high cost.

3. NUCLEON CATALYSIS

GUT monopoles may catalyze a baryon-violating process in nuclei with a strong interaction cross section; the Rubakov-Callan effect [9]. Thus, a monopole's mean free path would be small compared with the dimensions of a nucleon-stability detector, where its passage could be inferred from a line of decays. Since proton decay experiments have not detected a single event, taken together they set a limit on the catalyzing GUT monopole flux of 10^{-16}/cm^2/sr/s, which is somewhat dependent on the assumed cross section for the catalysis process.

An indirect, but more stringent, limit on the monopole flux has been set by searching for the possible effects of monopole catalysis on the luminosity distributions of neutron stars [10]. From the diffuse ultra-violet and X-ray emissions, a monopole flux limit of about 10^{-22}/cm²/sr/s is obtained. So either monopoles are incredibly rare or they are do not catalyze baryon decay.

4. IONIZATION EXPERIMENTS

After much uncertainly about the slow ($\beta \sim .001$) monopole energy loss, a consensus appears to have developed [11]. A similar theory for protons [12] was recently tested down to β=.0009 where measurable amounts of light were still emitted [13]. This essential experiment now gives confidence that comparably slow monopoles will produce detectable ionization. The totality of null results from a variety of large scintillation detectors [2] places a limit on the monopole flux of 10^{-15}/cm²/sr/s.

Monopoles have a second ionization energy-loss mechanism [14] in which helium is Zeeman excited into a metastable state. This excitation is transferred by the Penning process to a methane molecule which immediately ionizes. The experimental limit on the monopole flux using this effect is 10^{-13}/cm²/sr/s for monopoles with $\beta \gtrsim .0007$ [15].

Etching of certain materials which can be permanently radiation damaged by the passage of highly ionizing radiation define another class of ionization detectors [16] with which to search for magnetic monopoles. The most sensitive experiment to use these techniques looked at ancient mica samples whose age was calibrated using heavy ions [17]. It limits the flux to less than 10^{-16}/cm²/sr/s, but under a special (improbable?) scenario: a bare monopole enters the earth, captures an aluminium nucleus which remains bound and uncatalyzed throughout the mica encounter.

5. THE MONOPOLE, ASTROPHYSICS, AND COSMIC-RAY OBSERVATORY (MACRO)

An ambitious ionization detector, MACRO, is under construction by an Italian-American collaboration in the underground Gran Sasso Laboratory [18]. MACRO will serve as an observatory for monopoles, a variety of astrophysical phenomena as will as cosmic rays. Its principal motivation is to detect monopoles by a variety of redundant proven techniques: tracking by helium-n-pentane filled limited streamer tubes (9 layers), liquid scintillator log layers (2 layers), plastics (3 kinds, 5 layers), and interleaved concrete absorber (sufficient to range-out charged particles from neutrinos with energies up to 3 GeV). MACRO will not only be sensitive to bare monopoles, but also to monopoles with an attached fermion nucleus (e.g. proton, aluminium, magnesium etc.) Specifically it is designed to detect monopoles at a tenth of the Parker limit [19], $10^{-16}/cm^2/sr/s$.

MACRO is also ideally equipped to do high energy neutrino astronomy and cosmic-ray muon physics. It will distinguish upward from downward moving muons produced by TeV neutrinos interacting in the rocks and so should be able to see high energy gamma ray sources in the southern hemisphere like Vela X-1 with heretofore unachieved sensitivity. With single muons MACRO can search for point sources like the recently proposed Cygnus X-3. Multi-muon events can be used to study the composition of the primary cosmic rays in the PeV region. Finally, MACRO should be able to detect neutrinos from stellar collapse with sufficient lead time to alert optical and radio astronomers. In addition it will be able to study neutrino oscillations, if such occur. However the most interesting non-monopole physics which it can accomplish has yet to be imagined.

MACRO is physically large (111x12x6 m³) with a comparably large acceptance (~12,000 m²-sr) and is located deep underground (4 kmwe) to reduce the background. It will occupy an impressively large hall with easy roadway access. The Gran Sasso Laboratory itself has the basic facilities typically of accelerator laboratories. The first nineth of the detector should be operating in early 1986 and the complete MACRO is scheduled to be turned on two years later.

6. CONCLUSION

The importance to physics of the discovery of magnetic monpoles transcends any particular model which includes them. However, even a null search with the sensitivity of MACRO would be valuable as it would exclude GUT monopoles as the primary component of the dark matter in the universe.

REFERENCES

1. P.A.M. Dirac, Proc. Roy. Soc. London, Ser. A, <u>133</u>, 60 (1931).

2. M. Aguilar-Benitez et al. [Particle Data Group], Phys. Lett. B170, 1 (1986).

3. B. Cabrera, Phys. Rev. Lett. 48, 1378 (1982).

4. B. Cabrera, M. Taber, M. Gardner, and J. Bourg Phys. Rev. Lett. 51, 1933 (1983).

5. B. Cabrera (private communication).

6. L.W. Alvarez, Lawrence Radiation Laboratory Physics Note 470, 1963 (unpublished).

7. B. Cabrera, Ph.D. thesis, Stanford University, 1975.

8. B. Cabrera and W.P. Trower, Found. Phys. 13, 13 (1983).

9. V. Rubakov JETP Lett. 33, 644 (1981) and Nucl. Phys. B203, 311 (1982); and C.G. Callan, Phys. Rev. D25, 2141 (1982) and Nucl. Phys. B203, 311 (1982).

10. E.W. Kolb, S.A. Colgate, and J.A. Harvey, Phys. Rev. Lett. 49, 1373 (1982); and S. Dimopoulos, J. Preskill, and F. Wilczek, Phys. Lett. B119, 320 (1982).

11. S.P. Ahlen and G. Tarle, Phys. Rev. D27, 688 (1983).

12. J. Lindhard, Mat. Fys. Medd. Dan. Vid. Selsk. 28, 8 (1954).

13. S.P. Ahlen, T.M. Liss, C. Lane, and G. Liu, Phys. Rev. Lett. 55, 181 (1985).

14. S.D. Drell, N.M. Kroll, M.T. Mueller, S.J. Parke, and M.A. Ruderman, Phys. Rev. Lett. 50, 644 (1983).

15. T. Hara et al., Phys. Rev. Lett. 56, 553 (1986).

16. R.L. Fleischer, P.B. Price, and R.M. Walker, Nuclear Tracks in Solids, (Universilty of California Press, Berkeley, 1975).

17. P.B. Price, S-L. Guo, S.P. Ahlen, and R.L. Fleischer, Phys. Rev. Lett. 52, 1265 (1984).

18. C. De Marzo et al. CALT-68-1224a (1984) pp.141.

19. M.S. Turner, E.N. Parker and T.J. Bogdan, Phys. Rev. D26, 1296 (1982).

LIST OF PARTICIPANTS

BATAKIS, N.,
University of Ioannina

BÉG, M.A.B.,
Rockefeller University

BERTOLAMI, O.,
Oxford University

BILIĆ, N.,
Rudjer Bošković Institute

BRANDENBERGER, R.,
University of California

CHRYSICOPOULOU, P.,
University of Michigan

DENMAT, LE, G.,
Institute H. Poincaré

EVERETT, A.,
Tufts University

GANOULIS, N.,
University of Thessaloniki

GEORGI, H.,
Harvard University

GEORGIOPOULOS, C.,
Florida State University

GOGGI, G.,
University of Pavia

GRILLO, A.,
Lab. Naz. INFN, Frascati

GROSS, D.J.,
Princeton University

HATZIS, M.,
University of Thessaloniki

HOLDOM, B.,
University of Toronto

HU, B.,
University of Maryland

KANGAS, T.,
Indiana University

KARAPIPERIS, T.,
University of Thessaloniki

KHAN, I.,
International Centre for Theoretical Physics, Trieste

KIBBLE, T.W.B.,
Imperial College

KOLASSIS, C.,
Institute H. Poincaré

KOLB, E.,
Fermi National Accelerator Laboratory

KTORIDES, X.,
University of Athens

LAZARIDES, G.,
University of Thessaloniki

LEONTARIS, G.,
University of Ioannina

LÜST, D.,
University of Munich

MAEDA, K.,
International Centre for Theoretical Physics, Trieste

MANKA, R.,
Silesian University

MAYROMATOS, N.,
University of Oxford

MORGAN, T.,
Massachusetts Institute of Technology

NAPPI, C.R.,
Princeton University

NARAIN, K.,
Rutherford Appleton Laboratory

NIKOLAIDIS, A.,
University of Thessaloniki

NIŽIC, B.,
Rudjer Bošković

ORITO, S.,
University of Tokyo

OVRUT, B.,
Rockefeller University

PANAGIOTAKOPOULOS, C.,
Rockefeller University

PAPAIOANNOU, G.,
N.R.C. Demokritos

PASHALIS, J.,
University of Thessaloniki

PATI, J.,
University of Maryland

PERANTONIS, S.,
University of Oxford

PETRIDOU, C.,
Brookhaven National Laboratory

PIAZZOLI, A.,
University Di Pavia

POLYCHRONAKOS, A.,
CALTECH

RONGA, F.,
Lab. Naz. INFN, Frascati

SALAMON, M.,
University of California at Berkeley

SARANTAKOS, S.,
University of Thessaloniki

SASKI, M.,
Kyoto University

SENJANOVIĆ, G.,
Brookhaven National Laboratory

SHAFI, Q.,
University of Delaware

SIKIVIE, P.,
University of Florida

SIOPSIS, G.,
CALTECH

SOTO, J.,
University of Barcelona

STOGIANNIDOU, A.,
University of Ioannina

STONE, J.,
University of Michigan

TRAKAS, N.,
Nat. Tech. University of Athens

TROWER, W.P.,
Virginia Polytechnic Institute

TSOUBELIS, D.,
University of Ioannina

TUROK, N.,
Santa Barbara University

VAYONAKIS,
International Centre for Theoretical Physics, Trieste

VERGADOS, J.D.,
University of Ioannina

VILENKIN, A.,
Tufts University

VLACHOS, N.,
University of Crete

WARR, B.,
CALTECH

WEILER, T.,
Vanderbilt University

WETTERICH, C.,
CERN

WILLIS, S.,
University of Oklahoma

ZEE, A.,
University of Washington

AUTHOR INDEX

BÉG, M.A.B., 61
BRANDENBERGER, R., 151

GEORGI, H.M., 79
GOGGI, G., 271
GROSS, D.J., 1

HOLDOM, B., 133
HU, B.L., 169

KIBBLE, T.W.B., 177
KOLB, E.W., 247

LAZARIDES, G., 223
LÜST, D., 93

MAEDA, K., 239
NAPPI, C.R., 29

NARAIN, K.S., 263

OVRUT, B.A., 21

PANAGIOTAKOPOULOS, C., 139
PATI, J.C., 39

SENJANOVIĆ, G., 103
SIKIVIE, P., 201

TROWER, W.P., 287
TUROK, N., 189

VERGADOS, J.D., 115
VILENKIN, A., 143

WEILER, T.J., 215
WETTERICH, C., 231

ZEE, A., 257
ZOUPANOS, G., 93

RAYMOND H. FOGLER LIBRARY
DATE DUE

BOOKS ARE SUBJECT TO RECALL AFTER TWO WEEKS

JAN 1 4 1987

MAR 3 0 1987

FEB 1 2 1988